Dynamical Systems and Linear Algebra

Dynamical Systems and Linear Algebra

Fritz Colonius
Wolfgang Kliemann

Graduate Studies
in Mathematics

Volume 158

American Mathematical Society
Providence, Rhode Island

EDITORIAL COMMITTEE

Dan Abramovich
Daniel S. Freed
Rafe Mazzeo (Chair)
Gigliola Staffilani

2010 *Mathematics Subject Classification.* Primary 15-01, 34-01, 37-01, 39-01, 60-01, 93-01.

For additional information and updates on this book, visit
www.ams.org/bookpages/gsm-158

Library of Congress Cataloging-in-Publication Data

Colonius, Fritz.
 Dynamical systems and linear algebra / Fritz Colonius, Wolfgang Kliemann.
  pages cm. – (Graduate studies in mathematics ; volume 158)
 Includes bibliographical references and index.
 ISBN 978-0-8218-8319-8 (alk. paper)
 1. Algebras, Linear. 2. Topological dynamics. I. Kliemann, Wolfgang. II. Title.

QA184.2.C65 2014
512'.5–dc23
                                                                        2014020316

Copying and reprinting. Individual readers of this publication, and nonprofit libraries acting for them, are permitted to make fair use of the material, such as to copy a chapter for use in teaching or research. Permission is granted to quote brief passages from this publication in reviews, provided the customary acknowledgment of the source is given.

Republication, systematic copying, or multiple reproduction of any material in this publication is permitted only under license from the American Mathematical Society. Requests for such permission should be addressed to the Acquisitions Department, American Mathematical Society, 201 Charles Street, Providence, Rhode Island 02904-2294 USA. Requests can also be made by e-mail to reprint-permission@ams.org.

© 2014 by the American Mathematical Society. All rights reserved.
The American Mathematical Society retains all rights
except those granted to the United States Government.
Printed in the United States of America.

∞ The paper used in this book is acid-free and falls within the guidelines
established to ensure permanence and durability.
Visit the AMS home page at http://www.ams.org/

10 9 8 7 6 5 4 3 2 1   19 18 17 16 15 14

This book is dedicated to the Institut für Dynamische Systeme at Universität Bremen, which had a lasting influence on our mathematical thinking, as well as to our students. This book would not have been possible without the interaction at the Institut and the graduate programs in our departments.

Contents

Introduction	xi
Notation	xv

Part 1. Matrices and Linear Dynamical Systems

Chapter 1.	Autonomous Linear Differential and Difference Equations	3
§1.1.	Existence of Solutions	3
§1.2.	The Real Jordan Form	6
§1.3.	Solution Formulas	10
§1.4.	Lyapunov Exponents	12
§1.5.	The Discrete-Time Case: Linear Difference Equations	18
§1.6.	Exercises	24
§1.7.	Orientation, Notes and References	27
Chapter 2.	Linear Dynamical Systems in $\mathbb{R}^d$	29
§2.1.	Continuous-Time Dynamical Systems or Flows	29
§2.2.	Conjugacy of Linear Flows	33
§2.3.	Linear Dynamical Systems in Discrete Time	38
§2.4.	Exercises	43
§2.5.	Orientation, Notes and References	43
Chapter 3.	Chain Transitivity for Dynamical Systems	47
§3.1.	Limit Sets and Chain Transitivity	47
§3.2.	The Chain Recurrent Set	54

§3.3.	The Discrete-Time Case	59
§3.4.	Exercises	63
§3.5.	Orientation, Notes and References	65

Chapter 4. Linear Systems in Projective Space — 67
- §4.1. Linear Flows Induced in Projective Space — 67
- §4.2. Linear Difference Equations in Projective Space — 75
- §4.3. Exercises — 78
- §4.4. Orientation, Notes and References — 78

Chapter 5. Linear Systems on Grassmannians — 81
- §5.1. Some Notions and Results from Multilinear Algebra — 82
- §5.2. Linear Systems on Grassmannians and Volume Growth — 86
- §5.3. Exercises — 94
- §5.4. Orientation, Notes and References — 95

Part 2. Time-Varying Matrices and Linear Skew Product Systems

Chapter 6. Lyapunov Exponents and Linear Skew Product Systems — 99
- §6.1. Existence of Solutions and Continuous Dependence — 100
- §6.2. Lyapunov Exponents — 106
- §6.3. Linear Skew Product Flows — 113
- §6.4. The Discrete-Time Case — 118
- §6.5. Exercises — 121
- §6.6. Orientation, Notes and References — 123

Chapter 7. Periodic Linear Differential and Difference Equations — 127
- §7.1. Floquet Theory for Linear Difference Equations — 128
- §7.2. Floquet Theory for Linear Differential Equations — 136
- §7.3. The Mathieu Equation — 144
- §7.4. Exercises — 151
- §7.5. Orientation, Notes and References — 153

Chapter 8. Morse Decompositions of Dynamical Systems — 155
- §8.1. Morse Decompositions — 155
- §8.2. Attractors — 159
- §8.3. Morse Decompositions, Attractors, and Chain Transitivity — 164
- §8.4. Exercises — 166

§8.5.	Orientation, Notes and References	167
Chapter 9.	Topological Linear Flows	169
§9.1.	The Spectral Decomposition Theorem	170
§9.2.	Selgrade's Theorem	178
§9.3.	The Morse Spectrum	184
§9.4.	Lyapunov Exponents and the Morse Spectrum	192
§9.5.	Application to Robust Linear Systems and Bilinear Control Systems	197
§9.6.	Exercises	207
§9.7.	Orientation, Notes and References	208
Chapter 10.	Tools from Ergodic Theory	211
§10.1.	Invariant Measures	211
§10.2.	Birkhoff's Ergodic Theorem	214
§10.3.	Kingman's Subadditive Ergodic Theorem	217
§10.4.	Exercises	220
§10.5.	Orientation, Notes and References	221
Chapter 11.	Random Linear Dynamical Systems	223
§11.1.	The Multiplicative Ergodic Theorem (MET)	224
§11.2.	Some Background on Projections	233
§11.3.	Singular Values, Exterior Powers, and the Goldsheid-Margulis Metric	237
§11.4.	The Deterministic Multiplicative Ergodic Theorem	242
§11.5.	The Furstenberg-Kesten Theorem and Proof of the MET in Discrete Time	252
§11.6.	The Random Linear Oscillator	263
§11.7.	Exercises	266
§11.8.	Orientation, Notes and References	268
Bibliography		271
Index		279

Introduction

Background

Linear algebra plays a key role in the theory of dynamical systems, and concepts from dynamical systems allow the study, characterization and generalization of many objects in linear algebra, such as similarity of matrices, eigenvalues, and (generalized) eigenspaces. The most basic form of this interplay can be seen as a quadratic matrix A gives rise to a discrete time dynamical system $x_{k+1} = Ax_k$, $k = 0, 1, 2, \ldots$ and to a continuous time dynamical system via the linear ordinary differential equation $\dot{x} = Ax$.

The (real) Jordan form of the matrix A allows us to write the solution of the differential equation $\dot{x} = Ax$ explicitly in terms of the matrix exponential, and hence the properties of the solutions are intimately related to the properties of the matrix A. Vice versa, one can consider properties of a linear flow in $\mathbb{R}^d$ and infer characteristics of the underlying matrix A. Going one step further, matrices also define (nonlinear) systems on smooth manifolds, such as the sphere $\mathbb{S}^{d-1}$ in $\mathbb{R}^d$, the Grassmannian manifolds, the flag manifolds, or on classical (matrix) Lie groups. Again, the behavior of such systems is closely related to matrices and their properties.

Since A.M. Lyapunov's thesis [97] in 1892 it has been an intriguing problem how to construct an appropriate linear algebra for time-varying systems. Note that, e.g., for stability of the solutions of $\dot{x} = A(t)x$ it is not sufficient that for all $t \in \mathbb{R}$ the matrices $A(t)$ have only eigenvalues with negative real part (see, e.g., Hahn [61], Chapter 62). Classical Floquet theory (see Floquet's 1883 paper [50]) gives an elegant solution for the periodic case, but it is not immediately clear how to build a linear algebra around Lyapunov's 'order numbers' (now called Lyapunov exponents) for more general time dependencies. The key idea here is to write the time dependency as a dynamical system with certain recurrence properties. In this way, the multiplicative ergodic theorem of Oseledets from 1968 [**109**] resolves the basic issues for measurable linear systems with stationary time dependencies, and the Morse spectrum together with Selgrade's theorem [**124**] goes a long way in describing the situation for continuous linear systems with chain transitive time dependencies.

A third important area of interplay between dynamics and linear algebra arises in the linearization of nonlinear systems about fixed points or arbitrary trajectories. Linearization of a differential equation $\dot y = f(y)$ in $\mathbb{R}^d$ about a fixed point $y_0 \in \mathbb{R}^d$ results in the linear differential equation $\dot x = f'(y_0)x$ and theorems of the type Grobman-Hartman (see, e.g., Bronstein and Kopanskii [**21**]) resolve the behavior of the flow of the nonlinear equation locally around y_0 up to conjugacy, with similar results for dynamical systems over a stochastic or chain recurrent base.

These observations have important applications in the natural sciences and in engineering design and analysis of systems. Specifically, they are the basis for stochastic bifurcation theory (see, e.g., Arnold [**6**]), and robust stability and stabilizability (see, e.g., Colonius and Kliemann [**29**]). Stability radii (see, e.g., Hinrichsen and Pritchard [**68**]) describe the amount of perturbation the operating point of a system can sustain while remaining stable, and stochastic stability characterizes the limits of acceptable noise in a system, e.g., an electric power system with a substantial component of wind or wave based generation.

Goal

This book provides an introduction to the interplay between linear algebra and dynamical systems in continuous time and in discrete time. There are a number of other books emphasizing these relations. In particular, we would like to mention the book [**69**] by M.W. Hirsch and S. Smale, which always has been a great source of inspiration for us. However, this book restricts attention to autonomous equations. The same is true for other books like M. Golubitsky and M. Dellnitz [**54**] or F. Lowenthal [**96**], which is designed to serve as a text for a first course in linear algebra, and the relations to linear autonomous differential equations are established on an elementary level only.

Our goal is to review the autonomous case for one $d \times d$ matrix A via induced dynamical systems in $\mathbb{R}^d$ and on Grassmannians, and to present the main nonautonomous approaches for which the time dependency $A(t)$ is given via skew-product flows using periodicity, or topological (chain recurrence) or ergodic properties (invariant measures). We develop generalizations of (real parts of) eigenvalues and eigenspaces as a starting point for a linear algebra for classes of time-varying linear systems, namely periodic, random, and perturbed (or controlled) systems. Several examples of (low-dimensional) systems that play a role in engineering and science are presented throughout the text.

Originally, we had also planned to include some basic concepts for the study of genuinely nonlinear systems via linearization, emphasizing invariant manifolds and Grobman-Hartman type results that compare nonlinear behavior locally to the behavior of associated linear systems. We decided to skip this discussion, since it would increase the length of this book considerably and, more importantly, there are excellent treatises of these problems available in the literature, e.g., Robinson [**117**] for linearization at fixed points, or the work of Bronstein and Kopanskii [**21**] for more general linearized systems.

Another omission is the rich interplay with the theory of Lie groups and semigroups where many concepts have natural counterparts. The monograph [**48**] by R. Feres provides an excellent introduction. We also do not treat nonautonomous differential equations via pullback or other fiberwise constructions; see, e.g., Crauel and Flandoli [**37**], Schmalfuß [**123**], and Rasmussen [**116**]; our emphasis is on the treatment of families of nonautonomous equations. Further references are given at the end of the chapters.

Finally, it should be mentioned that all concepts and results in this book can be formulated in continuous and in discrete time. However, sometimes results in discrete time may be easier to state and to prove than their analogues in continuous time, or vice versa. At times, we have taken the liberty to pick one convenient setting, if the ideas of a result and its proof are particularly intuitive in the corresponding setup. For example, the results in Chapter 5 on induced systems on Grassmannians are formulated and derived only in continuous time. More importantly, the proof of the multiplicative ergodic theorem in Chapter 11 is given only in discrete time (the formulation and some discussion are also given in continuous time). In contrast, Selgrade's Theorem for topological linear dynamical systems in Chapter 9 and the results on Morse decompositions in Chapter 8, which are used for its proof, are given only in continuous time.

Our aim when writing this text was to make 'time-varying linear algebra' in its periodic, topological and ergodic contexts available to beginning graduate students by providing complete proofs of the major results in at least one typical situation. In particular, the results on the Morse spectrum in Chapter 9 and on multiplicative ergodic theory in Chapter 11 have detailed proofs that, to the best of our knowledge, do not exist in the current literature.

Prerequisites

The reader should have basic knowledge of real analysis (including metric spaces) and linear algebra. No previous exposure to ordinary differential equations is assumed, although a first course in linear differential equations certainly is helpful. Multilinear algebra shows up in two places: in Section 5.2 we discuss how the volumes of parallelepipeds grow under the flow of a linear autonomous differential equation, which we relate to chain recurrent sets of the induced flows on Grassmannians. The necessary elements of multilinear algebra are presented in Section 5.1. In Chapter 11 the proof of the multiplicative ergodic theorem requires further elements of multilinear algebra which are provided in Section 11.3. Understanding the proofs in Chapter 10 on ergodic theory and Chapter 11 on random linear dynamical systems also requires basic knowledge of σ-algebras and probability measures (actually, a detailed knowledge of Lebesgue measure should suffice). The results and methods in the rest of the book are independent of these additional prerequisites.

Acknowledgements

The idea for this book grew out of the preparations for Chapter 79 in the *Handbook of Linear Algebra* [71]. Then WK gave a course "Dynamics and Linear Algebra" at the Simposio 2007 of the Sociedad de Matemática de Chile. FC later taught a course on the same topic at Iowa State University within the 2008 Summer Program for Graduate Students of the Institute of Mathematics and Its Applications, Minneapolis. Parts of the manuscript were also used for courses at the University of Augsburg in the summer semesters 2010 and 2013 and at Iowa State University in Spring 2011. We gratefully acknowledge these opportunities to develop our thoughts, the feedback from the audiences, and the financial support.

Thanks for the preparation of figures are due to: Isabell Graf (Section 4.1), Patrick Roocks (Section 5.2); Florian Ecker and Julia Rudolph (Section 7.3.); and Humberto Verdejo (Section 11.6). Thanks are also due to Philipp Düren, Julian Braun, and Justin Peters. We are particularly indebted to Christoph Kawan who has read the whole manuscript and provided us with long lists of errors and inaccuracies. Special thanks go to Ina Mette of the AMS for her interest in this project and her continuous support during the last few years, even when the text moved forward very slowly.

The authors welcome any comments, suggestions, or corrections you may have.

Fritz Colonius	Wolfgang Kliemann
Institut für Mathematik	Department of Mathematics
Universität Augsburg	Iowa State University

Notation

Throughout this text we will use the following notation:

$A \triangle B$	$= (A \setminus B) \cup (B \setminus A)$, the symmetric difference of sets
$f^{-1}(E)$	$= \{x \mid f(x) \in E\}$ for a map $f : X \to Y$ and $E \subset Y$
E^c	the complement of a subset $E \subset X$, $E^c = X \setminus E$
$\mathbb{I}_E$	the characteristic function of a set E, $\mathbb{I}_E(x) := 1$ if $x \in E$ and $\mathbb{I}_E(x) := 0$ elsewhere
$f^+(x)$	$= \max(f(x), 0)$, the positive part of $f : X \to \mathbb{R}$
$\log^+ x$	$\log^+ x := \log x$ for $x \geq 1$ and $\log^+ x := 0$ for $x \leq 1$
$gl(d, \mathbb{R}), gl(d, \mathbb{C})$	the set of real (complex) $d \times d$ matrices
$Gl(d, \mathbb{R}), Gl(d, \mathbb{C})$	the set of invertible real (complex) $d \times d$ matrices
$A^\top$	the transpose of a matrix $A \in gl(d, \mathbb{R})$
$\|\cdot\|$	a norm on $\mathbb{R}^d$ or an induced matrix norm
$\mathrm{spec}(A)$	the set of eigenvalues $\mu \in \mathbb{C}$ of a matrix A
$\mathrm{im} A, \mathrm{tr} A$	the image and the trace of a linear map A, resp.
$\limsup, \liminf$	limit superior, limit inferior
$\mathbb{N}, \mathbb{N}_0$	the set of natural numbers excluding and including 0
$\imath$	$\imath = \sqrt{-1}$
$\overline{z}$	the complex conjugate of $z \in \mathbb{C}$
$\overline{A}$	$\overline{A} = (\overline{a_{ij}})$ for $A = (a_{ij}) \in gl(d, \mathbb{C})$
$\mathbb{P}^{d-1}$	the real projective space $\mathbb{P}^{d-1} = \mathbb{RP}^{d-1}$
$\mathbb{G}_k(d)$	the kth Grassmannian of $\mathbb{R}^d$
$L(\lambda)$	the Lyapunov space associated with a Lyapunov exponent λ
$\mathbb{E}$	expectation (relative to a probability measure P)

For points x and nonvoid subsets E of a metric space X with metric d:

$N(x, \varepsilon)$	$= \{y \in X \mid d(x, y) < \varepsilon\}$, the ε-neighborhood of x
$\mathrm{diam}\, E$	$= \sup\{d(x, y) \mid x, y \in E\}$, the diameter of E
$\mathrm{dist}(x, E)$	$= \inf\{d(x, y) \mid y \in E\}$, the distance of x to E
$\mathrm{cl}\, E, \mathrm{int}\, E$	the topological closure and interior of E, resp.

Part 1

Matrices and Linear Dynamical Systems

Chapter 1

Autonomous Linear Differential and Difference Equations

An autonomous linear differential equation or difference equation is determined by a fixed matrix. Here linear algebra can directly be used to derive properties of the corresponding solutions. This is due to the fact that these equations can be solved explicitly if one knows the eigenvalues and a basis of eigenvectors (and generalized eigenvectors, if necessary). The key idea is to use the (real) Jordan form of a matrix. The real parts of the eigenvectors determine the exponential growth behavior of the solutions, described by the Lyapunov exponents and the corresponding Lyapunov subspaces. In this chapter we recall the necessary concepts and results from linear algebra and linear differential and difference equations. Section 1.1 establishes existence and uniqueness of solutions for initial value problems for linear differential equations and shows continuous dependence of the solution on the initial value and the matrix. Section 1.2 recalls the Jordan normal form over the complex numbers and derives the Jordan normal form over the reals. This is used in Section 1.3 to write down explicit formulas for the solutions. Section 1.4 introduces Lyapunov exponents and relates them to eigenvalues. Finally, Section 1.5 presents analogous results for linear difference equations.

1.1. Existence of Solutions

This section presents basic results on existence of solutions for linear autonomous differential equations and their continuous dependence on the initial value and the matrix.

A linear differential equation (with constant coefficients) is given by a matrix $A \in gl(d, \mathbb{R})$ via $\dot{x}(t) = Ax(t)$, where $\dot{x}$ denotes differentiation with respect to t. Any differentiable function $x : \mathbb{R} \longrightarrow \mathbb{R}^d$ such that $\dot{x}(t) = Ax(t)$ for all $t \in \mathbb{R}$, is called a solution of $\dot{x} = Ax$.

An initial value problem for a linear differential equation $\dot{x} = Ax$ consists in finding for a given point $x_0 \in \mathbb{R}^d$ a solution $x(\cdot, x_0)$ that satisfies the initial condition $x(0, x_0) = x_0$. The differential equation $\dot{x} = Ax$ is also called a time invariant or autonomous linear differential equation. The parameter $t \in \mathbb{R}$ is interpreted as time.

A description of the solutions is based on the matrix exponential: For a matrix $A \in gl(d, \mathbb{R})$ the exponential $e^A \in Gl(d, \mathbb{R})$ is defined by $e^A = \sum_{n=0}^{\infty} \frac{1}{n!} A^n \in Gl(d, \mathbb{R})$. This series converges absolutely, i.e.,

$$\sum_{n=0}^{\infty} \frac{1}{n!} \|A^n\| \leq \sum_{n=0}^{\infty} \frac{1}{n!} \|A\|^n = e^{\|A\|} < \infty,$$

and the convergence is uniform for bounded matrices A, i.e., for every $M > 0$ and all $\varepsilon > 0$ there is $\delta > 0$ such that for all $A, B \in gl(d, \mathbb{R})$ with $\|A\|, \|B\| \leq M$

$$\|A - B\| < \delta \text{ implies } \|e^A - e^B\| < \varepsilon.$$

This follows since the power series of the scalar exponential function is uniformly convergent for bounded arguments. Furthermore, the inverse of e^A is e^{-A} and the matrices A and e^A commute, i.e., $Ae^A = e^A A$. Note also that for an invertible matrix $S \in Gl(d, \mathbb{R})$,

$$e^{SAS^{-1}} = Se^A S^{-1}.$$

The same properties hold for matrices with complex entries.

The solutions of $\dot{x} = Ax$ satisfy the following properties.

Theorem 1.1.1. *(i) For each $A \in gl(d, \mathbb{R})$ the solutions of $\dot{x} = Ax$ form a d-dimensional vector space $sol(A) \subset C^{\infty}(\mathbb{R}, \mathbb{R}^d)$ over $\mathbb{R}$, where $C^{\infty}(\mathbb{R}, \mathbb{R}^d) = \{f : \mathbb{R} \longrightarrow \mathbb{R}^d \mid f \text{ is infinitely often differentiable}\}$.*

(ii) For each initial value problem the solution $x(\cdot, x_0)$ is unique and given by $x(t, x_0) = e^{At} x_0, t \in \mathbb{R}$.

(iii) For a basis $v_1, \ldots, v_d$ of $\mathbb{R}^d$, the functions $x(\cdot, v_1), \ldots, x(\cdot, v_d)$ form a basis of the solution space $sol(A)$. The matrix function

$$X(\cdot) := [x(\cdot, v_1), \ldots, x(\cdot, v_d)]$$

is called a fundamental solution of $\dot{x} = Ax$ and $\dot{X}(t) = AX(t), t \in \mathbb{R}$.

Proof. Using the series expression $e^{tA} = \sum_{n=0}^{\infty} \frac{A^n}{n!} t^n$, one finds that the matrix e^{tA} satisfies (in $\mathbb{R}^{d \cdot d}$) $\frac{d}{dt} e^{tA} = A e^{tA}$. Hence $e^{At} x_0$ is a solution of the initial value problem. To see that the solution is unique, let $x(t)$ be any

1.1. Existence of Solutions

solution of the initial value problem and put $y(t) = e^{-tA}x(t)$. Then by the product rule,

$$\dot{y}(t) = \left(\frac{d}{dt}e^{-tA}\right)x(t) + e^{-tA}\dot{x}(t)$$
$$= -Ae^{-tA}x(t) + e^{-tA}Ax(t) = e^{-tA}(-A+A)x(t) = 0.$$

Therefore $y(t)$ is a constant. Setting $t = 0$ shows $y(t) = x_0$, and uniqueness follows. The claims on the solution space follow by noting that for every $t \in \mathbb{R}$ the map

$$x_0 \longmapsto x(t, x_0) : \mathbb{R}^d \to \mathbb{R}^d$$

is a linear isomorphism. Hence also the map

$$x_0 \longmapsto x(\cdot, x_0) : \mathbb{R}^d \to sol(A)$$

is a linear isomorphism. $\square$

Let $\mathbf{e}_1 = (1, 0, \ldots, 0)^\top, \ldots, \mathbf{e}_d = (0, 0, \ldots, 1)^\top$ be the standard basis of $\mathbb{R}^d$. Then $x(\cdot, \mathbf{e}_1), \ldots, x(\cdot, \mathbf{e}_d)$ is a basis of the solution space $sol(A)$ and the corresponding fundamental solution is e^{At}. Note that the solutions of $\dot{x} = Ax$ are even real analytic, i.e., given by a convergent power series in t. As an easy consequence of the solution formula one obtains the following continuity properties.

Corollary 1.1.2. *(i) The solution map $(t, x_0) \mapsto x(t, x_0) = e^{At}x_0 : \mathbb{R} \times \mathbb{R}^d \to \mathbb{R}^d$ is continuous.*

(ii) For every $M > 0$ the map $A \mapsto x(t, x_0) = e^{At}x_0 : \{A \in gl(d, \mathbb{R}) \mid \|A\| \le M\} \to \mathbb{R}^d$ is uniformly continuous for $x_0 \in \mathbb{R}^d$ with $\|x_0\| \le M$ and t in a compact time interval $[a, b], a < b$.

Proof. (i) This follows, since for $x_0, y_0 \in \mathbb{R}^d$ and $s, t \in \mathbb{R}$,

$$\|e^{As}x_0 - e^{At}y_0\| \le \|e^{As} - e^{At}\|\|x_0\| + \|e^{At}\|\|x_0 - y_0\|.$$

(ii) Let $\|A\|, \|B\| \le M$. One has

$$\|e^{At}x_0 - e^{Bt}x_0\| \le \|e^{At} - e^{Bt}\|\|x_0\| = \left\|\sum_{n=0}^{\infty}(A^n - B^n)\frac{t^n}{n!}\right\|\|x_0\|.$$

For $\varepsilon > 0$ there is $N \in \mathbb{N}$ such that for all $t \in [a, b]$,

$$\left\|\sum_{n=N+1}^{\infty}(A^n - B^n)\frac{t^n}{n!}\right\| \le \sum_{n=N+1}^{\infty}(\|A\|^n + \|B\|^n)\frac{|t|^n}{n!} \le \sum_{n=N+1}^{\infty}\frac{2|Mt|^n}{n!} < \varepsilon.$$

Hence N may be chosen independent of A, B. Then the assertion follows, since there is $\delta > 0$ such that for $\|A - B\| < \delta$,

$$\left\|\sum_{n=0}^{\infty}(A^n - B^n)\frac{t^n}{n!}\right\| \le \left\|\sum_{n=0}^{N}(A^n - B^n)\frac{t^n}{n!}\right\| + \left\|\sum_{n=N+1}^{\infty}(A^n - B^n)\frac{t^n}{n!}\right\| < 2\varepsilon.$$

$\square$

1.2. The Real Jordan Form

The key to obtaining explicit solutions of linear, time-invariant differential equations $\dot{x} = Ax$ are the eigenvalues, eigenvectors, and the real Jordan form of the matrix A. Here we show how to derive it from the well-known complex Jordan form: Recall that any matrix $A \in gl(d, \mathbb{C})$ is similar over the complex numbers to a matrix in Jordan normal form, i.e., there is a matrix $S \in Gl(d, \mathbb{C})$ such that $J^{\mathbb{C}} = S^{-1}AS$ has block diagonal form

$$J^{\mathbb{C}} = \text{blockdiag}[J_1, \ldots, J_s]$$

with Jordan blocks J_i given with $\mu \in \text{spec}(A)$, the set of eigenvalues or the spectrum of A, by

(1.2.1) $$J_i = \begin{bmatrix} \mu & 1 & . & . & . & 0 \\ 0 & \mu & 1 & & & . \\ . & & . & . & & . \\ . & & & . & . & . \\ . & & & 0 & \mu & 1 \\ 0 & & & & 0 & \mu \end{bmatrix}.$$

For an eigenvalue μ, the dimensions of the Jordan blocks (with appropriate ordering) are unique and for every $m \in \mathbb{N}$ the number of Jordan blocks of size equal to or less than $m \times m$ is determined by the dimension of the kernel $\ker(A - \mu I)^m$. The complex generalized eigenspace of an eigenvalue $\mu \in \mathbb{C}$ can be characterized as $\ker(A - \mu I)^n$, where n is the dimension of the largest Jordan block for μ (thus it is determined by the associated map and independent of the matrix representation.) The space $\mathbb{C}^d$ is the direct sum of the generalized eigenspaces. Furthermore, the eigenspace of an eigenvalue μ is the subspace of all eigenvectors of μ; its dimension is given by the number of Jordan blocks for μ. The subspace corresponding to a Jordan block (1.2.1) of size $m \times m$ intersects the eigenspace in the multiples of $(1, 0, \ldots, 0)^\top \in \mathbb{C}^m$.

We begin with the following lemma on similarity over $\mathbb{R}$ and over $\mathbb{C}$.

Lemma 1.2.1. *Let $A, B \in gl(d, \mathbb{R})$ and suppose that there is $S \in Gl(d, \mathbb{C})$ with $B = S^{-1}AS$. Then there is also a matrix $T \in Gl(d, \mathbb{R})$ with $B = T^{-1}AT$.*

1.2. The Real Jordan Form

An easy consequence of this lemma is: If $J^{\mathbb{C}}$ only has real entries, i.e., if all eigenvalues of A are real, then A is similar over the reals to $J^{\mathbb{C}}$. Hence we will only have to deal with complex eigenvalues.

Proof of Lemma 1.2.1. By assumption there is an invertible matrix $S \in Gl(d, \mathbb{C})$ with $SB = AS$. Decompose S into its real and imaginary entries, $S = S_1 + \imath S_2$, where S_1 and S_2 have real entries. Then $S_1 B = AS_1$ and $S_2 B = AS_2$, hence

$$(S_1 + xS_2)B = A(S_1 + xS_2) \text{ for all } x \in \mathbb{R}.$$

It remains to show that there is $x \in \mathbb{R}$ such that $S_1 + xS_2$ is invertible. This follows, since the polynomial $\det(S_1 + xS_2)$ in x, which has real coefficients, evaluated in $x = \imath$ satisfies $\det S = \det(S_1 + \imath S_2) \neq 0$. Thus it is not the zero polynomial, and hence there is also $x \in \mathbb{R}$ with $\det(S_1 + xS_2) \neq 0$ as claimed. $\square$

For a real matrix $A \in gl(d, \mathbb{R})$, proper complex eigenvalues always occur in complex-conjugate pairs, since the eigenvalues are the zeros of the real polynomial $p(x) = \det(A - xI) = (-1)^n x^n + q_{n-1} x^{n-1} + \ldots + q_1 x + q_0$. Thus if $\mu = \lambda + \imath\nu, \lambda, \nu \in \mathbb{R}$, is an eigenvalue of A, then also $\bar{\mu} = \lambda - \imath\nu$.

In order to motivate the real Jordan form, consider first one-dimensional complex Jordan blocks for $\mu, \bar{\mu} \in \sigma(A)$ (the result for arbitrary dimensions will not depend on this proposition.)

Proposition 1.2.2. *There is a matrix $S \in gl(2, \mathbb{C})$ with*

$$\begin{bmatrix} \lambda & -\nu \\ \nu & \lambda \end{bmatrix} = S \begin{bmatrix} \lambda + \imath\nu & 0 \\ 0 & \lambda - \imath\nu \end{bmatrix} S^{-1}.$$

Proof. Define

$$S := \begin{bmatrix} -\imath & 1 \\ -1 & \imath \end{bmatrix} \text{ with } S^{-1} = \frac{1}{2} \begin{bmatrix} \imath & -1 \\ 1 & -\imath \end{bmatrix}.$$

Then one computes

$$S \begin{bmatrix} \lambda + \imath\nu & 0 \\ 0 & \lambda - \imath\nu \end{bmatrix} S^{-1} = S \begin{bmatrix} \frac{\imath}{2}\lambda - \frac{1}{2}\nu & -\frac{\lambda}{2} - \frac{\imath}{2}\nu \\ \frac{\lambda}{2} - \frac{\imath}{2}\nu & -\frac{\imath}{2}\lambda - \frac{\nu}{2} \end{bmatrix} = \begin{bmatrix} \lambda & -\nu \\ \nu & \lambda \end{bmatrix}. \quad \square$$

A consequence of Lemma 1.2.1 and Proposition 1.2.2 is that for every matrix $A \in gl(2, \mathbb{R})$ there is a real matrix $T \in gl(2, \mathbb{R})$ with $T^{-1}AT = J \in gl(2, \mathbb{R})$, where either J is a Jordan matrix for real eigenvalues or $J = \begin{bmatrix} \lambda & -\nu \\ \nu & \lambda \end{bmatrix}$ for a complex conjugate eigenvalue pair $\mu, \bar{\mu} = \lambda \pm \imath\nu$ of A.

The following theorem describes the real Jordan normal form $J^{\mathbb{R}}$ for matrices in $gl(d, \mathbb{R})$ with arbitrary dimension d.

Theorem 1.2.3. *For every real matrix $A \in gl(d, \mathbb{R})$ there is an invertible real matrix $S \in Gl(d, \mathbb{R})$ such that $J^{\mathbb{R}} = S^{-1}AS$ is a block diagonal matrix,*

$$J^{\mathbb{R}} = \text{blockdiag}(J_1, \ldots, J_l),$$

with real Jordan blocks given for $\mu \in \text{spec}(A) \cap \mathbb{R}$ by (1.2.1) and for $\mu, \bar{\mu} = \lambda \pm \imath \nu \in \text{spec}(A), \nu > 0$, by

$$(1.2.2) \qquad J_i = \begin{bmatrix} \lambda & -\nu & 1 & 0 & & & & & & 0 \\ \nu & \lambda & 0 & 1 & \cdot & & \cdot & & & \\ & & \lambda & -\nu & & & & & & \\ & 0 & \nu & \lambda & & & & & & \cdot \\ & & & & \cdot & & \cdot & & & \\ \cdot & & & & & \cdot & & & & \cdot \\ \cdot & & & & \cdot & & & \lambda & -\nu & 1 & 0 \\ & & & & & & & \nu & \lambda & 0 & 1 \\ & & & & & & & & & \lambda & -\nu \\ & 0 & & & & & 0 & & & \nu & \lambda \end{bmatrix}$$

Proof. The matrix $A \in gl(d, \mathbb{R}) \subset gl(d, \mathbb{C})$ defines a linear map $F : \mathbb{C}^d \to \mathbb{C}^d$. Then one can for $\mu \in \text{spec}(F) \cap \mathbb{R}$ find a basis such that the restriction of F to the subspace for a Jordan block has the matrix representation (1.2.1). Hence it suffices to consider Jordan blocks for complex-conjugate pairs $\mu \neq \bar{\mu}$ in $\text{spec}(A)$. First we show that the complex Jordan blocks for μ and $\bar{\mu}$ have the same dimensions (with appropriate ordering). In fact, if there is $S \in Gl(d, \mathbb{C})$ with $J^{\mathbb{C}} = S^{-1}AS$, then the conjugate matrices satisfy

$$\overline{J^{\mathbb{C}}} = \overline{S^{-1}AS} = \bar{S}^{-1}A\bar{S}.$$

Now uniqueness of the complex Jordan normal form implies that $J^{\mathbb{C}}$ and $\overline{J^{\mathbb{C}}}$ are distinguished at most by the order of the blocks.

If J is an m-dimensional Jordan block of the form (1.2.1) corresponding to the eigenvalue μ, then $\bar{J}$ is an m-dimensional Jordan block corresponding to $\bar{\mu}$. Let $z_j = a_j + \imath b_j \in \mathbb{C}^m, j = 1, \ldots, m$, be the basis vectors corresponding to J with $a_j, b_j \in \mathbb{R}^m$. This means that $F(z_1)$ has the coordinate μ with respect to z_1 and, for $j = 2, \ldots, m$, the image $F(z_j)$ has the coordinate 1 with respect to z_{j-1} and μ with respect to z_j; all other coordinates vanish. Thus $F(z_1) = \mu z_1$ and $F(z_j) = z_{j-1} + \mu z_j$ for $j = 2, \ldots, m$. Then $F(\bar{z_1}) = A\bar{z_1} = \overline{Az_1} = \bar{\mu}\,\bar{z_1}$ and $F(\bar{z_j}) = A\bar{z_j} = \overline{Az_j} = \overline{z_{j-1}} + \bar{\mu}\,\bar{z_j}, j = 2, \ldots, m$, imply that $\bar{z_1}, \ldots, \bar{z_m}$ is a basis corresponding to $\bar{J}$. Define for $j = 1, \ldots, m$,

$$x_j = \frac{1}{\sqrt{2}}(z_j + \bar{z_j}) = \sqrt{2}\, a_j \in \mathbb{R}^m \text{ and } y_j = \frac{1}{\imath\sqrt{2}}(z_j - \bar{z_j}) = \sqrt{2}\, b_j \in \mathbb{R}^m.$$

1.2. The Real Jordan Form

Thus, up to a factor, these are the real and the imaginary parts of the generalized eigenvectors z_j. Then one computes for $j = 2, \ldots, m$,

$$\begin{aligned}
F(x_j) &= \frac{1}{\sqrt{2}} \left(F(z_j) + F(\overline{z_j}) \right) = \frac{1}{\sqrt{2}} \left(\mu z_j + \bar{\mu} \overline{z_j} + z_{j-1} + \overline{z_{j-1}} \right) \\
&= \frac{1}{2}(\mu + \bar{\mu}) \frac{z_j + \overline{z_j}}{\sqrt{2}} + \frac{\iota}{2}(\mu - \bar{\mu}) \frac{z_j - \overline{z_j}}{\iota \sqrt{2}} + x_{j-1} \\
&= (\operatorname{Re} \mu) x_j - (\operatorname{Im} \mu) y_j + x_{j-1} \\
&= \lambda x_j - \nu y_j + x_{j-1}, \\
F(y_j) &= \frac{1}{\iota \sqrt{2}} \left(F(z_j) - F(\overline{z_j}) \right) = \frac{1}{\iota \sqrt{2}} \left(\mu z_j - \bar{\mu} \overline{z_j} + z_{j-1} - \overline{z_{j-1}} \right) \\
&= \frac{1}{\iota \sqrt{2}} \left[(\mu - \bar{\mu}) \frac{1}{2}(z_j + \overline{z_j}) + \frac{1}{2}(\mu + \bar{\mu})(z_j - \overline{z_j}) \right] + y_{j-1} \\
&= (\operatorname{Im} \mu) x_j + (\operatorname{Re} \mu) y_j + y_{j-1} \\
&= \nu x_j + \lambda y_j + y_{j-1}.
\end{aligned}$$

In the case $j = 1$ the last summands to the right are skipped. We may identify the vectors $x_j, y_j \in \mathbb{R}^m \subset \mathbb{C}^m$ with elements of $\mathbb{C}^{2m}$ by adding 0's below and above, respectively. Then they form a basis of $\mathbb{C}^{2m}$, since they are obtained from $z_1, \ldots, z_m, \overline{z_1}, \ldots, \overline{z_m}$ by an invertible transformation (every element of $\mathbb{C}^{2m}$ is obtained as a linear combination of these real vectors with complex coefficients). This shows that on the corresponding subspace the map F has the matrix representation (1.2.2) with respect to this basis. Thus the matrix A is similar over $\mathbb{C}$ to a matrix with blocks given by (1.2.1) and (1.2.2). Finally, Lemma 1.2.1 shows that it is also similar over $\mathbb{R}$ to this real matrix. $\square$

Consider the basis of $\mathbb{R}^d$ corresponding to the real Jordan form $J^{\mathbb{R}}$. The real generalized eigenspace of a real eigenvalue $\mu \in \mathbb{R}$ is the subspace generated by the basis elements corresponding to the Jordan blocks for μ (This is $\ker(A - \mu I)^n$, where n is the dimension of the largest Jordan block for μ.) The real generalized eigenspace for a complex-conjugate pair of eigenvalues $\mu, \bar{\mu}$ is the subspace generated by the basis elements corresponding to the real Jordan blocks for $\mu, \bar{\mu}$ (See Exercise 1.6.9 for characterization which is independent of a basis.) Analogously we define real eigenspaces. Next we fix some notation.

Definition 1.2.4. For $A \in gl(d, \mathbb{R})$ let $\mu_k, k = 1, \ldots, r_1$, be the distinct real eigenvalues and $\mu_k, \overline{\mu_k}, k = 1, \ldots, r_2$, the distinct complex-conjugate eigenvalue pairs with $r := r_1 + 2r_2 \leq d$. The real generalized eigenspaces are denoted by $E(A, \mu_k) \subset \mathbb{R}^d$ or simply E_k for $k = 1, \ldots, r$.

Sometimes, we call the elements of the real generalized eigenspaces also generalized eigenvectors. For a real eigenvalue μ its algebraic multiplicity coincides with the dimension $n_k = \dim E_k$ of the corresponding real generalized eigenspace, for a complex-conjugate eigenvalue pair $\mu, \bar{\mu}$ the dimension of the corresponding real generalized eigenspace is given by $n_k = \dim E_k = 2m$ where m is the algebraic multiplicity of μ. It follows that $\bigoplus_{k=1}^{r} E_k = \mathbb{R}^d$, i.e., every matrix has a set of generalized eigenvectors that form a basis of $\mathbb{R}^d$.

1.3. Solution Formulas

The real Jordan form derived in the previous section will allow us to write down explicit formulas for the solution of the differential equation $\dot{x} = Ax$. This is due to the fact that for matrices in real Jordan normal form it is possible to compute the matrix exponential. The price to pay is that first the matrix A has to be transformed into real Jordan normal form. Thus we begin by noting what happens to the solutions under linear transformations.

Proposition 1.3.1. *Let $A, B \in gl(d, \mathbb{R})$ with $B = S^{-1}AS$ for some $S \in Gl(d, \mathbb{R})$. Then the solutions $y(t, y_0)$ of $\dot{y} = By$ and $x(t, x_0)$ of $\dot{x} = Ax$ are related by*
$$y(t, y_0) = S^{-1}x(t, Sy_0), t \in \mathbb{R}.$$

Proof. One computes for $t \in \mathbb{R}$,
$$y(t, y_0) = e^{Bt}y_0 = e^{S^{-1}ASt}y_0 = S^{-1}e^{At}Sy_0 = S^{-1}x(t, Sy_0). \qquad \square$$

A consequence of this proposition is that for the computation of exponentials of matrices it is sufficient to know the exponentials of Jordan form matrices and the transformation matrix. The following results for the solutions of $\dot{x} = Ax$ are obtained with the notation above from the properties of the real Jordan normal form.

Proposition 1.3.2. *Let $A \in gl(d, \mathbb{R})$ with real generalized eigenspaces $E_k = E_k(\mu_k)$ with dimensions $n_k = \dim E_k, k = 1, \ldots, r$.*

(i) Let $v_1, \ldots, v_d$ be a basis of $\mathbb{R}^d$, e.g., consisting of generalized real eigenvectors of A. If $x_0 = \sum_{i=1}^{d} \alpha_i v_i$, then $x(t, x_0) = \sum_{i=1}^{d} \alpha_i x(t, v_i)$ for all $t \in \mathbb{R}$.

(ii) Each generalized real eigenspace E_k is invariant for the linear differential equation $\dot{x} = Ax$, i.e., $x(t, x_0) \in E_k$ for all $t \in \mathbb{R}$, if $x_0 \in E_k$.

If A is a diagonal matrix, the solutions are easily obtained.

Example 1.3.3. Let $D = \text{diag}(\mu_1, \ldots, \mu_d)$ be a diagonal matrix. Then the standard basis $\mathbf{e}_1, \ldots, \mathbf{e}_d$ of $\mathbb{R}^d$ consists of eigenvectors of D and the solution

1.3. Solution Formulas

of the linear differential equation $\dot{x} = Dx$ with initial value $x_0 \in \mathbb{R}^d$ is given by

$$(1.3.1) \qquad e^{Dt} x_0 = \begin{bmatrix} e^{\mu_1 t} & & \\ & \ddots & \\ & & e^{\mu_d t} \end{bmatrix} x_0.$$

More generally, suppose that $A \in gl(d, \mathbb{R})$ is diagonalizable, i.e., there exists a transformation matrix $T \in Gl(d, \mathbb{R})$ and a diagonal matrix $D \in gl(d, \mathbb{R})$ with $A = T^{-1}DT$. Then the solution of the linear differential equation $\dot{x} = Ax$ with initial value $x_0 \in \mathbb{R}^d$ is given by $x(t, x_0) = T^{-1} e^{Dt} T x_0$, where e^{Dt} is given in (1.3.1).

By Proposition 1.3.1 the solution determined by a Jordan block J of a matrix A determines $e^{At} x_0$ in the corresponding subspace. Next we give explicit formulas for various Jordan blocks.

Example 1.3.4. Let J be a Jordan block of dimension m associated with the real eigenvalue μ of a matrix $A \in gl(d, \mathbb{R})$. Then

$$J = \begin{bmatrix} \mu & 1 & & & \\ & \ddots & \ddots & & \\ & & \ddots & \ddots & \\ & & & \ddots & 1 \\ & & & & \mu \end{bmatrix} \text{ and } e^{Jt} = e^{\mu t} \begin{bmatrix} 1 & t & \frac{t^2}{2!} & \cdots & \cdots & \frac{t^{m-1}}{(m-1)!} \\ & \ddots & \ddots & \ddots & & \vdots \\ & & \ddots & \ddots & \ddots & \vdots \\ & & & \ddots & \ddots & \frac{t^2}{2!} \\ & & & & \ddots & t \\ & & & & & 1 \end{bmatrix}.$$

In other words, for $x_0 = [x_1, \ldots, x_m]^\top$ the jth component of the solution of $\dot{x} = Jx$ reads

$$(1.3.2) \qquad x_j(t, x_0) = e^{\mu t} \sum_{k=j}^{m} \frac{t^{k-j}}{(k-j)!} x_k.$$

Example 1.3.5. Let $J = \begin{bmatrix} \lambda & -\nu \\ \nu & \lambda \end{bmatrix}$ be a real Jordan block associated with a complex eigenvalue pair $\mu = \lambda \pm i\nu$ of the matrix $A \in gl(d, \mathbb{R})$. Let x_0 be in the corresponding real eigenspace of μ. Then the solution $x(t, x_0)$ of $\dot{x} = Jx$ is given by

$$x(t, x_0) = e^{\lambda t} R(t) x_0 \text{ with } R(t) := \begin{bmatrix} \cos \nu t & -\sin \nu t \\ \sin \nu t & \cos \nu t \end{bmatrix}.$$

This can be proved by verifying directly that this yields solutions of the differential equation. Alternatively, one may use the series expansions of sin and cos.

Example 1.3.6. Let J be a real Jordan block of dimension $2m$ associated with the complex eigenvalue pair $\mu = \lambda \pm i\nu$ of a matrix $A \in gl(d, \mathbb{R})$. With

$$D := \begin{bmatrix} \lambda & -\nu \\ \nu & \lambda \end{bmatrix}, R := R(t) := \begin{bmatrix} \cos \nu t & -\sin \nu t \\ \sin \nu t & \cos \nu t \end{bmatrix}, \text{ and } I := \begin{bmatrix} 1 & 0 \\ 0 & 1 \end{bmatrix},$$

one obtains for

$$J = \begin{bmatrix} D & I & & & \\ & \ddots & \ddots & & \\ & & \ddots & \ddots & \\ & & & I \\ & & & & D \end{bmatrix} \text{ that } e^{Jt} = e^{\lambda t} \begin{bmatrix} R & tR & \frac{t^2}{2!}R & \cdots & \frac{t^{m-1}}{(m-1)!}R \\ & \ddots & \ddots & & \vdots \\ & & \ddots & \ddots & \frac{t^2}{2!}R \\ & & & \ddots & tR \\ & & & & R \end{bmatrix}$$

In other words, for $x_0 = [x_1, y_1, \ldots, x_m, y_m]^\top \in \mathbb{R}^{2m}$ the solution of $\dot{x} = Jx$ is given with $j = 1, \ldots, m$, by

(1.3.3a) $$x_j(t, x_0) = e^{\lambda t} \sum_{k=j}^{m} \frac{t^{k-j}}{(k-j)!} (x_k \cos \nu t - y_k \sin \nu t),$$

(1.3.3b) $$y_j(t, y_0) = e^{\lambda t} \sum_{k=j}^{m} \frac{t^{k-j}}{(k-j)!} (x_k \sin \nu t + y_k \cos \nu t).$$

Remark 1.3.7. Consider a Jordan block for a real eigenvalue μ as in Example 1.3.4. Then the k-dimensional subspace generated by the first k canonical basis vectors $(1, 0, \ldots, 0)^\top, \ldots, (0, \ldots 0, 1, 0, \ldots, 0)^\top, 1 \leq k \leq m$, is invariant under e^{Jt}. For a complex-conjugate eigenvalue pair $\mu, \bar{\mu}$ as in Example 1.3.6 the subspace generated by the first $2k$ canonical basis vectors is invariant.

Remark 1.3.8. Consider a solution $x(t), t \in \mathbb{R}$, of $\dot{x} = Ax$. Then the chain rule shows that the function $y(t) := x(-t), t \in \mathbb{R}$, satisfies

$$\frac{d}{dt} y(t) = \frac{d}{dt} x(-t) = -Ax(-t) = -Ay(t), t \in \mathbb{R}.$$

Hence we call $\dot{x} = -Ax$ the time-reversed equation.

1.4. Lyapunov Exponents

The asymptotic behavior of the solutions $x(t, x_0) = e^{At} x_0$ of the linear differential equation $\dot{x} = Ax$ plays a key role in understanding the connections between linear algebra and dynamical systems. For this purpose, we introduce Lyapunov exponents, a concept that is fundamental for this book, since it also applies to time-varying systems.

Definition 1.4.1. Let $x(\cdot, x_0)$ be a solution of the linear differential equation $\dot{x} = Ax$. Its Lyapunov exponent or exponential growth rate for $x_0 \neq 0$

1.4. Lyapunov Exponents

is defined as $\lambda(x_0, A) = \limsup_{t\to\infty} \frac{1}{t} \log \|x(t, x_0)\|$, where log denotes the natural logarithm and $\|\cdot\|$ is any norm in $\mathbb{R}^d$.

Let $E_k = E(\mu_k), k = 1, \ldots, r$, be the real generalized eigenspaces, denote the distinct real parts of the eigenvalues μ_k by λ_j, and order them as $\lambda_1 > \ldots > \lambda_\ell, 1 \leq \ell \leq r$. Define the Lyapunov space of λ_j as $L_j = L(\lambda_j) := \bigoplus E_k$, where the direct sum is taken over all generalized real eigenspaces associated to eigenvalues with real part equal to λ_j.

Note that
$$\mathbb{R}^d = L(\lambda_1) \oplus \ldots \oplus L(\lambda_\ell).$$

When the considered matrix A is clear from the context, we write the Lyapunov exponent as $\lambda(x_0)$. We will clarify in a moment the relation between Lyapunov exponents, eigenvalues, and Lyapunov spaces. It is helpful to look at the scalar case first: For $\dot{x} = \lambda x, \lambda \in \mathbb{R}$, the solutions are $x(t, x_0) = e^{\lambda t} x_0$. Hence the Lyapunov exponent is a limit (not just a limit superior) and

$$\lim_{t\to\pm\infty} \frac{1}{t} \log \left|e^{\lambda t} x_0\right| = \lim_{t\to\pm\infty} \frac{1}{t} \log \left(e^{\lambda t}\right) + \lim_{t\to\pm\infty} \frac{1}{t} \log |x_0| = \lambda.$$

Thus $-\lambda$ is the Lyapunov exponent of the time-reversed equation $\dot{x} = -\lambda x$. First we state that the Lyapunov exponents do not depend on the norm and that they remain constant under similarity transformations of the matrix.

Lemma 1.4.2. *(i) The Lyapunov exponent does not depend on the norm in $\mathbb{R}^d$ used in its definition.*

(ii) Let $A, B \in gl(d, \mathbb{R})$ with $B = S^{-1}AS$ for some $S \in Gl(d, \mathbb{R})$. Then the Lyapunov exponents $\lambda(x_0, A)$ and $\lambda(y_0, B)$ of the solutions $x(t, x_0)$ of $\dot{x} = Ax$ and $y(t, y_0)$ of $\dot{y} = By$, respectively, are related by

$$\lambda(y_0, B) = \lambda(Sy_0, A).$$

Proof. (i) This is left as Exercise 1.6.10. (ii) Using Proposition 1.3.1 we find

$$\lambda(y_0, B) = \limsup_{t\to\infty} \frac{1}{t} \log \|y(t, y_0)\| = \limsup_{t\to\infty} \frac{1}{t} \log \left\|S^{-1} x(t, Sy_0)\right\|$$

$$\leq \limsup_{t\to\infty} \frac{1}{t} \log \left\|S^{-1}\right\| + \limsup_{t\to\infty} \frac{1}{t} \log \|x(t, Sy_0)\| = \lambda(Sy_0, A).$$

Writing $x_0 = S\left(S^{-1} x_0\right)$, one obtains also the converse inequality. □

The following result clarifies the relationship between the Lyapunov exponents of $\dot{x} = Ax$ and the real parts of the eigenvalues of A. It is the main result of this chapter concerning systems in continuous time and explains the relation between Lyapunov exponents for $\dot{x} = Ax$ and the matrix A, hence establishes a first relation between dynamical systems and linear algebra.

Theorem 1.4.3. *For $\dot{x} = Ax, A \in gl(d, \mathbb{R})$, there are exactly ℓ Lyapunov exponents $\lambda(x_0)$, the distinct real parts λ_j of the eigenvalues of A. For a solution $x(\cdot, x_0)$ (with $x_0 \neq 0$) one has $\lambda(x_0) = \lim_{t \to \pm\infty} \frac{1}{t} \log \|x(t, x_0)\| = \lambda_j$ if and only if $x_0 \in L(\lambda_j)$.*

Proof. Using Lemma 1.4.2 we may assume that A is given in real Jordan form. Then the assertions of the theorem can be derived from the solution formulas in the generalized eigenspaces. For a Jordan block for a real eigenvalue $\mu = \lambda_j$, formula (1.3.2) yields for every component $x_i(t, x_0)$ of the solution $x(t, x_0)$ and $|t| \geq 1$,

$$\log |x_i(t, x_0)| = \mu t + \log \left| \sum_{k=i}^{m} \frac{t^{k-i}}{(k-i)!} x_k \right| \leq \mu t + m \log |t| + \log \max_k |x_k|,$$

$$\log |x_i(t, x_0)| \geq \mu t - m \log |t| - \log \max_k |x_k|.$$

Since $\frac{1}{t} \log |t| \to 0$ for $t \to \pm\infty$, it follows that $\lim_{t \to \pm\infty} \frac{1}{t} \log |x_i(t, x_0)| = \mu = \lambda_j$. With a bit more writing effort, one sees that this is also valid for every component of x_0 in a subspace for a Jordan block corresponding to a complex-conjugate pair of eigenvalues with real part equal to λ_j: By (1.3.3) one obtains the product of $e^{\lambda_j t}$ with a polynomial in t and sin and cos functions. The logarithm of the second factor, divided by t, converges to 0 for $t \to \pm\infty$. The Lyapunov space $L(\lambda_j)$ is obtained as the direct sum of such subspaces and every component of a corresponding solution has exponential growth rate λ_j. Since we may take the maximum-norm on $\mathbb{R}^d$, this shows that every solution starting in $L(\lambda_j)$ has exponential growth rate λ_j for $t \to \pm\infty$.

The only if part will follow from Theorem 1.4.4. $\square$

We emphasize that a characterization of the Lyapunov spaces $L(\lambda_j)$ via the exponential growth rates of solutions needs the limits for $t \to +\infty$ and for $t \to -\infty$. In other words, if one only considers the exponential growth rate for positive time, i.e., for time tending to $+\infty$, one cannot characterize the Lyapunov spaces. Nevertheless, we can extend the result of Theorem 1.4.3 by describing the exponential growth rates for positive and negative times and arbitrary initial points.

Theorem 1.4.4. *Consider the linear autonomous differential equation $\dot{x} = Ax$ with $A \in gl(d, \mathbb{R})$ and corresponding Lyapunov spaces $L_j := L(\lambda_j), \lambda_1 > \ldots > \lambda_\ell$. Let $V_{\ell+1} = W_0 := \{0\}$ and for $j = 1, \ldots, \ell$ define*

(1.4.1) $\qquad V_j := L_\ell \oplus \ldots \oplus L_j$ *and* $W_j := L_j \oplus \ldots \oplus L_1.$

1.4. Lyapunov Exponents

Then a solution $x(\cdot, x_0)$ with $x_0 \neq 0$ satisfies

$$\lim_{t \to +\infty} \frac{1}{t} \log \|x(t, x_0)\| = \lambda_j \text{ if and only if } x_0 \in V_j \setminus V_{j+1},$$

$$\lim_{t \to -\infty} \frac{1}{|t|} \log \|x(t, x_0)\| = -\lambda_j \text{ if and only if } x_0 \in W_j \setminus W_{j-1}.$$

In particular, $\lim_{t \to \pm\infty} \frac{1}{|t|} \log \|x(t, x_0)\| = \pm\lambda_j$ if and only if $x_0 \in L(\lambda_j) = V_j \cap W_j$.

Proof. This follows using the arguments in the proof of Theorem 1.4.3: For every j one has $V_{j+1} \subset V_j$ and for a point $x_0 \in V_j \setminus V_{j+1}$ the component in $L(\lambda_j)$ is nonzero and hence the exponential term $e^{\lambda_j t}$ determines the exponential growth rate for $t \to \infty$. By definition,

$$L_j = V_j \cap W_j, j = 1, \ldots, \ell.$$

Note further that $-\lambda_\ell > \ldots > -\lambda_1$ are the real parts of the eigenvalues $-\mu$ of $-A$, where μ are the eigenvalues of A. $\square$

Strictly increasing sequences of subspaces as in (1.4.1),

$$\{0\} = V_{\ell+1} \subset V_\ell \subset \ldots \subset V_1 = \mathbb{R}^d, \{0\} = W_0 \subset W_1 \subset \ldots \subset W_\ell = \mathbb{R}^d,$$

are called flags of subspaces. By definition $V_\ell = L_\ell$ is the Lyapunov space corresponding to the smallest Lyapunov exponent, and V_j is the sum of the Lyapunov spaces corresponding to the j smallest Lyapunov exponents.

Stability

Using the concept of Lyapunov exponents and Theorem 1.4.4 we can describe the behavior of solutions of linear differential equations $\dot{x} = Ax$ as time tends to infinity. By definition, a solution with negative Lyapunov exponent tends to the origin and a solution with positive Lyapunov exponent becomes unbounded (the converse need not be true, as we will see in a moment).

It is appropriate to formulate the relevant stability concepts not just for linear differential equations, but for general nonlinear differential equations of the form $\dot{x} = f(x)$ where $f : \mathbb{R}^d \to \mathbb{R}^d$. We assume that there are unique solutions $\varphi(t, x_0), t \geq 0$, of every initial value problem of the form

(1.4.2) $$\dot{x} = f(x), \ x(0) = x_0 \in \mathbb{R}^d.$$

Remark 1.4.5. Suppose that f is a locally Lipschitz continuous vector field, i.e., for every $x \in \mathbb{R}^d$ there are an ε-neighborhood $N(x, \varepsilon) = \{y \in \mathbb{R}^d \mid \|y - x\| < \varepsilon\}$ with $\varepsilon > 0$ and a Lipschitz constant $L > 0$ such that

$$\|f(y) - f(x)\| \leq L \|y - x\| \text{ for all } y \in N(x, \varepsilon).$$

Then the initial value problems above have unique solutions which, in general, are only defined on open intervals containing $t = 0$ (we will see an example for this in the beginning of Chapter 4). If the solutions remain bounded on bounded intervals, one can show that they are defined on $\mathbb{R}$.

Various stability concepts characterize the asymptotic behavior of the solutions $\varphi(t, x_0)$ for $t \to \pm\infty$.

Definition 1.4.6. Let $x^* \in \mathbb{R}^d$ be a fixed point of the differential equation $\dot{x} = f(x)$, i.e., the solution $\varphi(t, x^*)$ with initial value $\varphi(0, x^*) = x^*$ satisfies $\varphi(t, x^*) \equiv x^*$. Then the point x^* is called:

stable if for all $\varepsilon > 0$ there exists a $\delta > 0$ such that $\varphi(t, x_0) \in N(x^*, \varepsilon)$ for all $t \geq 0$ whenever $x_0 \in N(x^*, \delta)$;

asymptotically stable if it is stable and there exists a $\gamma > 0$ such that $\lim_{t \to \infty} \varphi(t, x_0) = x^*$ whenever $x_0 \in N(x^*, \gamma)$;

exponentially stable if there exist $\alpha \geq 1$ and $\beta, \eta > 0$ such that for all $x_0 \in N(x^*, \eta)$ the solution satisfies $\|\varphi(t, x_0) - x^*\| \leq \alpha \|x_0 - x^*\| e^{-\beta t}$ for all $t \geq 0$;

unstable if it is not stable.

It is immediate to see that a point x^* is a fixed point of $\dot{x} = Ax$ if and only if $x^* \in \ker A$, the kernel of A. The origin $0 \in \mathbb{R}^d$ is a fixed point of any linear differential equation.

Definition 1.4.7. The stable, center, and unstable subspaces associated with the matrix $A \in gl(d, \mathbb{R})$ are defined as

$$L^- = \bigoplus_{\lambda_j < 0} L(\lambda_j),\, L^0 = L(0), \text{ and } L^+ = \bigoplus_{\lambda_j > 0} L(\lambda_j).$$

The following theorem characterizes asymptotic and exponential stability of the origin for $\dot{x} = Ax$ in terms of the eigenvalues of A.

Theorem 1.4.8. For a linear differential equation $\dot{x} = Ax$ in $\mathbb{R}^d$ the following statements are equivalent:

(i) The origin $0 \in \mathbb{R}^d$ is asymptotically stable.

(ii) The origin $0 \in \mathbb{R}^d$ is exponentially stable.

(iii) All Lyapunov exponents (i.e., all real parts of the eigenvalues) are negative.

(iv) The stable subspace L^- satisfies $L^- = \mathbb{R}^d$.

Proof. First observe that by linearity, asymptotic and exponential stability of the fixed point $x^* = 0 \in \mathbb{R}^d$ in a neighborhood $N(x^*, \gamma)$ implies asymptotic and exponential stability, respectively, for all points $x_0 \in \mathbb{R}^d$. In fact,

1.4. Lyapunov Exponents

suppose exponential stability holds in $N(0, \gamma)$. Then for $x_0 \in \mathbb{R}^d$ the point $x_1 := \frac{\gamma}{2} \frac{x_0}{\|x_0\|} \in N(0, \gamma)$, and hence

$$\|\varphi(t, x_0)\| = \|e^{At} x_0\| = \left\| \frac{2\|x_0\|}{\gamma} e^{At} \frac{\gamma}{2} \frac{x_0}{\|x_0\|} \right\| = \frac{2\|x_0\|}{\gamma} \|\varphi(t, x_1)\|$$
$$\leq \frac{2\|x_0\|}{\gamma} \alpha \|x_1\| e^{-\beta t} = \alpha \|x_0\| e^{-\beta t},$$

and analogously for asymptotic stability. Clearly, properties (ii), (iii) and (iv) are equivalent and imply (i). Conversely, suppose that one of the Lyapunov exponents is nonnegative. Thus one of the eigenvalues, say μ, has nonnegative real part. If μ is real, i.e., $\mu \geq 0$, the solution corresponding to an eigenvector in $\mathbb{R}^d$ does not tend to the origin as time tends to infinity (if $\mu = 0$, all corresponding solutions are fixed points.) If μ is not real, consider the solution (1.3.3) in the two-dimensional eigenspace corresponding to the complex eigenvalue pair $\mu, \bar{\mu}$. This solution also does not tend to the origin as time tends to infinity. Hence (i) implies (iii). $\square$

Remark 1.4.9. In particular, the proof above shows that for linear systems the existence of a neighborhood $N(0, \gamma)$ with $\lim_{t \to \infty} \varphi(t, x_0) = x^*$ whenever $x_0 \in N(x^*, \gamma)$ implies that one may replace $N(0, \gamma)$ by $\mathbb{R}^d$. In this sense 'local stability = global stability' here.

It remains to characterize stability of the origin. If for an eigenvalue μ all complex Jordan blocks are one-dimensional, i.e., a complete set of eigenvectors exists, it is called semisimple. Equivalently, the corresponding real Jordan blocks are one-dimensional if μ is real, and two-dimensional if $\mu, \bar{\mu} \in \mathbb{C} \setminus \mathbb{R}$.

Theorem 1.4.10. *The origin $0 \in \mathbb{R}^d$ is stable for the linear differential equation $\dot{x} = Ax$ if and only if all Lyapunov exponents (i.e., all real parts of eigenvalues) are nonpositive and the eigenvalues with real part zero are semisimple.*

Proof. We only have to discuss eigenvalues with zero real part. Suppose first that $\lambda = 0 \in \text{spec}(A)$. Then the solution formula (1.3.2) shows that an eigenvector yields a stable solution. For a Jordan block of size $m > 1$, consider $y_0 = [y_1, \ldots, y_m]^\top = [0, \ldots, 0, 1]^\top$. Then stability does not hold, since

$$y_1(t, y_0) = e^{\lambda t} \sum_{k=1}^{m} \frac{t^{k-1}}{(k-1)!} y_k = \frac{t^{m-1}}{(m-1)!} \to \infty \text{ for } t \to \infty.$$

Similarly, one argues for a complex-conjugate pair of eigenvalues. $\square$

This result shows, in particular, that Lyapunov exponents alone do not allow us to characterize stability for linear systems. They are related to exponential or, equivalently, to asymptotic stability, and they do not detect polynomial instabilities.

The following example is a damped linear oscillator.

Example 1.4.11. The second order differential equation

$$\ddot{x} + 2b\dot{x} + x = 0$$

is equivalent to the system

$$\begin{bmatrix} \dot{x}_1 \\ \dot{x}_2 \end{bmatrix} = \begin{bmatrix} 0 & 1 \\ -1 & -2b \end{bmatrix} \begin{bmatrix} x_1 \\ x_2 \end{bmatrix}.$$

The eigenvalues are $\mu_{1,2} = -b \pm \sqrt{b^2 - 1}$. For $b > 0$ the real parts of the eigenvalues are negative and hence the stable subspace coincides with $\mathbb{R}^2$. Hence b is called a damping parameter. Note also that for every solution $x(\cdot)$ the function $y(t) := e^{bt}x(t), t \in \mathbb{R}$, is a solution of the equation $\ddot{y} + (1-b^2)y = 0$.

1.5. The Discrete-Time Case: Linear Difference Equations

This section discusses solution formulas and stability properties, in particular, Lyapunov exponents for autonomous linear difference equations. In this discrete-time case, the time domain $\mathbb{R}$ is replaced by $\mathbb{N}_0$ or $\mathbb{Z}$, and one obtains equations of the form

(1.5.1) $$x_{n+1} = Ax_n,$$

where $A \in gl(d, \mathbb{R})$. By induction, one sees that for positive time the solutions $\varphi(n, x_0)$ are given by

(1.5.2) $$\varphi(n, x_0) = x_n = A^n x_0, n \in \mathbb{N}.$$

If A is invertible, i.e., if 0 is not an eigenvalue of A, this formula holds for all $n \in \mathbb{Z}$, and $A^n, n \in \mathbb{Z}$, forms a fundamental solution. It is an important feature of the discrete time case that solutions may not exist for negative times. In fact, if A is not of full rank, only for points in the range of A there is x_{-1} with $x_0 = Ax_{-1}$ and the point x_{-1} is not unique. For simplicity, we will restrict the discussion in the present section and in the rest of this book to the invertible case $A \in Gl(d, \mathbb{R})$ where solutions are defined on $\mathbb{Z}$.

Existence of solutions and continuous dependence on initial values is clear from (1.5.2). Similarly, the result on the solution space, Theorem 1.1.1, is immediate: If $A \in Gl(d, \mathbb{R})$, the set of solutions $(\varphi(n, x_0))_{n \in \mathbb{Z}}$ forms a d-dimensional linear space (with pointwise addition and multiplication by scalars). When it comes to the question of solution formulas, the real

1.5. The Discrete-Time Case: Linear Difference Equations

Jordan form presented in Theorem 1.2.3 provides the following analogues to Propositions 1.3.1 and 1.3.2.

Theorem 1.5.1. Let $A \in Gl(d, \mathbb{R})$ with real generalized eigenspaces $E_k = E(\mu_k) k = 1, \ldots, r$, for the eigenvalues $\mu_1, \ldots, \mu_r \in \mathbb{C}$ with $n_k = \dim E_k$.

(i) If $A = T^{-1} J^{\mathbb{R}} T$, then $A^n = T^{-1} \left(J^{\mathbb{R}}\right)^n T$, i.e., for the computation of powers of matrices it is sufficient to know the powers of Jordan form matrices.

(ii) Let $v_1, \ldots, v_d$ be a basis of $\mathbb{R}^d$, e.g., consisting of generalized real eigenvectors of A. If $x_0 = \sum_{i=1}^{d} \alpha_i v_i$, then $\varphi(n, x_0) = \sum_{i=1}^{d} \alpha_i \varphi(n, v_i)$ for all $n \in \mathbb{Z}$.

(iii) Each real generalized eigenspace E_k is invariant for the linear difference equation $x_n = A^n x_0$, i.e., for $x_0 \in E_k$ it holds that the corresponding solution satisfies $\varphi(n, x_0) \in E_k$ for all $n \in \mathbb{Z}$.

This theorem shows that for explicit solution formulas iterates of Jordan blocks have to be computed. We consider the cases of real and complex eigenvalues.

Example 1.5.2. Let J be a Jordan block of dimension m associated with the real eigenvalue μ of a matrix $A \in gl(d, \mathbb{R})$. Then

$$J = \begin{bmatrix} \mu & 1 & & \\ & \ddots & \ddots & \\ & & \ddots & 1 \\ 0 & & & \mu \end{bmatrix} = \mu \begin{bmatrix} 1 & 0 & & \\ & \ddots & \ddots & \\ & & \ddots & 0 \\ & & & 1 \end{bmatrix} + \begin{bmatrix} 0 & 1 & & \\ & \ddots & \ddots & \\ & & \ddots & 1 \\ & & & 0 \end{bmatrix}.$$

Thus J has the form $J = \mu I + N$ with $N^m = 0$, hence N is a nilpotent matrix. Then one computes for $n \in \mathbb{N}$,

$$J^n = (\mu I + N)^n = \sum_{i=0}^{n} \binom{n}{i} \mu^{n-i} N^i.$$

Note that for $n \geq m - 1$,

$$(1.5.3) \qquad \varphi(n, y_0) = J^n y_0 = \mu^{n+1-m} \sum_{i=0}^{m-1} \binom{n}{i} \mu^{m-1-i} N^i y_0.$$

For $y_0 = [y_1, \ldots, y_m]^\top$ the j-th component of the solution $\varphi(n, y_0)$ of $y_{n+1} = J y_n$ reads

$$\varphi_j(n, y_0) = \binom{n}{0} \mu^n y_j + \binom{n}{1} \mu^{n-1} y_{j+1} + \ldots + \binom{n}{m-j} \mu^{n-(m-j)} y_n.$$

Example 1.5.3. Let J be a real Jordan block of dimension $2m$ associated with the pair of complex eigenvalues $\mu = \alpha \pm \imath\beta$ of a matrix $A \in gl(d, \mathbb{R})$. With $D = \begin{bmatrix} \alpha & -\beta \\ \beta & \alpha \end{bmatrix}$ and $I_2 = \begin{bmatrix} 1 & 0 \\ 0 & 1 \end{bmatrix}$, one obtains that J has the form

$$\begin{bmatrix} D & I_2 & & \\ & \ddots & \ddots & \\ & & \ddots & I_2 \\ & & & D \end{bmatrix} = \begin{bmatrix} D & 0 & & \\ & \ddots & \ddots & \\ & & \ddots & 0 \\ & & & D \end{bmatrix} + \begin{bmatrix} 0 & I_2 & & \\ & \ddots & \ddots & \\ & & \ddots & I_2 \\ & & & 0 \end{bmatrix}.$$

Thus J is the sum of a block-diagonal matrix $\tilde{D}$ with blocks D and a nilpotent matrix N with $N^{m-1} = 0$. Then, observing that $\tilde{D}$ and N commute, i.e., $\tilde{D}N = N\tilde{D}$, one computes for $n \geq m-1$,

$$J^n = (\tilde{D} + N)^n = \sum_{i=0}^{m-2} \binom{n}{i} \tilde{D}^{n-i} N^i$$

$$= \binom{n}{0} \tilde{D}^n I + \binom{n}{1} \tilde{D}^{n-1} N + \ldots + \binom{n}{m-2} \tilde{D}^{n-(m-2)} N^{m-1}.$$

Note that $|\mu| = \sqrt{\alpha^2 + \beta^2}$, hence with $\varphi \in [0, 2\pi)$ determined by $\cos\varphi = \frac{\alpha}{\sqrt{\alpha^2 + \beta^2}}$ (thus $\mu = |\mu| e^{\imath\varphi}$) one can write the matrix D as

$$D = \begin{bmatrix} \alpha & -\beta \\ \beta & \alpha \end{bmatrix} = |\mu| R \text{ with } R := \begin{bmatrix} \cos\varphi & -\sin\varphi \\ \sin\varphi & \cos\varphi \end{bmatrix}.$$

Thus D describes a rotation by the angle φ followed by multiplication by $|\mu|$. Write $\tilde{R}$ for the block diagonal matrix with blocks R. One obtains for $n \geq m-2$ the solution formula

$$\varphi(n, y_0) = J^n y_0 = \sum_{i=0}^{m-2} \binom{n}{i} |\mu|^{n-i} \tilde{R}^{n-i} N^i y_0$$

(1.5.4)
$$= |\mu|^{n+2-m} \sum_{i=0}^{m-2} \binom{n}{i} |\mu|^{m-2-i} \tilde{R}^{n-i} N^i y_0.$$

As in the continuous-time case, the asymptotic behavior of the solutions $\varphi(n, x_0) = A^n x_0$ of the linear difference equation $x_{n+1} = A x_n$ plays a key role in understanding the connections between linear algebra and dynamical systems. For this purpose, we introduce Lyapunov exponents.

Definition 1.5.4. Let $\varphi(n, x_0) = A^n x_0, n \in \mathbb{Z}$, be a solution of the linear difference equation $x_{n+1} = A x_n$. Its Lyapunov exponent or exponential growth rate for $x_0 \neq 0$ is defined as $\lambda(x_0) = \limsup_{n \to \infty} \frac{1}{n} \log \|\varphi(n, x_0)\|$, where log denotes the natural logarithm and $\|\cdot\|$ is any norm in $\mathbb{R}^d$.

1.5. The Discrete-Time Case: Linear Difference Equations

Let $E_k = E(\mu_k), k = 1, \ldots, r$, be the real generalized eigenspaces, and suppose that there are $1 \leq \ell \leq d$ distinct moduli $|\mu_k|$ of the eigenvalues μ_k. Let $\lambda_1 > \ldots > \lambda_\ell$ be the logarithms of these moduli and order them as $\lambda_1 > \ldots > \lambda_\ell, 1 \leq \ell \leq r$. Define the Lyapunov space of λ_j as $L_j = L(\lambda_j) := \bigoplus E_k$, where the direct sum is taken over all real generalized eigenspaces E_k associated to eigenvalues μ_k with $\lambda_j = \log |\mu_k|$. Note that $\bigoplus_{j=1}^\ell L(\lambda_j) = \mathbb{R}^d$.

A matrix A is invertible if and only if 0 is not an eigenvalue. Hence $\log |\mu|$ is finite for every eigenvalue of $A \in Gl(d, \mathbb{R})$. As in the continuous time case, it is helpful to look at the scalar case first: For $x_{n+1} = \lambda x_n, 0 \neq \lambda \in \mathbb{R}$, the solutions are $\varphi(n, x_0) = \lambda^n x_0$. Hence the Lyapunov exponent is a limit (not just a limit superior) and is given by

$$\lim_{n \to \pm\infty} \frac{1}{t} \log |\lambda^n x_0| = \lim_{n \to \pm\infty} \frac{1}{n} \log (|\lambda|^n) + \lim_{n \to \pm\infty} \frac{1}{n} \log |x_0| = |\lambda|.$$

Furthermore, the Lyapunov exponent of the time-reversed equation $x_{n+1} = \lambda^{-1} x_n$ is $|\lambda|^{-1}$.

Remark 1.5.5. Consider a solution $x_n, n \in \mathbb{Z}$, of $x_{n+1} = Ax_n$. Then the function $y_n := x_{-n}, n \in \mathbb{Z}$, satisfies

$$y_{n+1} = x_{-(n+1)} = A^{-1} x_{-n} = A^{-1} y_n, n \in \mathbb{Z}.$$

Hence we call $x_{n+1} = A^{-1} x_n, n \in \mathbb{Z}$, the time-reversed equation.

One finds that the Lyapunov exponents do not depend on the norm and that they remain unchanged under similarity transformations of the matrix. For arbitrary dimension d, the following result clarifies the relationship between the Lyapunov exponents of $x_{n+1} = Ax_n$ and the moduli of the eigenvalues of $A \in Gl(d, \mathbb{R})$. Furthermore, it shows that they are associated with the decomposition of the state space into the Lyapunov spaces.

Theorem 1.5.6. *Consider the linear difference equation $x_{n+1} = Ax_n$ with $A \in Gl(d, \mathbb{R})$. Then the state space $\mathbb{R}^d$ can be decomposed into the Lyapunov spaces*

$$\mathbb{R}^d = L(\lambda_1) \oplus \ldots \oplus L(\lambda_\ell),$$

and the Lyapunov exponents $\lambda(x_0), x_0 \in \mathbb{R}^d$, are given by the logarithms λ_j of the moduli of the eigenvalues of A. For a solution $\varphi(\cdot, x_0)$ (with $x_0 \neq 0$) one has $\lambda(x_0) = \lim_{n \to \pm\infty} \frac{1}{n} \log \|\varphi(n, x_0)\| = \lambda_j$ if and only if $x_0 \in L(\lambda_j)$.

Proof. For any matrix A there is a matrix $T \in Gl(d, \mathbb{R})$ such that $A = T^{-1} J^\mathbb{R} T$, where $J^\mathbb{R}$ is the real Jordan form of A. Hence we may assume that A is given in real Jordan form. Then the assertions of the theorem can be derived from the solution formulas in the generalized eigenspaces. For example, formula (1.5.3) yields for $n \geq m-1$,

(1.5.5)
$$\frac{1}{n}\log\|\varphi(n,x_0)\| = \frac{n+1-m}{n}\log|\mu| + \frac{1}{n}\log\left\|\sum_{i=0}^{m-1}\binom{n}{i}\mu^{m-1-i}N^i x_0\right\|.$$

One estimates
$$\left|\log\left\|\sum_{i=0}^{m-1}\binom{n}{i}\mu^{m-1-i}N^i x_0\right\|\right|$$
$$\leq \log m + \max_i \log\binom{n}{i} + \max_i \log\left(|\mu|^{m-1-i}\|N^i x_0\|\right),$$

where the maxima are taken over $i = 0, 1, \ldots, m-1$. For every i and $n \to \infty$ one can further estimate

$$\frac{1}{n}\log\binom{n}{i} = \frac{1}{n}\log\frac{n(n-1)\ldots(n-i+1)}{i!}$$
$$\leq \max_{i\in\{0,1,\ldots,m-1\}}\frac{i}{n}(\log n - \log i!) \to 0.$$

Hence taking the limit for $n \to \infty$ in (1.5.5), one finds that the Lyapunov exponent equals $\log|\mu|$ for every initial value x_0 in this generalized eigenspace. The same arguments work for complex conjugate pairs of eigenvalues and in the sum of the generalized eigenspaces corresponding to eigenvalues with equal moduli. Finally, for initial values in the sum of generalized eigenspaces for eigenvalues with different moduli, the largest modulus determines the Lyapunov exponent. Similarly, one argues for $n \to -\infty$.

The 'only if' part will follow from Theorem 1.5.8. □

Remark 1.5.7. As in continuous time, we note that a characterization of the Lyapunov spaces by the dynamic behavior of solutions needs the limits for $n \to +\infty$ and for $n \to -\infty$. For example, every initial value $x_0 = x_1 + x_2$ with $x_i \in L(\lambda_i), \lambda_1 < \lambda_2$ and $x_2 \neq 0$, has Lyapunov exponent $\lambda(x_0) = \lambda_2$.

We can sharpen the results of Theorem 1.5.6 in order to characterize the exponential growth rates for positive and negative times and arbitrary initial points. This involves flags of subspaces.

Theorem 1.5.8. *Consider the linear difference equation $x_{n+1} = Ax_n$ with $A \in Gl(d,\mathbb{R})$ and corresponding Lyapunov spaces $L_j := L(\lambda_j), \lambda_1 > \ldots > \lambda_\ell$. Let $V_{\ell+1} = W_0 := \{0\}$ and for $j = 1, \ldots, \ell$ define*

$$V_j := L_\ell \oplus \ldots \oplus L_j \text{ and } W_j := L_j \oplus \ldots \oplus L_1.$$

1.5. The Discrete-Time Case: Linear Difference Equations

Then a solution $\varphi(\cdot, x_0)$ with $x_0 \neq 0$ satisfies

$$\lim_{n \to +\infty} \frac{1}{n} \log \|\varphi(n, x_0)\| = \lambda_j \text{ if and only if } x_0 \in V_j \setminus V_{j+1},$$

$$\lim_{n \to -\infty} \frac{1}{|n|} \log \|\varphi(n, x_0)\| = -\lambda_j \text{ if and only if } x_0 \in W_j \setminus W_{j-1}.$$

In particular, $\lim_{n \to \pm\infty} \frac{1}{|n|} \log \|\varphi(n, x_0)\| = \pm\lambda_j$ if and only if $x_0 \in L(\lambda_j) = V_j \cap W_j$.

Proof. This follows using the solution formulas and the arguments given in the proof of Theorem 1.5.6. Here the time-reversed equation has the form

$$x_{n+1} = A^{-1} x_n, n \in \mathbb{Z}.$$

The eigenvalues of A^{-1} are given by the inverses of the eigenvalues μ of A and the Lyapunov exponents are $-\lambda_\ell > \ldots > -\lambda_1$, since

$$\log\left(|\mu^{-1}|\right) = -\log|\mu|.$$

The generalized eigenspaces and hence the Lyapunov spaces $L(-\lambda_j)$ coincide with the corresponding generalized eigenspaces and Lyapunov spaces $L(\lambda_j)$, respectively. $\square$

Stability

Using the concept of Lyapunov exponents and Theorem 1.5.8 we can describe the behavior of solutions of linear difference equations $x_{n+1} = Ax_n$ as time tends to infinity. By definition, a solution with negative Lyapunov exponent tends to the origin and a solution with positive Lyapunov exponent becomes unbounded.

It is appropriate to formulate the relevant stability concepts not just for linear differential equations, but for general nonlinear difference equations of the form

$$x_{n+1} = f(x_n),$$

where $f : \mathbb{R}^d \to \mathbb{R}^d$. In general, the solutions $\varphi(n, x_0)$ are only defined for $n \geq 0$. Various stability concepts characterize the asymptotic behavior of $\varphi(n, x_0)$ for $n \to \infty$.

Definition 1.5.9. Let $x^* \in \mathbb{R}^d$ be a fixed point of the difference equation $x_{n+1} = f(x_n)$, i.e., the solution $\varphi(n, x^*)$ with initial value $\varphi(0, x^*) = x^*$ satisfies $\varphi(n, x^*) \equiv x^*$. Then the point x^* is called:

stable if for all $\varepsilon > 0$ there exists a $\delta > 0$ such that $\varphi(n, x_0) \in N(x^*, \varepsilon)$ for all $n \in \mathbb{N}$ whenever $x_0 \in N(x^*, \delta)$;

asymptotically stable if it is stable and there exists a $\gamma > 0$ such that $\lim_{n \to \infty} \varphi(n, x_0) = x^*$ whenever $x_0 \in N(x^*, \gamma)$;

exponentially stable if there exist $\alpha \geq 1, \eta > 0$ and $\beta \in (0,1)$ such that for all $x_0 \in N(x^*, \eta)$ the solution satisfies $\|\varphi(n, x_0) - x^*\| \leq \alpha \beta^n \|x_0 - x^*\|$ for all $n \in \mathbb{N}$;

unstable if it is not stable.

The origin $0 \in \mathbb{R}^d$ is a fixed point of any linear difference equation. The following definition referring to the Lyapunov spaces will be useful.

Definition 1.5.10. The stable, center, and unstable subspaces associated with the matrix $A \in Gl(d, \mathbb{R})$ are defined as

$$L^- = \bigoplus_{\lambda_j < 0} L(\lambda_j), L^0 = L(0), \text{ and } L^+ = \bigoplus_{\lambda_j > 0} L(\lambda_j).$$

One obtains a characterization of asymptotic and exponential stability of the origin for $x_{n+1} = Ax_n$ in terms of the eigenvalues of A.

Theorem 1.5.11. *For a linear difference equation $x_{n+1} = Ax_n$ in $\mathbb{R}^d$ the following statements are equivalent:*

(i) The origin $0 \in \mathbb{R}^d$ is asymptotically stable.

(ii) The origin $0 \in \mathbb{R}^d$ is exponentially stable.

(iii) All Lyapunov exponents are negative (i.e., all moduli of the eigenvalues are less than 1.)

(iv) The stable subspace L^- satisfies $L^- = \mathbb{R}^d$.

Proof. The proof is completely analogous to the proof for differential equations; see Theorem 1.4.8. $\square$

It remains to characterize stability of the origin.

Theorem 1.5.12. *The origin $0 \in \mathbb{R}^d$ is stable for the linear difference equation $x_{n+1} = Ax_n$ if and only if all Lyapunov exponents are nonpositive and the eigenvalues with modulus equal to 1 are semisimple.*

Proof. The proof is completely analogous to the proof for differential equations; see Theorem 1.4.10. $\square$

Again we see that Lyapunov exponents alone do not allow us to characterize stability. They are related to exponential stability or, equivalently, to asymptotic stability and they do not detect polynomial instabilities.

1.6. Exercises

Exercise 1.6.1. One can draw the solutions $x(t, x_0) \in \mathbb{R}^2$ of $\dot{x} = Ax$ with $A \in gl(2, \mathbb{R})$ either componentwise as functions of $t \in \mathbb{R}$ or as single parametrized curves in $\mathbb{R}^2$. The latter representation of all solutions is an example of a phase portrait of a differential equation. Observe that by uniqueness of solutions, these parametrized curves cannot intersect. Describe the solutions as functions of time $t \in \mathbb{R}$ and the phase portraits in the plane $\mathbb{R}^2$ of $\dot{x} = Ax$ for

$$A = \begin{bmatrix} -1 & 0 \\ 0 & -3 \end{bmatrix}, \quad A = \begin{bmatrix} 1 & 0 \\ 0 & -1 \end{bmatrix}, \quad A = \begin{bmatrix} 3 & 0 \\ 0 & 1 \end{bmatrix}.$$

What are the relations between the corresponding solutions?

Exercise 1.6.2. Describe the solutions as functions of time $t \in \mathbb{R}$ and the phase portraits in the plane $\mathbb{R}^2$ of $\dot{x} = Ax$ for

$$A = \begin{bmatrix} 1 & 1 \\ 0 & 1 \end{bmatrix}, \quad A = \begin{bmatrix} 0 & 1 \\ 0 & 0 \end{bmatrix}, \quad A = \begin{bmatrix} -1 & 1 \\ 0 & -1 \end{bmatrix}.$$

Exercise 1.6.3. Describe the solutions as functions of time $t \in \mathbb{R}$ and the phase portraits in the plane $\mathbb{R}^2$ of $\dot{x} = Ax$ for

$$A = \begin{bmatrix} 1 & -1 \\ 1 & 1 \end{bmatrix}, \quad A = \begin{bmatrix} 0 & -1 \\ 1 & 0 \end{bmatrix}, \quad A = \begin{bmatrix} -1 & -1 \\ 1 & -1 \end{bmatrix}.$$

Exercise 1.6.4. Describe all possible phase portraits in $\mathbb{R}^2$ taking into account the possible Jordan blocks.

Exercise 1.6.5. Compute the solutions of $\dot{x} = Ax$ for

$$A = \begin{bmatrix} 1 & 1 & 0 \\ 0 & 1 & 0 \\ 0 & 0 & 2 \end{bmatrix}.$$

Exercise 1.6.6. Determine the stable, center, and unstable subspaces of $\dot{x} = Ax$ for

$$A = \begin{bmatrix} 3 & -4 & 1 \\ 1 & 0 & -1 \\ -1 & 4 & -3 \end{bmatrix}.$$

Exercise 1.6.7. Show that for a matrix $A \in gl(d, \mathbb{R})$ and $T > 0$ the spectrum $\mathrm{spec}(e^{AT})$ is given by $\{e^{\lambda T} \mid \lambda \in \mathrm{spec}(A)\}$. Show also that the maximal dimension of a Jordan block for $\mu \in \mathrm{spec}(e^{AT})$ is given by the maximal dimension of a Jordan block of an eigenvalue $\lambda \in \mathrm{spec}(A)$ with $e^{\lambda T} = \mu$. Take into account that $e^{\nu T} = e^{\nu' T}$ for real ν, ν' does not imply $\nu = \nu'$. As an example, discuss the eigenspaces of A and the eigenspace for the eigenvalue 1 of e^{AT} with A given by

$$A = \begin{bmatrix} 0 & 0 & 0 \\ 0 & 0 & -1 \\ 0 & 1 & 0 \end{bmatrix}.$$

Exercise 1.6.8. Consider the differential equation
$$m\ddot{y}(t) + c\dot{y}(t) + ky(t) = 0,$$
where $m, c, k > 0$ are constants. Determine all solutions in the following three cases: (i) $c^2 - 4km > 0$; (ii) $c^2 - 4km = 0$; (iii) $c^2 - 4km < 0$. Show that in all cases all solutions tend to the origin as $t \to \infty$ (the system is asymptotically stable). Determine (in each of the cases (i) to (iii)) the solution $\varphi(\cdot)$ with $\varphi(0) = 1$ and $\dot{\varphi}(0) = 0$. Show that in case (iii) all solutions can be written in the form $\varphi(t) = Ae^{\alpha t} \cos(\beta t - \vartheta)$ and determine α and β. Discuss the differences in the phase portraits in cases (i) to (iii).

Exercise 1.6.9. As noted in Section 1.2, for a matrix $A \in gl(d, \mathbb{R})$ the complex generalized eigenspace of an eigenvalue $\mu \in \mathbb{C}$ can be characterized as $\ker(A - \mu I)^m$, where m is the dimension of the largest Jordan block for μ. Show that the real generalized eigenspace for $\mu, \bar{\mu}$ is the sum of the real and imaginary parts of the complex generalized eigenspace, i.e., it is given by
$$\{\operatorname{Re} v, \operatorname{Im} v \in \mathbb{R}^d \mid v \in \ker(A - \mu I)^m\}.$$
Hint: Find $S_2 = [v, \bar{v}, \ldots] \in gl(d, \mathbb{R})$ such that $S_2^{-1} A S_2 = \begin{bmatrix} \mu & 0 & 0 \\ 0 & \bar{\mu} & 0 \\ 0 & 0 & * \end{bmatrix}$. Then find $S_1 = \begin{bmatrix} -i & 1 & 0 \\ -1 & i & 0 \\ 0 & 0 & I \end{bmatrix}$ in $gl(d, \mathbb{R})$ (using Proposition 1.2.2) transforming A into a matrix of the form
$$S_1 S_2^{-1} A S_2 S_1^{-1} = \begin{bmatrix} \operatorname{Re} \mu & -\operatorname{Im} \mu & 0 \\ \operatorname{Im} \mu & \operatorname{Re} \mu & 0 \\ 0 & 0 & * \end{bmatrix}$$
and note that $S_1 S_2^{-1} = [\operatorname{Re} v, \operatorname{Im} v, \ldots]$. Show that $v \in \mathbb{C}^d$ satisfies $v \in \ker[A - \mu I]$ and $\bar{v} \in \ker[A - \bar{\mu} I]$ if and only if $\operatorname{Re} v, \operatorname{Im} v \in \ker[(A - \operatorname{Re} \mu I)^2 + (\operatorname{Im} \mu I)^2] = \ker(A - \mu I)(A - \bar{\mu} I)$. Generalize this discussion to m-dimensional Jordan blocks $J_\mu, J_{\bar{\mu}}$ and transformations
$$S_2 = [v_1, \ldots, v_n, \bar{v}_1, \ldots, \bar{v}_n, \ldots]$$
leading to $S_2^{-1} A S_2 = \begin{bmatrix} J_\mu & 0 & 0 \\ 0 & J_{\bar{\mu}} & 0 \\ 0 & 0 & * \end{bmatrix}$. Then use
$$S_1 := \begin{bmatrix} -iI_n & I_n & 0 \\ -I_n & iI_n & 0 \\ 0 & 0 & * \end{bmatrix} \text{ giving } S_1 S_2^{-1} A S_2 S_1^{-1} = \begin{bmatrix} R & 0 \\ 0 & * \end{bmatrix},$$
where R is a real Jordan block. Note that here
$$S_1 S_2^{-1} = [\operatorname{Re} v_1, \ldots, \operatorname{Re} v_n, \operatorname{Im} v_1, \ldots, \operatorname{Im} v_n, \ldots].$$

Exercise 1.6.10. Show that the Lyapunov exponents do not depend on the norm used in Definition 1.4.1.

Exercise 1.6.11. Let $A \in gl(d, \mathbb{R})$ be a matrix with all entries $a_{ij} > 0$. Suppose that $\lambda \geq 0$ is a real eigenvalue such that there is a corresponding eigenvector $w \in \mathbb{R}^d$ with all entries $w_j > 0$. Show that

$$\left(\sum_{i,j} (A^m)_{ij} \right)^{1/m} \to \lambda \text{ for } m \to \infty$$

and that $|\mu| \leq \lambda$ for all eigenvalues $\mu \in \mathbb{C}$ of A.

Hint: Suppose that there is an eigenvalue μ with $|\mu| > \lambda$ and consider a vector $v \neq 0$ in the Lyapunov space $L(\log |\mu|)$. Then discuss the time evolution of $w \pm \varepsilon v$.

1.7. Orientation, Notes and References

Orientation. This chapter has discussed linear autonomous systems in continuous and discrete time. The main results are Theorem 1.4.3 for differential equations and its counterpart for difference equations, Theorem 1.5.6. They explain the relation between Lyapunov exponents and corresponding initial points for $\dot{x} = Ax$ and $x_{n+1} = Ax_n$ on one hand and the matrix A, its eigenvalues and generalized eigenspaces on the other hand, hence they establish a first relation between dynamical systems and linear algebra. In the simple autonomous context here, this connection is deduced from explicit solution formulas. They are based on the Jordan canonical form, which is used to show that the Lyapunov exponents and the associated Lyapunov spaces are determined by the real parts of the eigenvalues and their absolute values, respectively, and the corresponding sums of generalized eigenspaces. The table below puts the results in continuous and discrete time in parallel.

Continuous time	Discrete time
$\dot{x} = Ax, \; A \in gl(d, \mathbb{R})$	$x_{n+1} = Ax_n, \; A \in Gl(d, \mathbb{R})$
Lyapunov exponents:	Lyapunov exponents:
$\lambda(x_0) = \limsup_{t \to \infty} \frac{1}{t} \log \|x(t, x_0)\|$	$\lambda(x_0) = \limsup_{n \to \infty} \frac{1}{n} \log \|\varphi(n, x_0)\|$
real parts of the eigenvalues μ:	moduli of the eigenvalues μ:
$\lambda_1 > \ldots > \lambda_\ell$	$\lambda_1 > \ldots > \lambda_\ell$
Lyapunov spaces:	Lyapunov spaces:
$L(\lambda_j) = \bigoplus_{\operatorname{Re} \mu_k = \lambda_j} E(\mu_k)$	$L(\lambda_j) = \bigoplus_{\log |\mu_k| = \lambda_j} E(\mu_k)$
$\mathbb{R}^d = L(\lambda_1) \oplus \ldots \oplus L(\lambda_\ell)$	$\mathbb{R}^d = L(\lambda_1) \oplus \ldots \oplus L(\lambda_\ell)$
Theorem 1.4.3: $x_0 \in L(\lambda_j) \Leftrightarrow$	**Theorem 1.5.6:** $x_0 \in L(\lambda_j) \Leftrightarrow$
$\lim_{t \to \pm\infty} \frac{1}{t} \log \|x(t, x_0)\| = \lambda_j$	$\lim_{n \to \pm\infty} \frac{1}{n} \log \|\varphi(n, x_0)\| = \lambda_j$

Chapter 2 gives further insight into the interplay between dynamical systems and linear algebra by discussing classification of dynamical systems versus classification of matrices.

Notes and references. The linear algebra and the theory of linear autonomous differential and difference equations presented here is standard and can be found in many textbooks in this area. For the derivation of the real Jordan form from the complex Jordan form, we have followed Kowalsky [**83**] and Kowalsky and Michler [**84**]. An alternative derivation is provided in Exercise 1.6.9. One may also give a proof based on irreducible polynomials; cf., e.g., Wonham [**143**, pp. 18-20] for an easily readable presentation. Introductions to difference equation are, e.g., Elaydi [**43**] and Krause and Nesemann [**85**]; in particular, the latter reference also analyzes stability of linear autonomous equations $x_{n+1} = Ax_n$ for noninvertible matrices A.

Chapter 2

Linear Dynamical Systems in $\mathbb{R}^d$

In this chapter we will introduce the general notion of dynamical systems in continuous and in discrete time and discuss some of their properties. Linear autonomous differential equations $\dot x = Ax$ where $A \in gl(d, \mathbb{R})$ generate linear continuous-time dynamical systems or flows in $\mathbb{R}^d$. Similarly, linear autonomous difference equations $x_{n+1} = Ax_n$ generate linear discrete-time dynamical systems in $\mathbb{R}^d$. A standard concept for the classification of these dynamical systems are conjugacies mapping solutions into solutions. It turns out that these classifications of dynamical systems are closely related to classifications of matrices. For later purposes, the definitions for dynamical systems are given in the abstract setting of metric spaces, instead of $\mathbb{R}^d$.

Section 2.1 introduces continuous-time dynamical systems on metric spaces and conjugacy notions. Section 2.2 determines the associated equivalence classes for linear flows in $\mathbb{R}^d$ in terms of matrix classifications. Section 2.3 presents analogous results in discrete time.

2.1. Continuous-Time Dynamical Systems or Flows

In this section we introduce general continuous-time dynamical systems or flows on metric spaces and the notions of conjugacy.

Recall that a metric space is a set X with a distance function $d : X \times X \to [0, \infty)$ satisfying $d(x, y) = 0$ if and only if $x = y$, $d(x, y) = d(y, x)$, and $d(x, z) \le d(x, y) + d(y, z)$ for all $x, y, z \in X$.

Definition 2.1.1. A continuous dynamical system or flow over the 'time set' $\mathbb{R}$ with state space X, a metric space, is defined as a continuous map $\Phi : \mathbb{R} \times X \longrightarrow X$ with the properties

(i) $\Phi(0, x) = x$ for all $x \in X$,

(ii) $\Phi(s + t, x) = \Phi(s, \Phi(t, x))$ for all $s, t \in \mathbb{R}$ and all $x \in X$.

For each $x \in X$ the set $\{\Phi(t, x) \mid t \in \mathbb{R}\}$ is called the orbit (or trajectory) of the system through the point x. For each $t \in \mathbb{R}$ the time-t map is defined as $\varphi_t = \Phi(t, \cdot) : X \longrightarrow X$. Using time-$t$ maps, properties (i) and (ii) above can be restated as $\varphi_0 = id$, the identity map on X, and $\varphi_{s+t} = \varphi_s \circ \varphi_t$ for all $s, t \in \mathbb{R}$.

More precisely, a system Φ as above is a continuous dynamical system in continuous time. For simplicity, we just talk about continuous dynamical systems in the following. Note that we have defined a dynamical system over the (two-sided) time set $\mathbb{R}$. This immediately implies invertibility of the time-t maps.

Proposition 2.1.2. *Each time-t map φ_t has the inverse $(\varphi_t)^{-1} = \varphi_{-t}$, and $\varphi_t : X \longrightarrow X$ is a homeomorphism, i.e., a continuous bijective map with continuous inverse. Denote the set of time-t maps again by $\Phi = \{\varphi_t \mid t \in \mathbb{R}\}$. A dynamical system is a group in the sense that $(\Phi, \circ)$, with $\circ$ denoting composition of maps, satisfies the group axioms, and $\varphi : (\mathbb{R}, +) \longrightarrow (\Phi, \circ)$, defined by $\varphi(t) = \varphi_t$ is a group homomorphism.*

Systems defined over the one-sided time set $\mathbb{R}^+ := \{t \in \mathbb{R} \mid t \geq 0\}$ satisfy the corresponding semigroup property and their time-t maps need not be invertible. Standard examples for continuous dynamical systems are given by solutions of differential equations.

Example 2.1.3. For $A \in gl(d, \mathbb{R})$ the solutions of a linear differential equation $\dot{x} = Ax$ form a continuous dynamical system with time set $\mathbb{R}$ and state space $X = \mathbb{R}^d$. Here $\Phi : \mathbb{R} \times \mathbb{R}^d \longrightarrow \mathbb{R}^d$ is defined by $\Phi(t, x_0) = x(t, x_0) = \mathrm{e}^{At} x_0$. This follows from Corollary 1.1.2.

Also, many nonlinear differential equations define dynamical systems. Since we do not need this general result, we just state it.

Example 2.1.4. Suppose that the function f in the initial value problem (1.4.2) is locally Lipschitz continuous and for all $x_0 \in \mathbb{R}^d$ there are solutions $\varphi(t, x_0)$ defined for all $t \in \mathbb{R}$. Then $\Phi(t, x_0) = \varphi(t, x_0)$ defines a dynamical system $\Phi : \mathbb{R} \times \mathbb{R}^d \longrightarrow \mathbb{R}^d$. In this case, the vector field f is called complete.

Two specific types of orbits will play an important role in this book, namely fixed points and periodic orbits.

Definition 2.1.5. A fixed point (or equilibrium) of a dynamical system Φ is a point $x \in X$ with the property $\Phi(t,x) = x$ for all $t \in \mathbb{R}$.

A solution $\Phi(t,x), t \in \mathbb{R}$, of a dynamical system Φ is called periodic if there exists $S > 0$ such that $\Phi(S+s,x) = \Phi(s,x)$ for all $s \in \mathbb{R}$. The infimum T of the numbers S with this property is called the period of the solution and the solution is called T-periodic.

Since a solution is continuous in t, the period T satisfies $\Phi(T+s,x) = \Phi(s,x)$ for all $s \in \mathbb{R}$. Note that a solution of period 0 is a fixed point. For a periodic solution we also call the orbit $\{\Phi(t,x) \,|\, t \in \mathbb{R}\}$ periodic. If the system is given by a differential equation as in Example 2.1.4, the fixed points are easily characterized, since we assume that solutions of initial value problems are unique: A point $x_0 \in X$ is a fixed point of the dynamical system Φ associated with a differential equation $\dot{x} = f(x)$ if and only if $f(x_0) = 0$.

For linear differential equations as in Example 2.1.3 we can say a little more.

Proposition 2.1.6. *(i) A point $x_0 \in \mathbb{R}^d$ is a fixed point of the dynamical system Φ associated with the linear differential equation $\dot{x} = Ax$ if and only if $x_0 \in \ker A$, the kernel of A.*

(ii) The solution for $x_0 \in \mathbb{R}^d$ is T-periodic if and only if x_0 is in the eigenspace of the eigenvalue 1 of e^{AT}. This holds, in particular, if x_0 is in the eigenspace of an imaginary eigenvalue pair $\pm\imath\nu \neq 0$ of A and $T = \frac{2\pi}{\nu}$.

Proof. Assertion (i) and the first assertion in (ii) are obvious from direct constructions of solutions. The second assertion in (ii) follows from Example 1.3.6 and the fact that the eigenvalues $\pm\imath\nu$ of A are mapped onto the eigenvalue 1 of $e^{A\frac{2\pi}{\nu}}$. The reader is asked to prove this in detail in Exercise 2.4.5. □

Example 2.1.7. The converse of the second assertion in (ii) is not true: Consider the matrix

$$A = \begin{bmatrix} 0 & 0 & 0 \\ 0 & 0 & -1 \\ 0 & 1 & 0 \end{bmatrix}$$

with eigenvalues $\{0, \pm\imath\}$. The initial value $x_0 = (1,1,0)^\top$ (which is not in an eigenspace of A) leads to the 2π-periodic solution $e^{At}x_0 = (1, \cos t, \sin t)^\top$.

Conjugacy

A fundamental topic in the theory of dynamical systems concerns comparison of two systems, i.e., how can we tell that two systems are 'essentially the same'? In this case, they should have similar properties. For example, fixed points and periodic solutions should correspond to each other. This idea can be formalized through conjugacies, which we define next.

Definition 2.1.8. (i) Two continuous dynamical systems $\Phi, \Psi : \mathbb{R} \times X \longrightarrow X$ on a metric space X are called C^0-conjugate or topologically conjugate if there exist a homeomorphism $h : X \to X$ such that

$$(2.1.1) \qquad h(\Phi(t,x)) = \Psi(t, h(x)) \text{ for all } x \in X \text{ and } t \in \mathbb{R}.$$

(ii) Let $\Phi, \Psi : \mathbb{R} \times X \longrightarrow X$ be C^k-maps, $k \geq 1$, on an open subset X of $\mathbb{R}^d$. They are called C^k-**conjugate** if there exists a C^k diffeomorphism $h : X \to X$ with (2.1.1). Then h is also called a smooth conjugacy.

The conjugacy property (2.1.1) can be illustrated by the following commutative diagram

$$\begin{array}{ccc} X & \xrightarrow{\Phi(t,\cdot)} & X \\ h\downarrow & & \downarrow h \\ X & \xrightarrow{\Phi(t,\cdot)} & X \end{array}.$$

Note that while this terminology is standard in dynamical systems, the term *conjugate* is used differently in linear algebra. (Smooth) *conjugacy* as used here is related to matrix similarity (compare Theorem 2.2.1), not to matrix conjugacy. Topological conjugacies preserve many properties of dynamical systems. The next proposition shows some of them.

Proposition 2.1.9. *Let $h : X \to X$ be a topological conjugacy of two dynamical systems $\Phi, \Psi : \mathbb{R} \times X \longrightarrow X$ on a metric state space X. Then*

(i) the point $p \in X$ is a fixed point of Φ if and only if $h(p)$ is a fixed point of Ψ;

(ii) the solution $\Phi(\cdot, p)$ is T-periodic if and only if $\Psi(\cdot, h(p))$ is T-periodic.

(iii) Let, in addition, $g : Y \to Y$ be a topological conjugacy of two dynamical systems $\Phi_1, \Psi_1 : \mathbb{R} \times Y \longrightarrow Y$ on a metric space Y. Then the product flows $\Phi \times \Phi_1$ and $\Psi \times \Psi_1$ on $X \times Y$ are topologically conjugate via $h \times g : X \times Y \longrightarrow X \times Y$.

Proof. The proof of assertions (i) and (ii) is deferred to Exercise 2.4.1. Assertion (iii) follows, since $h \times g$ is a homeomorphism and for $x \in X, y \in Y$, and $t \in \mathbb{R}$ one has

$$(h \times g)(\Phi \times \Phi_1)(x,y) = (h(\Phi(t,x)), g(\Phi_1(t,y))) = (\Psi(t, h(x)), \Psi_1(t, g(y)))$$
$$= (\Psi \times \Psi_1)(t, (h \times g)(x,y)). \qquad \square$$

2.2. Conjugacy of Linear Flows

For linear flows associated with linear differential equations as introduced in Example 2.1.3, conjugacy can be characterized directly in terms of the matrix A. We start with smooth conjugacies.

Theorem 2.2.1. *For two linear flows Φ (associated with $\dot{x} = Ax$) and Ψ (associated with $\dot{x} = Bx$) in $\mathbb{R}^d$, the following are equivalent:*

(i) Φ and Ψ are C^k-conjugate for $k \geq 1$,

(ii) Φ and Ψ are linearly conjugate, i.e., the conjugacy map h is an invertible linear operator on $\mathbb{R}^d$,

(iii) A and B are similar, i.e., $A = SBS^{-1}$ for some $S \in Gl(d, \mathbb{R})$.

Each of these statements is equivalent to the property that A and B have the same Jordan form. Thus the C^k-conjugacy classes are exactly the real Jordan form equivalence classes in $gl(d, \mathbb{R})$.

Proof. Properties (ii) and (iii) are obviously equivalent and imply (i). Suppose that (i) holds, and let $h : \mathbb{R}^d \to \mathbb{R}^d$ be a C^k-conjugacy. Thus for all $x \in \mathbb{R}^d$ and $t \in \mathbb{R}$,

$$h(\Phi(t, x)) = h(\mathrm{e}^{At} x) = \mathrm{e}^{Bt} h(x) = \Psi(h(x)).$$

Differentiating with respect to x and using the chain rule we find

$$Dh(\mathrm{e}^{At} x)\mathrm{e}^{At} = \mathrm{e}^{Bt} Dh(x).$$

Evaluating this at $x = 0$ we get with $H := Dh(0)$,

$$H\mathrm{e}^{At} = \mathrm{e}^{Bt} H \text{ for all } t \in \mathbb{R}.$$

Differentiation with respect to t in $t = 0$ finally gives $HA = BH$. Since h is a diffeomorphism, the linear map $H = Dh(0)$ is invertible and hence defines a linear conjugacy. $\square$

In particular we obtain conjugacy for the linear dynamical systems induced by a matrix A and its real Jordan form $J^{\mathbb{R}}$.

Corollary 2.2.2. *For each matrix $A \in gl(d, \mathbb{R})$ its associated linear flow in $\mathbb{R}^d$ is C^k-conjugate for all $k \geq 1$ to the dynamical system associated with the Jordan form $J^{\mathbb{R}}$.*

Theorem 2.2.1 clarifies the structure of two matrices that give rise to conjugate flows under C^k-diffeomorphisms with $k \geq 1$. The eigenvalues and the dimensions of the Jordan blocks remain invariant, while the eigenspaces and generalized eigenspaces are mapped onto each other.

For homeomorphisms, i.e., for $k = 0$, the situation is quite different and somewhat surprising. To explain the corresponding result we first need to introduce the concept of hyperbolicity.

Definition 2.2.3. The matrix $A \in gl(d, \mathbb{R})$ is hyperbolic if it has no eigenvalues on the imaginary axis.

The set of hyperbolic matrices in $gl(d, \mathbb{R})$ is rather 'large' in $gl(d, \mathbb{R})$ (which may be identified with $\mathbb{R}^{d^2}$ and hence carries the corresponding topology.)

Proposition 2.2.4. *The set of hyperbolic matrices is open and dense in $gl(d, \mathbb{R})$ and for every hyperbolic matrix A there is a neighborhood $U \subset gl(d, \mathbb{R})$ of A such that the dimension of the stable subspace is constant for $B \in U$.*

Proof. Let A be hyperbolic. Then also for all matrices in a neighborhood of A all eigenvalues have nonvanishing real parts, since the eigenvalues depend continuously on the matrix entries. Hence openness and the last assertion follow. Concerning density, transform an arbitrary matrix $A \in gl(d, \mathbb{R})$ via a matrix T into a matrix $T^{-1}AT$ in Jordan normal form. For such a matrix it is clear that one finds arbitrarily close matrices B which are hyperbolic. Transforming them back into TBT^{-1} one obtains hyperbolic matrices arbitrarily close to A. □

With these preparations we can formulate the characterization of C^0-conjugacies of linear flows:

Theorem 2.2.5. *(i) If A and B are hyperbolic, then the associated linear flows Φ and Ψ in $\mathbb{R}^d$ are topologically conjugate if and only if the dimensions of the stable subspaces (and hence the dimensions of the unstable subspaces) of A and B agree.*

(ii) A matrix A is hyperbolic if and only if its flow is structurally stable, i.e., there exists a neighborhood $U \subset gl(d, \mathbb{R})$ of A such that for all $B \in U$ the associated linear flows are conjugate to the flow of A.

Observe that assertion (ii) is an immediate consequence of (i) and Proposition 2.2.4. The proof of assertion (i) is complicated and needs some preparation. Consider first asymptotically stable differential equations $\dot{x} = Ax$ and $\dot{x} = Bx$. In our construction of a topological conjugacy h, we will first consider the unit spheres and then extend h to $\mathbb{R}^d$. This requires that trajectories intersect the unit sphere exactly once. In general, this is not true, since asymptotic stability only guarantees that for all $\kappa > \max\{\text{Re}\,\lambda \mid \lambda \in \text{spec}(A)\}$ there is $c \geq 1$ with

$$\|e^{At}x\| \leq ce^{\kappa t}\|x\| \text{ for all } x \in \mathbb{R}^d,\ t \geq 0.$$

In the following simple example the Euclidean norm, $\|x\|_2 = \sqrt{x_1^2 + \ldots + x_d^2}$, does not decrease monotonically along solutions.

2.2. Conjugacy of Linear Flows

Example 2.2.6. Consider in $\mathbb{R}^2$,
$$\dot{x} = -x - y, \ \dot{y} = 4x - y.$$
The eigenvalues of $A = \begin{bmatrix} -1 & -1 \\ 4 & -1 \end{bmatrix}$ are the zeros of $\det(\lambda I - A) = (\lambda + 1)^2 + 4$, hence they are equal to $-1 \pm 2\imath$. The solutions are
$$\begin{bmatrix} x(t) \\ y(t) \end{bmatrix} = e^{-t} \begin{bmatrix} \cos 2t & -\tfrac{1}{2}\sin 2t \\ 2\sin 2t & \cos 2t \end{bmatrix} \begin{bmatrix} x_0 \\ y_0 \end{bmatrix}.$$
The origin is asymptotically stable, but the Euclidean distance to the origin does not decrease monotonically.

The next proposition shows that monotonicity always holds in a norm which is adapted to the matrix A.

Proposition 2.2.7. *For $\dot{x} = Ax$ with $A \in gl(d,\mathbb{R})$ the following properties are equivalent:*

(i) For every eigenvalue μ of A one has $\operatorname{Re} \mu < 0$.

(ii) For every norm $\|\cdot\|$ on $\mathbb{R}^d$ there are $a > 0$ and $c \geq 1$ with
$$\left\|e^{At}x\right\| \leq c\, e^{-at} \|x\| \ \text{for } t \geq 0.$$

(iii) There is a norm $\|\cdot\|_A$ on $\mathbb{R}^d$, called an adapted norm, such that for some $a > 0$ and for all $x \in \mathbb{R}^d$,
$$\left\|e^{At}x\right\|_A \leq e^{-at} \|x\|_A \ \text{for } t \geq 0.$$

Proof. (iii) implies (ii), since all norms on $\mathbb{R}^d$ are equivalent. Property (ii) is equivalent to (i) by Theorem 1.4.4. It remains to show that (ii) implies (iii). Let $b \in (0, a)$. Then (ii) (with any norm) implies for $t \geq 0$,
$$\left\|e^{At}x\right\| \leq c\, e^{-at}\|x\| = c\, e^{(b-a)t} e^{-bt} \|x\|.$$
Hence there is $\tau > 0$ such that $c\, e^{(b-a)t} < 1$ for all $t \geq \tau$ and therefore

(2.2.1)
$$\left\|e^{At}x\right\| \leq e^{-bt} \|x\|.$$
Then
$$\|x\|_A := \int_0^\tau e^{bs} \left\|e^{As}x\right\| ds, \ x \in \mathbb{R}^d,$$
defines a norm, since $\|x\|_A = 0$ if and only if $e^{bs}\left\|e^{As}x\right\| = 0$ for $s \in [0,\tau]$ if and only if $x = 0$, and
$$\|x + y\|_A = \int_0^\tau e^{bs} \left\|e^{As}(x+y)\right\| ds \leq \|x\|_A + \|y\|_A.$$

This norm has the desired monotonicity property: For $t \geq 0$ write $t = n\tau + T$ with $0 \leq T < \tau$ and $n \in \mathbb{N}_0$. Then

$$\left\|e^{At}x\right\|_A = \int_0^\tau e^{bs} \left\|e^{As}e^{At}x\right\| ds$$

$$= \int_0^{\tau-T} e^{bs} \left\|e^{An\tau}e^{A(T+s)}x\right\| ds + \int_{\tau-T}^\tau e^{bs} \left\|e^{A(n+1)\tau}e^{A(T-\tau+s)}x\right\| ds$$

$$\leq \int_T^\tau e^{b(\sigma-T)} \left\|e^{An\tau}e^{A\sigma}x\right\| d\sigma + \int_0^T e^{b(\sigma-T+\tau)} \left\|e^{A(n+1)\tau}e^{A\sigma}x\right\| d\sigma$$

with $\sigma := T + s$ and $\sigma := T - \tau + s$, respectively. We can use (2.2.1) to estimate the second summand from above, since $(n+1)\tau \geq \tau$. If $n = 0$, we leave the first summand unchanged, otherwise we can also apply (2.2.1). In any case we obtain

$$\leq \int_T^\tau e^{b(\sigma-T-n\tau)} \left\|e^{A\sigma}x\right\| d\sigma + \int_0^T e^{b(\sigma-T+\tau-(n+1)\tau)} \left\|e^{A\sigma}x\right\| d\sigma$$

$$= e^{-bt} \int_0^\tau e^{b\sigma} \left\|e^{A\sigma}x\right\| d\sigma = e^{-bt} \|x\|_A$$

and hence (ii) implies (iii). $\square$

We show the assertion of Theorem 2.2.5 first in the asymptotically stable case.

Proposition 2.2.8. *Let $A, B \in gl(d, \mathbb{R})$. If all eigenvalues of A and of B have negative real parts, then the flows e^{At} and e^{Bt} are topologically conjugate.*

Proof. Let $\|\cdot\|_A$ and $\|\cdot\|_B$ be corresponding adapted norms. Hence with constants $a, b > 0$,

$$\left\|e^{At}x\right\|_A \leq e^{-at} \|x\|_A \text{ and } \left\|e^{Bt}x\right\|_B \leq e^{-bt} \|x\|_B \text{ for } t \geq 0 \text{ and } x \in \mathbb{R}^d.$$

Then for $t \leq 0$,

$$\left\|e^{At}x\right\|_A \geq e^{a(-t)} \|x\|_A \text{ and } \left\|e^{Bt}x\right\|_B \geq e^{b(-t)} \|x\|_B,$$

by applying the inequality above to $e^{At}x$ and $-t$. Thus backwards in time, the norms of the solutions are strictly increasing. Consider the corresponding unit spheres

$$S_A = \{x \in \mathbb{R}^d \mid \|x\|_A = 1\} \text{ and } S_B = \{x \in \mathbb{R}^n \mid \|x\|_B = 1\}.$$

They are fundamental domains of the flows e^{At} and e^{Bt}, respectively (every nontrivial trajectory intersects them). Define a homeomorphism $h_0 : S_A \to S_B$ by

$$h_0(x) := \frac{x}{\|x\|_B} \text{ with inverse } h_0^{-1}(y) = \frac{y}{\|y\|_A}.$$

In order to extend this map to $\mathbb{R}^d$ observe that (by the intermediate value theorem and by definition of the adapted norms) there is for every $x \neq 0$ a unique time $\tau(x) \in \mathbb{R}$ with $\left\|e^{A\tau(x)}x\right\|_A = 1$. This immediately implies $\tau(e^{At}x) = \tau(x) - t$. The map $x \mapsto \tau(x)$ is continuous: If $x_n \to x$, then the assumption that $\tau(x_{n_k}) \to \sigma \neq \tau(x)$ for a subsequence implies $\|\varphi(\sigma, x)\| = \|\varphi(\tau(x), x)\| = 1$ contradicting uniqueness of $\tau(x)$. Now define $h : \mathbb{R}^d \to \mathbb{R}^d$ by
$$h(x) = \begin{cases} e^{-B\tau(x)}h_0(e^{A\tau(x)}x) & \text{for } x \neq 0, \\ 0 & \text{for } x = 0. \end{cases}$$
Then h is a conjugacy, since
$$h(e^{At}x) = e^{-B\tau(e^{At}x)}h_0(e^{A\tau(e^{At}x)}e^{At}x) = e^{-B[\tau(x)-t]}h_0(e^{A[\tau(x)-t]}e^{At}x)$$
$$= e^{Bt}e^{-B\tau(x)}h_0(e^{A\tau(x)}x) = e^{Bt}h(x).$$

The map h is continuous in $x \neq 0$, since e^{At} and e^{Bt} as well as $\tau(x)$ are continuous. In order to prove continuity in $x = 0$, consider a sequence $x_j \to 0$. Then $\tau_j := \tau(x_j) \to -\infty$. Let $y_j := h_0(e^{\tau_j}x_j)$. Then $\|y_j\|_B = 1$ and hence
$$\|h(x_j)\|_B = \left\|e^{-B\tau_j}y_j\right\|_B \leq e^{b\tau_j} \to 0 \text{ for } j \to \infty.$$
The map is injective: Suppose $h(x) = h(z)$. The case $x = 0$ is clear. Hence suppose that $x \neq 0$. Then $h(x) = h(z) \neq 0$, and with $\tau := \tau(x)$ the conjugation property implies
$$h(e^{A\tau}x) = e^{B\tau}h(x) = e^{B\tau}h(z) = h(e^{A\tau}z).$$
Thus $h(e^{A\tau}z) = h(e^{A\tau}x) \in S_B$. Since h maps only S_A to S_B, it follows that $e^{A\tau}z \in S_A$ and hence $\tau = \tau(x) = \tau(z)$. By
$$h_0(e^{A\tau}x) = h(e^{A\tau}x) = h(e^{A\tau}z) = h_0(e^{A\tau}z)$$
and injectivity of h_0 we find
$$e^{A\tau}x = e^{A\tau}z, \text{ and hence } x = z.$$
Exchanging the roles of A and B we see that h^{-1} exists and is continuous. $\square$

Proof of Theorem 2.2.5. If the dimensions of the stable subspaces coincide, there are topological conjugacies
$$h^s : E_A^s \to E_B^s \text{ and } h^u : E_A^u \to E_B^u$$
between the restrictions to the stable and the unstable subspaces of e^{At} and e^{Bt}, respectively. With the projections
$$\pi^s : \mathbb{R}^n \to E_A^s \text{ and } \pi^u : \mathbb{R}^n \to E_A^u,$$
a topological conjugacy is defined by
$$h(x) := h^s(\pi^s(x)) + h^u(\pi^u(x)).$$

Conversely, any topological conjugacy homeomorphically maps the stable subspace onto the stable subspace. This implies that the dimensions coincide (invariance of domain theorem, Massey [**101**, Chapter VIII, Theorem 6.6 and Exercise 6.5]). $\square$

2.3. Linear Dynamical Systems in Discrete Time

The purpose of this section is to analyze properties of autonomous linear difference equations from the point of view of dynamical systems. First we define dynamical systems in discrete time, and then we classify the conjugacy classes of the dynamical systems generated by autonomous linear difference equations of the form $x_{n+1} = Ax_n$ with $A \in Gl(d, \mathbb{R})$.

In analogy to the continuous-time case, Definition 2.1.1, we define dynamical systems in discrete time in the following way.

Definition 2.3.1. A continuous dynamical system in discrete time over the 'time set' $\mathbb{Z}$ with state space X, a metric space, is defined as a continuous map $\Phi : \mathbb{Z} \times X \longrightarrow X$ with the properties

(i) $\Phi(0, x) = x$ for all $x \in X$;

(ii) $\Phi(m + n, x) = \Phi(m, \Phi(n, x))$ for all $m, n \in \mathbb{Z}$ and all $x \in X$.

For each $x \in X$ the set $\{\Phi(n, x) \mid n \in \mathbb{Z}\}$ is called the orbit (or trajectory) of the system through the point x. For each $n \in \mathbb{Z}$ the time-n map is defined as $\varphi_n = \Phi(n, \cdot) : X \longrightarrow X$. Using time-$n$ maps, properties (i) and (ii) above can be restated as $\varphi_0 = id$, the identity map on X, and $\varphi_{m+n} = \varphi_n \circ \varphi_m$ for all $m, n \in \mathbb{Z}$.

It is an immediate consequence of the definition, that a dynamical system in discrete time is completely determined by its time-1 map, also called its generator.

Proposition 2.3.2. *For every $n \in \mathbb{Z}$, the time-n map φ_n is given by $\varphi_n = (\varphi_1)^n$. In particular, each time-n map φ_n has an inverse $(\varphi_n)^{-1} = \varphi_{-n}$, and $\varphi_n : X \longrightarrow X$ is a homeomorphism. A dynamical system in discrete time is a group in the sense that $(\{\varphi_n \mid n \in \mathbb{Z}\}, \circ)$, with $\circ$ denoting composition of maps, satisfies the group axioms, and $\varphi : (\mathbb{Z}, +) \longrightarrow (\{\varphi_n \mid n \in \mathbb{Z}\}, \circ)$, defined by $\varphi(n) = \varphi_n$ is a group homomorphism.*

This proposition also shows that every homeomorphism f defines a continuous dynamical system in discrete time by $\varphi_n := f^n, n \in \mathbb{Z}$. In particular, this holds if f is given by a matrix $A \in Gl(d, \mathbb{R})$.

Remark 2.3.3. It is worth mentioning that a major difference to the continuous-time case comes from the fact that, contrary to e^{At}, the matrix A may not be invertible or, equivalently, that 0 may be an eigenvalue

of A. In this case, one only obtains a map
$$\Phi : \mathbb{N}_0 \times \mathbb{R}^d \to \mathbb{R}^d, \ \Phi(n,x) := A^n x,$$
which is linear in the second argument and satisfies property (ii) in Definition 2.3.1 only for $m, n \geq 0$. If A is invertible, the map Φ can be extended to a dynamical system $\Phi : \mathbb{Z} \times \mathbb{R}^d \to \mathbb{R}^d$. We only consider the invertible case with $A \in Gl(d, \mathbb{R})$. Naturally, one may also define general systems Φ over the one-sided time set $\mathbb{N}_0$ satisfying the corresponding semigroup property; their time-n maps need not be invertible.

Continuous dynamical systems in discrete time may be classified up to conjugacies, in analogy to the case in continuous time. Formally, we define the following.

Definition 2.3.4. Let $\Phi, \Psi : \mathbb{Z} \times X \to X$ be continuous dynamical systems generated by homeomorphisms $f, g : X \to X$, respectively. These systems (and also f and g) are called topologically conjugate if there exists a homeomorphism $h : X \to X$ such that $h(\Phi(n,x)) = \Psi(n, h(x))$ for all $n \in \mathbb{Z}$ and all $x \in X$.

By induction, one sees that this is equivalent to the requirement that $h \circ f = g \circ h$.

Remark 2.3.5. We remark that the smooth conjugacy problem for linear systems in discrete time is trivial: For a C^k-conjugacy h with $k \geq 1$ differentiation of the equation $h(Ax) = Bh(x)$ in $x = 0$ yields the linear conjugacy or matrix similarity
$$Dh(0)A = BDh(0).$$

Two dynamical systems Φ_A and Φ_B in discrete time generated by matrices $A, B \in Gl(d, \mathbb{R})$, respectively, are topologically conjugate, if there is a homeomorphism $h : \mathbb{R}^d \to \mathbb{R}^d$ such that for all $n \in \mathbb{Z}$ and all $x \in \mathbb{R}^d$ one has $h(Ax) = Bh(x)$. We will discuss the topological conjugacy classes using arguments which are analogous to the continuous-time case. As seen in Section 1.5, the stability properties of this dynamical system are again determined by the eigenvalues of the matrix A. Here the role of the imaginary axis in continuous time is taken over by the unit circle: For example, an eigenvector v for a real eigenvalue μ of A satisfies
$$A^n v = \mu^n v \to 0 \text{ if and only if } |\mu| < 1.$$

First we introduce adapted norms for discrete-time dynamical systems.

Proposition 2.3.6. *For $x_{n+1} = Ax_n$ with $A \in Gl(d, \mathbb{R})$ the following properties are equivalent:*

(i) There is a norm $\|\cdot\|_A$ on $\mathbb{R}^d$, called an adapted norm, such that for some $0 < a < 1$ and for all $x \in \mathbb{R}^d$,
$$\|A^n x\|_A \leq a^n \|x\|_A \text{ for all } n \geq 0.$$

(ii) For every norm $\|\cdot\|$ on $\mathbb{R}^d$ there are $0 < a < 1$ and $c \geq 1$ such that for all $x \in \mathbb{R}^d$,
$$\left\|A^k x\right\| \leq c\, a^n \|x\| \text{ for all } n \geq 0.$$

(iii) For every eigenvalue μ of A one has $|\mu| < 1$.

Proof. The proof is analogous to the continuous-time case, Proposition 2.2.7. Property (i) implies (ii), since all norms on $\mathbb{R}^d$ are equivalent. Property (ii) (with $a \in (\max |\mu|, 1)$) implies asymptotic stability which is equivalent to (iii), by Theorem 1.5.11. It remains to show that (ii) implies (i). Let $b \in (a, 1)$. Then (ii) implies for $n \geq 0$
$$\|A^n x\| \leq c\, a^n \|x\| = c \left(\frac{a}{b}\right)^n b^n \|x\|.$$

There is $N \in \mathbb{N}$ such that $c \left(\frac{a}{b}\right)^n < 1$ for all $n \geq N$, hence $\|A^n x\| \leq b^n \|x\|$. Then one shows that

(2.3.1) $$\|x\|_A = \sum_{j=0}^{N-1} b^{-j} \left\|A^j x\right\|, \; x \in \mathbb{R}^d,$$

defines an adapted norm (Exercise 2.4.4). □

A matrix A satisfying the properties in Proposition 2.3.6(i) or, equivalently, $|\mu| < 1$ for all eigenvalues, is called a linear contraction.

Suppose that all eigenvalues μ of A satisfy $|\mu| > 1$. Then all eigenvalues of A^{-1} have modulus less than 1, since they are the inverses of the eigenvalues of A. An adapted norm for A^{-1} yields for some $a \in (0,1)$, all $x \in \mathbb{R}^d$ and all $n \geq 0$
$$\|A^{-n} x\|_A \leq a^n \|x\|_A.$$
In particular, this holds for $x = A^n y, y \in \mathbb{R}^d$, and hence for all $n \geq 0$,
$$\|y\|_A = \|A^{-n} x\|_A \leq a^n \|x\|_A = a^n \|A^n y\|_A$$
implying
$$\|A^n y\|_A \geq b^n \|y\|_A \text{ with } b := a^{-1} > 1.$$
A matrix A with this property is called a linear expansion.

We turn to a result on topological conjugacy of linear contractions, i.e., of autonomous linear dynamical systems in discrete time which are asymptotically stable. Here the discrete-time case is more complicated than the continuous-time case treated in Proposition 2.2.8 for two reasons: The space $Gl(d, \mathbb{R})$ of invertible $d \times d$-matrices is not connected, since the image of the determinant, which is a continuous function, has two path connected components corresponding to the sign of the determinant (in fact, this determines the two connected components of $Gl(d, \mathbb{R})$; cf. Remark 2.3.8). A second difficulty in the proof occurs, since not every orbit intersects the unit sphere; this was an essential ingredient in the proof of Proposition 2.2.7. Hence we have to blow up the sphere to a ring (an 'annulus') in order to guarantee that every orbit intersects this ring.

Theorem 2.3.7. *Let $A, B \in Gl(d, \mathbb{R})$ be invertible linear contractions in the same path connected component of the set of linear contractions, i.e., one finds linear contractions $A_t, t \in [0, 1]$, depending continuously on t with $A_0 = B$ and $A_1 = A$. Then the generated dynamical systems Φ_A and Φ_B are topologically conjugate.*

Proof. Let $A_t, t \in [0, 1]$, be a curve in $Gl(d, \mathbb{R})$ connecting A and B, $A_0 = B$ and $A_1 = A$. For corresponding adapted norms $\|\cdot\|_A$ and $\|\cdot\|_B$ consider the unit disc and sphere,

$$D_A := \{x \in \mathbb{R}^d \mid \|x\|_A < 1\} \text{ and } S_A := \{x \in \mathbb{R}^d \mid \|x\|_A = 1\},$$

and analogously for B. The following rings or annuli

$$F_A := \operatorname{cl}(D_A \setminus AD_A) \text{ and } F_B = \operatorname{cl}(D_B \setminus BD_B)$$

are called fundamental domains for the associated dynamical systems, since for all $x \neq 0$ there is $j = j_A \in \mathbb{Z}$ with $A^j x \in F_A$. In fact, by the definition of adapted norms, if $\|x\|_A > 1$, there is $j \in \mathbb{N}$ with $\|A^{j-1} x\|_A > 1$ and $\|A^j x\|_A \leq 1$, hence $A^j x \in \operatorname{cl}(D_A \setminus AD_A)$. Observe also that the 'outer' boundary of F_A equals S_A and the 'inner' boundary equals AS_A. Analogous statements hold for B. First we will construct a conjugating homeomorphism $h_0 : F_A \to F_B$, hence $h_0(Ax) = Bh_0(x), x \in F_A$, and then extend it to $\mathbb{R}^d$. The idea for the construction is to map the outer and inner boundary of F_A to the outer and inner boundary of F_B, respectively. On the outer boundary, h_0 will be the radial projection of S_A to S_B, and on the inner boundary, h_0 will essentially be equal to BA^{-1} (plus radial projection to $B(S_B)$). Then it will be easy to see that h_0 becomes a conjugacy. This construction separates the radial component from the angular component in $\mathbb{S}^{d-1}$.

For the radial component we will first define h_A, h_B on the standard ring $[0, 1] \times \mathbb{S}^{d-1}$ with values in F_A and F_B, respectively. Then we define $H : [0, 1] \times \mathbb{S}^{d-1} \to [0, 1] \times \mathbb{S}^{d-1}$ using the path from B to A. Here the t-values remain preserved and on $\mathbb{S}^{d-1}$ we use $A_t A^{-1}$. This yields the identity for $t = 1$ and BA^{-1} for $t = 0$.

Let us make this program precise. Define maps

$$\tau_A, h_A : [0, 1] \times \mathbb{S}^{d-1} \to F_A \text{ by } h_A(t, x) = \tau_A(t, x)x,$$

where τ_A is the map which is affine in t and determined by $\tau_A(1, x) = \|x\|_A^{-1}$ and $\tau_A(0, x) = 1/\|A^{-1}x\|_A$. Then $h_A(1, x) = x/\|x\|_A \in S_A$, the outer boundary of F_A, and $h_A(0, x)$ is on the inner boundary of F_A, since

$$h_A(0, x) = \tau_A(0, x)x = x\|A^{-1}x\|_A = A\left(A^{-1}x/\|A^{-1}x\|_A\right) \in AS_A.$$

Since τ_A is affine in t, it follows that

$$\tau_A(t, x) = \frac{t}{\|x\|_A} + \frac{1-t}{\|A^{-1}x\|_A}, \quad t \in [0, 1].$$

We find for all $y \in S_A$ (with $x = Ay/\|Ay\|$),

$$h_A\left(0, \frac{Ay}{\|Ay\|}\right) = \frac{Ay}{\|Ay\|} \frac{\|Ay\|}{\|y\|_A} = \frac{Ay}{\|y\|_A} = Ay, \text{ hence } h_A^{-1}(Ay) = \left(0, \frac{Ay}{\|Ay\|}\right).$$

Analogously, define τ_B and $h_B : [0, 1] \times \mathbb{S}^{d-1} \to F_B$ to obtain

$$h_B\left(0, \frac{Bx}{\|Bx\|}\right) = \frac{Bx}{\|x\|_B} \text{ and } h_B\left(1, \frac{x}{\|x\|}\right) = \frac{x}{\|x\|_B}.$$

Now we use the path A_t in $Gl(d, \mathbb{R})$ with $A_0 = B$, $A_1 = A$ to construct $H : [0, 1] \times \mathbb{S}^{d-1} \to [0, 1] \times \mathbb{S}^{d-1}$ such that the 'radius' $t \in [0, 1]$ is preserved and the 'angle' in $\mathbb{S}^{d-1}$ changes continuously: Define

$$H(t, x) := \left(t, \frac{A_t A^{-1} x}{\|A_t A^{-1} x\|}\right).$$

Then $H(1, x) = (1, x)$ and

$$H(0, x) = \left(0, \frac{BA^{-1}x}{\|BA^{-1}x\|}\right), \text{ hence } H\left(0, \frac{Ax}{\|Ax\|}\right) = \left(0, \frac{Bx}{\|Bx\|}\right).$$

The map h_0 defined by

$$h_0 : F_A \to F_B, \quad h_0 := h_B \circ H \circ h_A^{-1},$$

is a composition of injective maps, hence injective. It is a conjugacy (here we only have to consider the case $x, Ax \in F_A$, in particular, $x \in S_A$), since

$$Bh_0(x) = Bh_B \circ H \circ h_A^{-1}(x) = Bh_B \circ H\left(1, \frac{x}{\|x\|}\right) = \frac{Bx}{\|x\|_B}$$

and

$$h_0(Ax) = h_B \circ H \circ h_A^{-1}(Ax) = h_B \circ H\left(0, \frac{Ay}{\|Ay\|}\right)$$
$$= h_B\left(0, \frac{Bx}{\|Bx\|}\right) = \frac{Bx}{\|x\|_B} = Bh_0(x).$$

Now extend h_0 to $\mathbb{R}^d$ by $h(0) := 0$ and $h(x) = B^{-j(x)} h_0(A^{j(x)} x)$ for $x \neq 0$, where $j(x) \in \mathbb{N}$ is taken such that $A^{j(x)} \in F_A$. If in addition to $A^j x \in F_A$ also $A^{j+1} x \in F_A$, the conjugation property implies

$$B^{-j-1} h_0(A^{j+1} x) = B^{-j-1} h_0(A \, A^j x) = B^{-j-1} B h_0(A^j x) = B^{-j} h_0(A^j x).$$

Thus the map h is well defined. The map h is obviously continuous on $\mathbb{R}^d \setminus \{0\}$. It is also continuous in 0, since $x_n \to 0$ implies $j(x_n) \to -\infty$; now use that B^{-1} is a linear expansion and $\|h_0(A^{j(x_n)}x_n)\| = 1$ to show

$$h(x_n) = B^{-j(x_n)}h_0(A^{j(x_n)}x_n) \to 0.$$

Exchanging the roles of A and B one finds a continuous inverse h^{-1} proving that h is a homeomorphism. □

Remark 2.3.8. Hyperbolic systems in discrete time are given by matrices which do not have an eigenvalue on the unit circle in $\mathbb{C}$. The statement of Theorem 2.2.5 for hyperbolic systems also holds in the discrete-time case, with an analogous proof. Additionally, one has to take into account that the subset of contractions in $Gl(d, \mathbb{R})$ has exactly two path connected components determined by $\det A < 0$ and $\det A > 0$, respectively. A proof is sketched in Robinson [**117**, Chapter IV, Theorem 9.6].

2.4. Exercises

Exercise 2.4.1. Prove parts (i) and (ii) of Proposition 2.1.9: Let $h : X \to X$ be a topological conjugacy for dynamical systems $\Phi, \Psi : \mathbb{R} \times X \longrightarrow X$ on a metric state space X. Then (i) the point $p \in X$ is a fixed point of Φ if and only if $h(p)$ is a fixed point of Ψ; (ii) the solution $\Phi(\cdot, p)$ is periodic with period T if and only if $\Psi(\cdot, h(p))$ is periodic with period T.

Exercise 2.4.2. Construct explicitly a topological conjugacy $h : \mathbb{R} \to \mathbb{R}$ between the systems $\dot{x} = -x$ and $\dot{y} = -2y$.

Exercise 2.4.3. Construct explicitly a topological conjugacy for the linear differential equations determined by

$$A = \begin{bmatrix} -1 & 0 \\ 0 & -1 \end{bmatrix} \text{ and } B = \begin{bmatrix} -1 & -1 \\ 1 & -1 \end{bmatrix}.$$

Exercise 2.4.4. Work out the details of the proof of Proposition 2.3.6 by showing that formula (2.3.1) defines an adapted norm in discrete time.

Exercise 2.4.5. Prove the second part of Proposition 2.1.6(ii): Suppose that x_0 is in the eigenspace of an imaginary eigenvalue pair $\pm i\nu \neq 0$ of A and $T = \frac{2\pi}{\nu}$. Then the solution for $x_0 \in \mathbb{R}^d$ is periodic with period T.

2.5. Orientation, Notes and References

Orientation. The linear dynamical systems considered in this chapter are generated by linear autonomous differential equations $\dot{x} = Ax$ or difference equations $x_{n+1} = Ax_n$. In continuous time one has flows $\Phi : \mathbb{R} \times \mathbb{R}^d \longrightarrow \mathbb{R}^d$ over the time domain $\mathbb{R}$ satisfying $\Phi(t, \alpha x + \beta y) = \alpha \Phi(t, x) + \beta \Phi(t, y)$ for all $x, y \in \mathbb{R}^d, \alpha, \beta \in \mathbb{R}$ and $t \in \mathbb{R}$, and analogously in discrete time over the time domain $\mathbb{Z}$. A natural question to ask is, which properties of the systems are preserved under transformations of the system, i.e., conjugacies. Theorem 2.2.1 and Theorem 2.3.7 show that C^k-conjugacies with $k \geq 1$ reduce to linear conjugacies, thus they preserve the Jordan normal form of the generator A. As seen in Chapter 1 this means that practically all dynamical properties are preserved. On the other hand, mere topological conjugacies only fix the dimensions of the stable and the unstable subspaces. Hence both classifications do not characterize the Lyapunov spaces which determine the exponential growth rates of the solutions. Smooth conjugacies are too fine if one is interested in exponential growth rates, and topological conjugacies are too rough. Hence important features of matrices and their associated linear differential or difference equations cannot be described by these conjugacies in $\mathbb{R}^d$.

Recall that the exponential growth rates and the associated Lyapunov spaces are determined by the real parts of the eigenvalues of the matrix generator A; cf. Definition 1.4.1 and Theorem 1.4.3 (or by the logarithms of the moduli of the eigenvalues) and the generalized eigenspaces. In Chapter 4 we will take a different approach by looking not at conjugacies in order to characterize the Lyapunov spaces. Instead we analyze induced nonlinear systems in projective space and analyze them topologically. The next chapter, Chapter 3, introduces some concepts and results necessary for the analysis of nonlinear dynamical systems. We will use them in Chapters 4 and 5 to characterize the Lyapunov spaces, hence obtain additional information on the connections between matrices and dynamical systems given by autonomous linear differential and difference equations.

Notes and references. The ideas and results of this chapter can be found, e.g., in Robinson [**117**]; in particular, our construction of conjugacies for linear systems follows the exposition in [**117**, Chapter 4]. Continuous dependence of eigenvalues on the matrix is proved, e.g., in Hinrichsen and Pritchard [**68**, Corollary 4.2.1] as well as in Kato [**74**] and Baumgärtel [**17**].

Example 2.1.4 can be generalized to differentiable manifolds: Suppose that X is a C^k-differentiable manifold and f a C^k-vector field on X such that the differential equation $\dot{x} = f(x)$ has unique solutions $x(t, x_0), t \in \mathbb{R}$, with $x(0, x_0) = x_0$ for all $x_0 \in X$. Then $\Phi(t, x_0) = x(t, x_0)$ defines a dynamical system $\Phi : \mathbb{R} \times X \longrightarrow X$. Similarly, C^k-conjugacies can be defined in this setting.

The characterization of matrices via invariance properties of the associated linear autonomous differential and difference equations under smooth and continuous conjugacies may be viewed as part of Klein's Erlanger Programm in the nineteenth century defining geometries by groups of transformations. This point of view is emphasized by McSwiggen and Meyer in

2.5. Orientation, Notes and References

[**105**] who also discuss invariance properties under Lipschitz and Hölder conjugacies; see also Kawan and Stender [**76**] for a classification under Lipschitz conjugacies. Conjugacies are not the only way to classify flows: If one looks at the trajectories, the parametrization by time does not play a role, except for the orientation. This leads to the notion of C^k-equivalence, $k \geq 0$. For $k \geq 1$, the flows for $\dot{x} = Ax$ and $\dot{y} = By$ are C^k-equivalent if and only if there are a real number $\alpha > 0$ and $T \in Gl(d, \mathbb{R})$ with $A = \alpha TBT^{-1}$; cf. Ayala, Kliemann, and Colonius [**12**] for a proof.

The topological conjugacy problem for nonhyperbolic systems in the continuous-time case is treated by Kuiper [**87**]; cf. Ladis [**92**] for topological equivalence. The discrete-time case is much more complicated; cf. Kuiper and Robbin [**89**] and the references given in Ayala and Kawan [**14**].

Chapter 3

Chain Transitivity for Dynamical Systems

This chapter introduces limit sets of dynamical systems and a related concept called chain transitivity. The framework is general dynamical systems in continuous and discrete time. The results will find immediate applications in the next two chapters, where dynamical systems generated by autonomous linear differential equations are analyzed. The concepts treated in the present chapter are fundamental for the global theory of general dynamical systems.

Section 3.1 considers flows and introduces limit sets and the notion of chain transitivity. Section 3.2 characterizes the chain recurrent set and its connected components; the results in this short section will be needed in Chapters 8 and 9. Section 3.3 discusses limit sets and chain transitivity for dynamical systems in discrete time.

3.1. Limit Sets and Chain Transitivity

Global analysis of dynamical systems starts with limit sets, i.e., with the question, where do the trajectories go for $t \to \pm\infty$. The following definition formalizes this question. We suppose that the underlying state space is a complete metric space. This means that every Cauchy sequence has a limit, i.e., every sequence (x_k) in X such that for any $\varepsilon > 0$ there is $n_0(\varepsilon) \in \mathbb{N}$ with $d(x_n, x_m) < \varepsilon$ for all $n, m \geq n_0(\varepsilon)$ converges to an element $x \in X$.

Definition 3.1.1. Let a dynamical system $\Phi : \mathbb{R} \times X \longrightarrow X$ on a complete metric space X be given. For a subset $Y \subset X$ the α-limit set is defined as $\alpha(Y) = \{z \in X \,|\, \text{there exist sequences } (x_n) \text{ in } Y \text{ and } t_n \to -\infty \text{ in } \mathbb{R}$

with $\lim_{n\to\infty} \Phi(t_n, x_n) = z\}$, and similarly the ω-limit set of Y is defined as $\omega(Y) = \{z \in X \mid \text{there exist sequences } (x_n) \text{ in } Y \text{ and } t_n \to \infty \text{ in } \mathbb{R} \text{ with } \lim_{n\to\infty} \Phi(t_n, x_n) = z\}$.

If the set Y consists of a single point x, we just write the limit sets as $\omega(x)$ and $\alpha(x)$, respectively. The limit set $\omega(Y)$ allows points in Y to vary along the sequence. Hence $\omega(Y)$ is larger, in general, than $\bigcup_{y \in Y} \omega(y)$; cf. Example 3.1.4. Note that if X is compact, then the α-limit sets and ω-limit sets are nonvoid for all $Y \subset X$. This need not be true in the noncompact case. We look at some examples of limit sets.

Example 3.1.2. Consider the linear autonomous differential equation on $\mathbb{R}^2$ given by

$$\begin{bmatrix} \dot{x}_1 \\ \dot{x}_2 \end{bmatrix} = \begin{bmatrix} 1 & 0 \\ 0 & -1 \end{bmatrix} \begin{bmatrix} x_1 \\ x_2 \end{bmatrix}, \begin{bmatrix} x_1(0) \\ x_2(0) \end{bmatrix} = \begin{bmatrix} x_1^0 \\ x_2^0 \end{bmatrix} = x^0$$

with solutions $x(t) = (e^t x_1^0, e^{-t} x_2^0), t \in \mathbb{R}$. Thus the α-limit sets $\alpha(x^0)$ are void for all initial points x^0 with $x_2^0 \neq 0$ and $\alpha(x_0) = \{(0,0)\}$ for all $x^0 = (x_1^0, 0)$. The ω-limit sets are void for all initial points x^0 with $x_1^0 \neq 0$ and $\omega(x_0) = \{(0,0)\}$ for all $x^0 = (0, x_2^0)$.

Now look at the projection $\pi : \mathbb{R}^2 \to \mathbb{S}^1$ to the unit sphere $\mathbb{S}^1 := \{x = (x_1, x_2) \mid \|x\|_2 = \sqrt{x_1^2 + x_2^2} = 1\}$ which is a compact metric space with the metric inherited from $\mathbb{R}^2$. The projections of the trajectories $s(t) := \pi(x(t)) = \frac{x(t)}{\|x(t)\|_2}$ define a flow on the unit sphere $\mathbb{S}^1$. The solution formula for the differential equation in $\mathbb{R}^2$ shows that the flow properties hold for the projections (take $s(0) \in \mathbb{S}^1$). Since $\mathbb{S}^1$ is compact, all α- and ω-limit sets are nonvoid. Inspection of the phase portrait shows that the α-limit sets of points on $\mathbb{S}^1$ are given by the points $(0, \pm 1)$ and the ω-limit sets are given by $(\pm 1, 0)$. Note that the eigenvalues are ± 1 and that the α-limit sets are the intersections of the Lyapunov space $L(-1) = \{0\} \times \mathbb{R}$ with the unit sphere and the ω-limit sets are the intersections of the Lyapunov space $L(1) = \mathbb{R} \times \{0\}$ with the unit sphere. In the next chapter we will extend this observation to a topological characterization of the Lyapunov spaces for general linear autonomous differential equations.

Example 3.1.3. Consider the following dynamical system Φ in $\mathbb{R}^2 \setminus \{0\}$, given by a differential equation in polar coordinates with radius $r > 0$ and angle $\theta \in [0, 2\pi)$. Let $a \neq 0$ and

$$\dot{r} = 1 - r, \ \dot{\theta} = a.$$

For each $x \in \mathbb{R}^2 \setminus \{0\}$ the ω-limit set is the circle $\omega(x) = \mathbb{S}^1 = \{(r, \theta) \mid r = 1, \theta \in [0, 2\pi)\}$. The state space $\mathbb{R}^2 \setminus \{0\}$ is not compact (and also not complete), and α-limit sets are nonvoid only for the points $x \in \mathbb{S}^1$; for these points $\alpha(x) = \mathbb{S}^1$.

Example 3.1.4. Consider the ordinary differential equation

$$\dot{x} = x(x-1)(x-2)^2(x-3)$$

on the compact interval $X := [0,3]$ with the metric from $\mathbb{R}$. The solutions $\varphi(t,x)$ of this equation with $\varphi(0,x) = x$ are unique and exist for all $t \in \mathbb{R}$. Hence they define a dynamical system $\Phi : \mathbb{R} \times [0,3] \longrightarrow [0,3]$ via $\Phi(t,x) := \varphi(t,x)$. The ω-limit sets of this system are of the following form: For points $x \in [0,3]$ we have

$$\omega(x) = \begin{cases} \{0\} & \text{for } x = 0, \\ \{1\} & \text{for } x \in (0,2), \\ \{2\} & \text{for } x \in [2,3), \\ \{3\} & \text{for } x = 3. \end{cases}$$

Limit sets for subsets of $[0,3]$ can be entire intervals, e.g., for $Y = [a,b]$ with $a \in (0,1]$ and $b \in [2,3)$ we have $\omega(Y) = [1,2]$, which can be seen as follows: Obviously, it holds that $1,2 \in \omega(Y)$. Let $x \in (1,2)$, then $\lim_{t \to -\infty} \Phi(t,x) = 2$. We define $t_n := n \in \mathbb{N}$ and $x_n := \varphi(-n,x) \in (1,2) \subset Y$. Then $\Phi(t_n, x_n) = \Phi(n, \Phi(-n,x)) = x$ for all $n \in \mathbb{N}$, which shows that $\omega(Y) \supset [1,2]$. For the reverse inclusion let $x \in (0,1)$. Note that $\lim_{t \to \infty} \Phi(t,a) = 1$ and for all $y \in [a,1)$ and all $t \geq 0$ we have $d(\Phi(t,y),1) \leq d(\Phi(t,a),1)$. Hence for any sequence y_n in $[a,1)$ and any $t_n \to \infty$ one sees that $d(\Phi(t_n, y_n), 1) \leq d(\Phi(t_n, a), 1)$ and therefore $\lim_{n \to \infty} d(\Phi(t_n, y_n), 1) \leq \lim_{n \to \infty} d(\Phi(t_n, a), 1) = 0$. This implies that no point $x \in (0,1]$ can be in $\omega(Y)$. The same argument applies to $x = 0$, and one argues similarly for $x \in (2,3]$. In particular, one finds $\omega([1,2]) = [1,2]$, while $\bigcup_{x \in [1,2]} \omega(x) = \{1,2\}$.

Furthermore, the limit set of a subset Y can strictly include Y, e.g., for $Y = (0,3)$ it holds that $\omega(Y) = [0,3]$: In order to show that $0 \in \omega(Y)$ let $x \in (0,1)$. Define $y_n := \Phi(-2n, x)$ and $x_n := \Phi(-n, x)$, then $\Phi(n, y_n) = \Phi(n, \Phi(-2n, x)) = \Phi(-n, x) = x_n$ and $\lim x_n = 0$. Hence with $t_n := n$ and y_n as above we have $\Phi(t_n, y_n) \to 0$. The argument is similar for proving that all points in $[0,3]$ are in $\omega(Y)$ and it is clear that $\omega(Y) \subset [0,3]$.

Here are some elementary properties of limit sets. We will use the following notion: A metric space X is called connected, if it cannot be written as the disjoint union of two nonvoid open sets $U, V \subset X$. Thus $X = U \cup V, U \cap V = \emptyset$ implies $U = \emptyset$ or $V = \emptyset$. The intersection of a decreasing sequence of compact connected sets is compact connected; see Exercise 3.4.1.

Proposition 3.1.5. *Let Φ be a continuous flow on a compact metric space X and let $x \in X$.*

(i) The ω-limit set $\omega(x)$ is a compact connected set which is invariant under the flow Φ, i.e.,

$$\Phi(t,y) \in \omega(x) \text{ for all } y \in \omega(x) \text{ and all } t \in \mathbb{R}.$$

(ii) The α-limit set $\alpha(x)$ is the ω-limit set of x under the time-reversed flow $(t,x) \mapsto \Phi^*(t,x) := \Phi(-t,x)$. In particular, $\alpha(x)$ is compact connected and invariant under Φ.

Proof. (i) Compactness follows from

$$\omega(x) = \bigcap_{T>0} \text{cl}\{\Phi(t,x) \mid t \geq T\}.$$

This also implies that $\omega(x)$ is connected, since it is the intersection of a decreasing sequence of compact connected sets. For invariance, note that for $y \in \omega(x)$ there are $t_k \to \infty$ with $\Phi(t_k, x) \to y$. Hence, by continuity, it follows for all $t \in \mathbb{R}$ that

$$\Phi(t_k + t, x) = \Phi(t, \Phi(t_k, x)) \to \Phi(t, y) \in \omega(x).$$

(ii) This is immediate from the definitions. □

Limit sets describe the dynamical behavior of the considered system. However, topologically, they may be rather complicated objects, since they need not be isolated from each other. Then it may be advantageous to introduce a more general notion. Before we define it, we modify Example 3.1.2 concerning linear autonomous differential equations.

Example 3.1.6. Consider the linear autonomous differential equation on $\mathbb{R}^2$ given by

$$\begin{bmatrix} \dot{x}_1 \\ \dot{x}_2 \end{bmatrix} = \begin{bmatrix} 1 & 0 \\ 0 & 1 \end{bmatrix} \begin{bmatrix} x_1 \\ x_2 \end{bmatrix}$$

with solutions $x(t) = (e^t x_1^0, e^t x_2^0), t \in \mathbb{R}$. For the induced dynamical system on the unit sphere obtained by projection, every point is an equilibrium, hence an ω-limit set. Observe that here the eigenspace for the eigenvalue 1 is the Lyapunov space $L(1) = \mathbb{R}^2$.

In order to overcome the difficulties connected with nonisolated limit sets we introduce the concept of chains, which admit small jumps. This generalization of trajectories leads us to the notion of chain transitive sets.

Definition 3.1.7. For a flow Φ on a complete metric space X with metric d, and $\varepsilon, T > 0$ an (ε, T)-chain ξ from $x \in X$ to $y \in X$ is given by

$$n \in \mathbb{N}, \ x_0 = x, \ldots, x_n = y, \ T_0, \ldots, T_{n-1} > T$$

with

$$d(\Phi(T_i, x_i), x_{i+1}) < \varepsilon \text{ for } i = 0, 1, \ldots, n-1.$$

The total time of ξ is $\tau(\xi) := \sum_{i=0}^{n-1} T_i$. A nonvoid set $K \subset X$ is chain transitive if for all $x, y \in K$ and all $\varepsilon, T > 0$ there is an (ε, T)-chain from x to y. A point x is called chain recurrent, if for all $\varepsilon, T > 0$ there is an (ε, T)-chain from x to x and the chain recurrent set $\mathcal{R}$ is the set of all chain recurrent points. The flow Φ is called chain transitive if X is a chain transitive set, and Φ is chain recurrent if $\mathcal{R} = X$.

We also say that a point y is chain reachable from x, if for all $\varepsilon, T > 0$ there is an (ε, T)-chain from x to y. A number of remarks on this definition may be helpful: Note that the number n of 'jumps' is not bounded. As the notation suggests, only small values of $\varepsilon > 0$ are of interest. In particular, also 'trivial jumps' where $x_{i+1} = \Phi(T_i, x_i)$ are allowed. Furthermore, the (ε, T)-chains used to characterize a chain transitive set K need not be contained in K. A set consisting of chain recurrent points need not be chain transitive.

Example 3.1.8. Simple examples of chain transitive sets are given by a fixed point or a periodic orbit; see Exercise 3.4.2.

In Example 3.1.6 the unit sphere is chain transitive: Given $\varepsilon, T > 0$ one can get from any point s_1 on the unit sphere to any other point s_2 by an (ε, T)-chain constructed as follows: Stay in s_1 for a time $T_1 > T$, then jump to another equilibrium with distance less than ε, stay there for a time $T_2 > T$, etc., until s_2 is reached. Thus, here, the nonisolated ω-limit sets form a single chain transitive set. One also sees that 'reachability of y from x by chains' may be very different from 'existence of trajectories from x to y'.

Proposition 3.1.9. *Every chain transitive set is contained in a maximal chain transitive set, called a chain component.*

Proof. For a chain transitive set K consider the union $\bigcup K'$ over all chain transitive sets K' containing K. We show that $\bigcup K'$ is chain transitive, hence it is the maximal chain transitive set containing K. In fact, consider $x, y \in \bigcup K'$ and let $\varepsilon, T > 0$. Then there are chain transitive sets K_x and K_y containing K with $x \in K_x$ and $y \in K_y$. Pick $z \in K$. Then there are (ε, T)-chains from x to z and from z to y. Hence there is an (ε, T)-chain from x to y, since the concatenation of (ε, T)-chains is again an (ε, T)-chain. $\square$

The following proposition shows that for chain transitivity it is not important to use chains with arbitrarily large jump times T_i.

Proposition 3.1.10. *Let Φ be a continuous flow on a compact metric space X. A set $K \subset X$ is chain transitive if and only if for all $x, y \in K$ and all $\varepsilon > 0$ there is an (ε, T)-chain from x to y with all jump times $T_i \in (1, 2]$.*

Proof. If $K \subset X$ is chain transitive, we may introduce 'artificial' trivial jumps (of size zero), thus reducing the jump times such that they are in $(1,2]$. For the converse, it suffices to show the following: Let $x, y \in K$ and let $\tau > 0$. If for every $\varepsilon > 0$ there exists an (ε, τ)-chain from x to y, then for every $\varepsilon, T > 0$ there exists an (ε, T)-chain from x to y. This, in turn, follows, if we can show that for every $\varepsilon > 0$ there is an $(\varepsilon, 2\tau)$-chain from x to y. By compactness of X the map Φ is uniformly continuous on $[0, 3\tau] \times X$. Hence there is $\delta \in \left(0, \frac{\varepsilon}{2}\right)$ such that for all $a, b \in X$ and $t \in [0, 3\tau]$:

$$d(a,b) < \delta \text{ implies } d(\Phi(t,a), \Phi(t,b)) < \frac{\varepsilon}{2}.$$

Now let a (δ, τ)-chain $x_0 = x, x_1, \ldots, x_n = y$ with times $\tau_0, \ldots, \tau_{n-1} \geq \tau$ be given. We may assume that $\tau_i \in [\tau, 2\tau]$. We also may assume that $n \geq 2$, because we may concatenate this chain with a chain from y to y. Thus there are $q \in \{0, 1, \ldots\}$ and $r \in \{2, 3\}$ with $n = 2q + r$. We obtain an $(\varepsilon, 2\tau)$-chain from x to y given by points

$$y_0 = x, \ y_1 = x_2, \ y_2 = x_4, \ldots, \ y_q = x_{2q}, \ y_{q+1} = x_n = y$$

with times

$$T_0 = \tau_0 + \tau_1, \ T_1 = \tau_2 + \tau_3, \ldots, \ T_q = \tau_{2q} + \tau_n.$$

This follows by the triangle inequality and the choice of δ. □

The following proposition strengthens the assertion of Proposition 3.1.10 in a special case. This will be used in Section 5.2 for systems on Grassmannians.

Proposition 3.1.11. *Let Φ be a continuous flow on a compact metric space X and consider a chain transitive set K containing an equilibrium or a periodic solution and let $x, y \in K$. Then for all $\varepsilon > 0$ there is an (ε, T)-chain from x to y with all jump times $T_i = 1$ and the number of jumps may be taken arbitrarily large.*

Proof. Suppose that e is an equilibrium in K. By Proposition 3.1.10, we find for every $\varepsilon > 0$ a chain from x to e with all jump times in the interval $(1,2]$. Then, using continuous dependence on initial values and uniform continuity as in the proof of Proposition 3.1.10, one can construct a chain with all jumps less than ε and all jump times T_i adjusted to 1, except possibly the last one, say T_n, which we may take in $(0,1)$. Since $d(\Phi(T_n, x_n), e)$ can be made arbitrarily small, we may assume that (using continuity again)

$$d(\Phi(1, x_n), e) = d(\Phi(1 - T_n, \Phi(T_n, x_n)), \Phi(1 - T_n, e)) < 2\varepsilon.$$

Thus, also, the last jump time may be taken equal to 1. In a similar vein, starting in the equilibrium e one may adjust all jump times of a chain to y equal to 1. Hence the points x and y can be connected by such chains by first going from x to the equilibrium, and from there to y. Finally, we may introduce arbitrarily many trivial jumps at the equilibrium e. If there is a periodic solution in K, one argues similarly by adjusting jump times around the period. □

Next we show that α-limit sets and ω-limit sets are chain transitive.

Proposition 3.1.12. *Let Φ be a continuous flow on a compact metric space X. Then for all $x \in X$ the limit set $\omega(x)$ is chain transitive.*

Proof. Let $y, z \in \omega(x)$ and fix $\varepsilon > 0$. By continuity, one finds $\delta > 0$ such that for all y_1 with $d(y_1, y) < \delta$ one has $d(\Phi(2, y_1), \Phi(2, y)) < \varepsilon$. By definition of $\omega(x)$ there are times $S > 0$ and $T > S + 3$ such that

$$d(\Phi(S, x), y) < \delta \text{ and } d(\Phi(T, x), z) < \varepsilon.$$

Thus the chain $y_0 = y$, $y_1 = \Phi(S + 2, x)$, $y_2 = z$ with jump times $T_0 := 2$ and $T_1 = T - (S + 2) > 1$ is an $(\varepsilon, 1)$-chain from y to z and the assertion follows from Proposition 3.1.10. □

Together with Proposition 3.1.5 this also implies that α-limit sets are chain transitive, since the next proposition shows that chain transitivity remains invariant under time reversal.

Proposition 3.1.13. *Let Φ be a continuous flow on a compact metric space X.*

(i) Let $x, y \in X$ and suppose that for all $\varepsilon, T > 0$ there is an (ε, T)-chain from x to y. Then for the time-reversed flow Φ^* has the property that for all $\varepsilon, T > 0$ there is an (ε, T)-chain from y to x.

(ii) A chain transitive set K for Φ is also chain transitive for the time-reversed flow.

Proof. Exercise 3.4.6. □

An example of a flow for which the union of the limits sets from points is strictly contained in the chain recurrent set can be obtained as follows:

Example 3.1.14. Let a continuous flow Φ on $X := [0, 1] \times [0, 1]$ be defined such that all points on the boundary are fixed points, and the orbits for points $(x, y) \in (0, 1) \times (0, 1)$ are straight lines $\Phi(\cdot, (x, y)) = \{(z_1, z_2) \mid z_1 = x, z_2 \in (0, 1)\}$ with $\lim_{t \to \pm\infty} \Phi(t, (x, y)) = (x, \pm 1)$. For this system, each point on the boundary is its own α- and ω-limit set. The α-limit sets for points in the interior $(x, y) \in (0, 1) \times (0, 1)$ are of the form $\{(x, -1)\}$, and the ω-limit sets are of the form $\{(x, 1)\}$. On the other hand, the whole space $X = [0, 1] \times [0, 1]$ is chain transitive.

The concepts of limit sets and chain transitive sets describe the qualitative behavior of a dynamical system. If these concepts describe intrinsic properties of a system that can be used for its characterization, they should survive under topological conjugacies. The next results show that this is actually true, thus extending the results of Proposition 2.1.9. A closed set Y is called minimal invariant if for every $y \in Y$ the closure of its orbit coincides with Y, i.e., $\mathrm{cl}\{\Phi(t,y) \mid t \in \mathbb{R}\} = Y$.

Proposition 3.1.15. *Let $h : X \to X$ be a topological conjugacy for two dynamical systems $\Phi, \Psi : \mathbb{R} \times X \longrightarrow X$ on a compact metric state space X.*

(i) For $Y \subset X$ the limit sets $\alpha(Y)$ and $\omega(Y)$ of Φ are mapped onto the limit sets $\alpha(h(Y))$ and $\omega(h(Y))$ of Ψ, respectively.

(ii) If $Y \subset X$ is an invariant set for Φ, then $h(Y)$ is an invariant set for Ψ. In particular, closed minimal invariant sets are mapped onto closed minimal invariant sets.

(iii) If $Y \subset X$ is a chain transitive set for Φ, then $h(Y)$ is chain transitive for Ψ. In particular, maximal chain transitive sets are mapped onto maximal chain transitive sets and the chain recurrent set of Φ is mapped onto the chain recurrent set of Ψ.

Proof. The proof of assertion (i) is left to the reader in Exercise 3.4.5. Assertion (ii) follows, since a topological conjugacy maps orbits onto orbits and closures of orbits onto closures of orbits. For (iii), it suffices to show that for a chain transitive set $Y \subset X$ of Φ the set $h(Y)$ is chain transitive for Ψ. The conjugacy h is a homeomorphism and X is compact by assumption. Hence for every $\varepsilon > 0$ there is $\delta > 0$ such that $d(x, x') < \delta$ implies $d(h(x), h(x')) < \varepsilon$ for all $x, x' \in X$. This shows that any (δ, T)-chain connecting points $x, y \in Y$ is mapped to an (ε, T)-chain connecting $h(x), h(y) \in h(Y)$, since for all i,

$$d(\Psi(T_i, h(x_i)), h(x_{i+1})) = d(h(\Phi(T_i, x_i)), h(x_{i+1})) < \varepsilon. \qquad \square$$

3.2. The Chain Recurrent Set

This section gives further insight into the global behavior of flows. The maximal chain transitive sets, i.e., the chain components, are the connected components of the chain recurrent set and all ω-limit sets $\omega(x)$ and all α-limit sets $\alpha(x)$ are contained in chain components. Technically, the results of this section will only be needed in Chapters 8 and 9, but they also help to appreciate the relevance and beauty of the concept of chain transitivity.

We start with two simple examples.

Example 3.2.1. Consider again the dynamical system Φ discussed in Example 3.1.4 on $X = [0, 3]$ given by

$$\dot{x} = x(x-1)(x-2)^2(x-3).$$

Obviously, the fixed points $x^* = 0, 1, 2, 3$ are chain recurrent. In this example, there are no other chain recurrent points, which can be seen as follows: Consider a point $x \in [0, 3]$ that is not a fixed point and let $\delta := \min |x - x^*|$, where x^* is a fixed point. Let $0 < \varepsilon \leq \delta/3$. Denote $a := \lim_{t \to \infty} \Phi(t, x)$ and let $T := \min\{t > 0 \,|\, |\Phi(t, x) - a| = \varepsilon\}$. We claim that there is no (ε, T)-chain from x to x. Consider an (ε, T)-chain starting in $x_0 = x$. Then for $T_0 > T$, one has $|\Phi(T_0, x_0) - a| \leq \varepsilon$, since convergence of $\Phi(t, x)$ to a is monotone. For x_1 with $|\Phi(T_0, x_0) - x_1| < \varepsilon$ it follows that $|x - x_1| > \varepsilon$ and there are two possibilities: (a) $x_1 \notin \{\Phi(t, x) \,|\, t \geq 0\}$, in this case $\inf\{|x - \Phi(t, x_1)| \,|\, t \geq 0\} \geq \delta \geq 3\varepsilon$. (b) $x_1 \in \{\Phi(t, x) \,|\, t \geq 0\}$, in this case $|a - \Phi(t, x_1)| \leq \varepsilon$ for all $t \geq T$ and hence $|x - \Phi(t, x_1)| > \varepsilon$. Repeating the construction for $\Phi(T_1, x_1)$ with $T_1 > T$ and x_2 with $|\Phi(T_1, x_1) - x_2| < \varepsilon$ we see that for all $n \in \mathbb{N}$ it holds that $|x_n - x| > \varepsilon$, and hence there is no (ε, T)-chain from x to x.

The key to Example 3.2.1 is that trajectories starting from x 'move away' and cannot return, even using jumps of size ε, to x because of the topology of the state space $[0, 3]$. This is different in the following example.

Example 3.2.2. Consider the compact metric space $\mathbb{S}^1$, the one-dimensional sphere, which we identify here with $\mathbb{R}/(2\pi\mathbb{R})$. On $\mathbb{S}^1$ the differential equation

$$\dot{x} = \sin^2 x$$

defines a dynamical system. In this case we have $\mathcal{R} = \mathbb{S}^1$, i.e., the entire circle is the chain recurrent set: Let $x \in \mathbb{S}^1$ and $\varepsilon, T > 0$ be given, assume without loss of generality that $x \in (0, \pi]$. Since $\lim_{t \to \infty} \Phi(t, x) = \pi$ there is $T_0 > T$ with $d(\Phi(T_0, x), \pi) < \frac{\varepsilon}{2}$. Pick $x_1 \in N(\pi, \frac{\varepsilon}{2}) \cap (\pi, 2\pi)$ (note that 2π is identified with 0). Because of $\lim_{t \to \infty} \Phi(t, x_1) = 0$ there is $T_1 > T$ with $d(\Phi(T_1, x_1), 0) < \frac{\varepsilon}{2}$. Furthermore, $\lim_{t \to -\infty} \Phi(t, x) = 0$ and hence there is $T_2 > T$ with $x_2 := \Phi(-T_2, x) \in N(0, \frac{\varepsilon}{2})$. Now $x = x_0$, x_1, x_2, $x_3 = x$ form an (ε, T)-chain from x to x. In a similar way one constructs for any $\varepsilon, T > 0$ an (ε, T)-chain from x to y for any two points $x, y \in \mathbb{S}^1$, showing that this dynamical system is chain transitive and hence chain recurrent on $\mathbb{S}^1$.

Next we discuss chain transitive sets which are maximal with respect to set inclusion. We will need the following result about compact metric spaces.

Theorem 3.2.3 (Blaschke). *The set of nonvoid closed subsets of a compact metric space becomes a compact metric space under the Hausdorff distance*

$$(3.2.1) \qquad d_H(A, B) = \max\left\{ \max_{a \in A}\left[\min_{b \in B} d(a,b)\right], \max_{b \in B}\left[\min_{a \in A} d(a,b)\right] \right\}.$$

In fact, one can verify that this space is complete and totally bounded and hence compact.

Proposition 3.2.4. *Let Φ be a flow on a compact metric space X.*

(i) Chain components, i.e., maximal chain transitive sets, are closed and invariant.

(ii) The flow restricted to a maximal chain transitive subset is chain transitive. In particular, the flow restricted to the chain recurrent set $\mathcal{R}$ is chain recurrent.

Proof. The proof uses repeatedly that the map $\Phi : [0,2] \times X \to X$ is uniformly continuous.

(i) In order to see closedness consider a sequence (y_n) in a maximal chain transitive set Y with $y_n \to y$. Then it is obvious that for all $z \in Y$ and $\varepsilon, T > 0$ there is an (ε, T)-chain from z to y. Conversely, chains from y_n to z lead to chains from y to z, using uniform continuity. For invariance, let $\tau \in \mathbb{R}$ and $y \in Y$. In order to show that the point $\Phi(\tau, y)$ is in Y, consider for $\varepsilon > 0$ and $T > 0$ an $(\varepsilon, T + |\tau|)$-chain with $y_0 = y, \ldots, y_n = y$ from y to itself. Then $\Phi(\tau, y), y_0, \ldots, y_n$ gives an (ε, T)-chain from $\Phi(\tau, y)$ to y. In order to construct an (ε, T)-chain from y to $\Phi(\tau, y)$, note that by continuity there is $\delta \in (0, \varepsilon)$ such that $d(x, x') < \delta$ implies $d(\Phi(\tau, x), \Phi(\tau, x')) < \varepsilon$. Then a $(\delta, T + |\tau|)$-chain $y_0 = y, \ldots, y_{n-1}, y$ gives rise to an (ε, T)-chain with $y_0 = y, \ldots, y_{n-1}, \Phi(\tau, y)$.

(ii) A maximal chain transitive set $\mathcal{M}$ and the chain recurrent set $\mathcal{R}$ are invariant. Hence it suffices to show that any two points in $\mathcal{M}$ can be connected by chains with jump points $x_i \in \mathcal{M}$. Let $y, y' \in \mathcal{M}$. For every $p \in \mathbb{N}$ there is an $(1/p, 1)$-chain in X from y to y', say with $x_0 = y, x_1, \ldots, x_m = y' \in X$ and times $T_0, \ldots, T_{m-1} \in (1, 2]$. Similarly, there is a $(1/p, 1)$-chain in X from y' to y which, for convenience, we denote by $x_m = y', \ldots, x_n = y$ with times $T_m, \ldots, T_{n-1} \in (1, 2]$ (everything depends on p, naturally). Define compact sets $K^p := \bigcup_{i=0}^{n} \Phi([0, T_i], x_i)$. By Blaschke's theorem, Theorem 3.2.3, there exists a subsequence of K^p converging in the Hausdorff metric d_H to some nonvoid compact subset $K \subset X$ with $y, y' \in K$.

Claim: For all $x, z \in K$ and all $q \in \mathbb{N}$ there is a $(1/q, 1)$-chain in K with times $\tau_0, \ldots, \tau_{r-1} \in (1, 2]$ from x to z.

If this claim is true, then, in particular, $y, y' \in K \cap \mathcal{M}$, which with maximality of the chain transitive set $\mathcal{M}$ implies $K \subset \mathcal{M}$ and hence it

3.2. The Chain Recurrent Set

follows that $y, y' \in \mathcal{M}$ can be connected by chains with points in $K \subset \mathcal{M}$ and the assertion follows.

The claim is proved as follows. Let $x, z \in K$ and $q \in \mathbb{N}$. By uniform continuity of Φ on X there is a number $\delta \in (0, \frac{1}{3q})$ such that

$$d(a, b) < \delta \text{ implies } d(\Phi(t, a), \Phi(t, b)) < 1/(6q) \text{ for all } t \in [0, 2], a, b \in X.$$

Choosing $p \in \mathbb{N}$ with $p > \max\{6q, \delta^{-1}\}$ and $d_H(K^p, K) < \delta$ one can construct a $(\frac{1}{q}, 1)$-chain from x to z in K as required. In fact, one finds for $x, z \in K$ points $\tilde{x} = \Phi(t_1, x_k)$ and $\tilde{z} = \Phi(t_2, x_l)$ in K^p with $d(x, \tilde{x}) < \delta, d(z, \tilde{z}) < \delta$ and $t_1 \in [0, T_k], t_2 \in [0, T_l]$, and $k, l \in \{0, 1, \ldots, n-1\}$. Without loss of generality, one may take $l \geq k + 3$ (otherwise, one follows the chain from $\tilde{x}$ to y, then back to x, and then to $\tilde{z}$). Then define the following auxiliary chain in K^p:

$$\xi_0 = \tilde{x}, \xi_1 := x_{k+2}, \xi_2 := x_{k+3}, \ldots, \xi_{l-k-1} := x_{l-1} = \tilde{z},$$
$$\tau_0 := T_k - t_1 + T_{k+1}, \tau_1 := T_{k+2}, \ldots, \tau_{l-k-2} := T_{l-1} + t_2.$$

All $\tau_i > 1$ and by the choice of δ and q one sees that

$$d(\Phi(\tau_0, \xi_0), \xi_1) < \frac{1}{3q} \text{ and } d(\Phi(\tau_{l-k-2}, \xi_{l-k-2}), \xi_{l-k-1}) < \frac{1}{3q}.$$

Furthermore, for the other jumps one has

$$d(\Phi(\tau_i, \xi_i), \xi_{i+1}) < \frac{1}{p} < \frac{1}{3q}.$$

Thus we have constructed a $(\frac{1}{3q}, 1)$-chain in K^p. Introducing, if necessary, trivial jumps, we may assume that all jump times $\tau_i \in (1, 2]$. Since $d_H(K^p, K) < \delta$, we find $\eta_i \in K$ with $d(\xi_i, \eta_i) < \delta$ for $i = 1, \ldots, l - k - 2$ and let $\eta_0 = x$ and $\eta_{l-k-1} = z$. Then it follows for all $i = 0, 1, \ldots, l-k-2$,

$$d(\Phi(\tau_i, \eta_i), \eta_{i+1}) \leq d(\Phi(\tau_i, \eta_i), \Phi(\tau_i, \xi_i)) + d(\Phi(\tau_i, \xi_i), \xi_{i+1}) + d(\xi_{i+1}, \eta_{i+1})$$
$$< \frac{1}{3q} + \frac{1}{3q} + \frac{1}{3q} = \frac{1}{q}.$$

This shows that we have constructed a $(\frac{1}{q}, 1)$-chain from x to z with all $\eta_i \in K$. □

We need the following properties of connected sets. A subset $A \subset X$ is connected if it cannot be written as the disjoint union of two (relative to A) open sets. The union of all connected sets $A \subset X$ containing a point $x \in X$ is again connected and called the connected component of x. In fact, the union of (an arbitrary family of) connected sets with nonvoid intersection is connected. This is seen as follows. X is not connected if and only if there is a continuous surjective map $f: X \to \{0, 1\}$; then one can define $U := f^{-1}(0)$ and $V := f^{-1}(1)$. Thus any continuous map $f: X \to \{0, 1\}$ is constant on a connected subset. If $Y = \bigcup_\alpha A_\alpha$ is the connected component of x, where the A_α denote the connected sets containing x, then a continuous function $f : Y \to \{0, 1\}$ is constant on every A_α, hence on Y. Similar arguments show that the closure of a connected set is connected, hence connected components are closed.

Proposition 3.2.5. *A closed subset Y of a compact metric space X is chain transitive if it is chain recurrent and connected. Conversely, if the flow on X is chain transitive, then X is connected.*

Proof. Suppose first that Y is chain recurrent and connected. Let $x, y \in Y$ and fix $\varepsilon, T > 0$. Cover Y by balls $N(y, \frac{\varepsilon}{4})$. By compactness there are finitely many points, say $y_1, \ldots, y_n \in Y$ such that for all $z \in Y$ there is y_i with $d(z, y_i) < \frac{\varepsilon}{4}$. Let the index set $I_1 := \{i \in \{1, \ldots, n\} \mid x \in N(y_i, \frac{\varepsilon}{4})\}$ and let inductively

$$I_{k+1} := \{i \notin I_1 \cup \ldots \cup I_k \mid d(y_i, y_j) < \varepsilon/4 \text{ for some } j \in I_1 \cup \ldots \cup I_k\}, k \geq 1.$$

Since Y is connected, the union of all I_k coincides with $\{1, \ldots, n\}$. Hence one can number the y_i such that $x \in N(y_1, \frac{\varepsilon}{4}), y \in N(y_K, \frac{\varepsilon}{4})$ for some K and such that for all i the distance between y_i and y_{i+1} is bounded above by $\frac{\varepsilon}{2}$. Now use that by chain recurrence of the flow there are $(\frac{\varepsilon}{4}, T)$-chains from y_i to y_i for $i = 0, 1, \ldots, n-1$. Appropriate concatenation of these chains leads to an (ε, T)-chain from x to y. Hence chain transitivity follows. Conversely, let the flow on X be chain transitive. If X is not connected, it can be written as the disjoint union of nonvoid open sets V and W. Then these sets are also closed, hence compact and

$$\varepsilon_0 := \inf \{d(v, w) \mid v \in V, w \in W\} > 0.$$

Hence for $\varepsilon < \varepsilon_0/2$ there cannot exist (ε, T)-chains from an element of V to an element of W. $\square$

The following theorem provides a characterization of the chain components. It gives fundamental information on the structure of flows on compact metric spaces.

Theorem 3.2.6. *Let Φ be a flow on a compact metric space X.*

(i) The connected components of the chain recurrent set $\mathcal{R}$ coincide with the maximal chain transitive subsets of $\mathcal{R}$, the chain components.

(ii) The flow restricted to a connected component of $\mathcal{R}$ is chain transitive.

(iii) Each ω-limit set $\omega(x)$ and each α-limit set $\alpha(x)$ is contained in a chain component.

Proof. By Proposition 3.2.4 we know that the flow restricted to a maximal chain transitive subset $\mathcal{R}_0$ of $\mathcal{R}$ is chain transitive. Hence, by the second part of Proposition 3.2.5, the set $\mathcal{R}_0$ is connected and thus contained in a connected component of $\mathcal{R}$. Conversely, the first part of Proposition 3.2.5 implies that every connected component of $\mathcal{R}$ is chain transitive, because it is closed, chain recurrent, and connected. Hence assertions (i) and (ii) follow. Assertion (iii) follows by Propositions 3.1.5 and 3.1.13. □

Example 3.2.7. In Example 3.2.1 the chain components are $\{0\}, \{1\}, \{2\}, \{3\} \subset X = [0, 3]$.

We also note the following simple lemma, which indicates a uniform upper bound for the total time needed to connect any two points in a chain component.

Lemma 3.2.8. *Let $\mathcal{M}$ be a chain component and fix $\varepsilon, T > 0$. Then there exists $\bar{T}(\varepsilon, T) > 0$ such that for all $x, y \in \mathcal{M}$ there is an (ε, T)-chain ξ from x to y with total time $\tau(\xi) \leq \bar{T}(\varepsilon, T)$.*

Proof. By assumption, one finds for all $x, y \in \mathcal{M}$ an $(\frac{\varepsilon}{2}, T)$-chain from x to y. Using continuous dependence on initial values and compactness, one finds finitely many (ε, T)-chains connecting every $x \in \mathcal{M}$ with a fixed $z \in \mathcal{M}$. One also finds finitely many (modulo their endpoints) (ε, T)-chains connecting z with arbitrary elements $y \in \mathcal{M}$. Thus one ends up with finitely many (ε, T)-chains connecting all points in $\mathcal{M}$. The maximum of their total times is an upper bound $\bar{T}(\varepsilon, T)$. □

3.3. The Discrete-Time Case

In this section, we introduce limit sets and chain transitivity for continuous dynamical systems in discrete time.

As noted in Section 2.3, every homeomorphism f on a complete metric space defines a continuous dynamical system in discrete time by $\varphi_n := f^n, n \in \mathbb{Z}$. We call this the dynamical system generated by f. Its global behavior is described by the following notions.

Definition 3.3.1. Let a dynamical system $\Phi : \mathbb{Z} \times X \longrightarrow X$ on a complete metric space X be given. For a subset $Y \subset X$ the α-limit set is defined as $\alpha(Y) := \{z \in X \mid \text{there exist sequences } (x_k) \text{ in } Y \text{ and } n_k \to -\infty \text{ in } \mathbb{Z}$ with $\lim_{k \to \infty} \Phi(n_k, x_k) = z\}$, and similarly the ω-limit set of Y is defined as $\omega(Y) := \{z \in X \mid \text{there exist sequences } (x_k) \text{ in } Y \text{ and } n_k \to \infty \text{ in } \mathbb{Z}$ with $\lim_{k \to \infty} \Phi(n_k, x_k) = z\}$.

If the set Y consists of a single point x, we just write the limit sets as $\omega(x)$ and $\alpha(x)$, respectively. Where appropriate, we also write $\omega(x, f), \alpha(x, f)$ if the considered dynamical system is generated by f. Note that if X is compact, the α-limit sets and ω-limit sets are nonvoid for all $Y \subset X$. If Φ is generated by f, then the dynamical system Φ^* generated by f^{-1} satisfies

$$\Phi^*(n,x) = \Phi(-n,x) \text{ for all } n \in \mathbb{Z} \text{ and all } x \in X.$$

Thus Φ^* is the time-reversed system. Here are some elementary properties of limit sets.

Proposition 3.3.2. *Let Φ be a continuous dynamical system generated by a homeomorphism f on a compact metric space X. Then for every $x \in X$ the following holds true.*

(i) The ω-limit set $\omega(x)$ is a compact set which is invariant under f, i.e.,

$$f(y), f^{-1}(y) \in \omega(x) \text{ for all } y \in \omega(x).$$

(ii) The α-limit set $\alpha(x)$ is the ω-limit set of x for the time-reversed system.

Proof. (i) Compactness follows from

$$\omega(x) = \bigcap_{N \in \mathbb{N}} \mathrm{cl}\{\Phi(n,x) \mid n \geq N\}.$$

For invariance, note that for $y \in \omega(x)$ there are $n_k \to \infty$ with $\Phi(n_k, x) \to y$. Hence, by continuity, it follows for every $n \in \mathbb{Z}$ (in particular, for $n = \pm 1$) that

$$\Phi(n_k + n, x) = \Phi(n, \Phi(n_k, x)) \to \Phi(n, y) \in \omega(x).$$

(ii) This is immediate from the definitions. $\square$

Next the concept of chains and chain transitivity is introduced.

Definition 3.3.3. For a dynamical system Φ generated by a homeomorphism f on a complete metric space X with metric d and $\varepsilon > 0$ an ε-chain from $x \in X$ to $y \in X$ is given by $x_0 = x, \ldots, x_n = y$ with

$$d(f(x_i), x_{i+1}) < \varepsilon \text{ for all } i.$$

A set $K \subset X$ is chain recurrent, if for all $x \in K$ and all $\varepsilon > 0$ there is an ε-chain from x to x. It is called chain transitive if for all $x, y \in K$ and all $\varepsilon > 0$ there is an ε-chain from x to y.

The chain recurrent set $\mathcal{R}$ is the set of all points that are chain recurrent, i.e., $\mathcal{R} = \{x \in X \mid \text{for all } \varepsilon > 0 \text{ there is an } \varepsilon\text{-chain from } x \text{ to } x\}$. The maximal chain transitive subsets are also called chain components or basic sets. The flow Φ is called chain transitive if X is a chain transitive set, and Φ is chain recurrent if $\mathcal{R} = X$.

3.3. The Discrete-Time Case

This definition of chains is considerably simpler than the definition in continuous time, Definition 3.1.7, since in discrete time all times are fixed to 1. See also Propositions 3.1.10 and 3.1.11 and the notes at the end of this chapter for a discussion.

Next we show that limit sets are chain transitive.

Proposition 3.3.4. *Let Φ be a dynamical system generated by a homeomorphism f on a compact metric space X. Then for every $x \in X$ the limit set $\omega(x)$ is chain transitive. In particular, the ω-limit sets $\omega(x)$ are contained in chain components.*

Proof. Let $y, z \in \omega(x)$ and fix $\varepsilon > 0$. By continuity, one finds $\delta > 0$ such that for all y_1 with $d(y_1, y) < \delta$ one has $d(f(y_1), f(y)) < \varepsilon$. By definition of $\omega(x)$ there are times $N \in \mathbb{N}$ and $K > N$ such that $d(f^N(x), y) < \delta$ and $d(f^K(x), z) < \varepsilon$. Thus the chain $x_0 = y$, $x_1 = f^{N+1}(x), \ldots, f^K(x), z$ is an ε-chain from y to z, and the assertion follows. $\square$

We note the following useful properties of chain transitive sets.

Proposition 3.3.5. *Consider a dynamical system Φ in discrete time on a compact metric space X.*

(i) Every chain transitive set K of Φ is also chain transitive for the time-reversed dynamical system Φ^*.

(ii) Let Y be a maximal chain transitive set, i.e., if $Y' \supset Y$ is chain transitive, then $Y' = Y$. If $K \cap Y \neq \varnothing$ for a chain transitive set K, then $K \subset Y$.

(iii) If a closed subset Y of a compact metric space X is chain recurrent and connected, then it is chain transitive.

Proof. For the proof of assertions (i) and (ii) see Exercise 3.4.7. For assertion (iii) suppose that Y is chain recurrent, connected, and closed. Let $x, y \in Y$ and fix $\varepsilon > 0$. Cover Y by balls of radius $\varepsilon/4$. By compactness there are finitely many points, say $y_1, \ldots, y_{n-1} \in Y$ such that for all $z \in Y$ there is y_i with $d(z, y_i) < \varepsilon/4$. Define $y_0 = x$ and $y_n = y$. Because Y is connected, one can choose the y_i such that the distance between y_i and y_{i+1} is bounded above by $\frac{\varepsilon}{2}$; see the proof of Proposition 3.2.5 for details. Now use that by chain recurrence of the flow there are $\varepsilon/4$-chains from y_i to y_i for $i = 0, 1, \ldots, n-1$. Appropriate concatenation of these chains leads to an ε-chain from x to y. Hence chain transitivity follows. $\square$

The converse of property (iii) in Proposition 3.3.5 is not true in the discrete-time case considered here: There are chain transitive sets that are not connected.

Proposition 3.3.6. *For a dynamical system in discrete time generated by a homeomorphism f on a compact metric space X the restriction to a chain component $\mathcal{M}$, i.e., a maximal chain transitive subset, is chain transitive. In particular, the flow restricted to the chain recurrent set $\mathcal{R}$ is chain recurrent.*

Proof. We have to show that any two points in $\mathcal{M}$ can be connected by chains with jump points $x_i \in \mathcal{M}$. Let $y, y' \in \mathcal{M}$. For every $p \in \mathbb{N}$ there is a $\frac{1}{p}$-chain in X from y to y', say with $x_0 = y, x_1, \ldots, x_m = y' \in X$. Similarly, there is a $\frac{1}{p}$-chain in X from y' to y which, for convenience, we denote by $x_m = y', \ldots, x_n = y$. Define compact sets $K^p := \{x_0, \ldots, x_n\}$ which consist of periodic $\frac{1}{p}$-chains, i.e., chains with coinciding initial and final point. By Blaschke's theorem, Theorem 3.2.3, there exists a subsequence of K^p converging in the Hausdorff metric d_H to some nonvoid compact subset $K \subset X$ with $y, y' \in K$.

Claim: For all $x, z \in K$ and all $q \in \mathbb{N}$ there is a $\frac{1}{q}$-chain in K from x to z.

Since $y, y' \in K \cap \mathcal{M}$, maximality of the chain transitive set $\mathcal{M}$ implies $K \subset \mathcal{M}$ by Proposition 3.3.5(ii), and hence $y, y' \in \mathcal{M}$ can be connected by chains with points $x_i \in K \subset \mathcal{M}$, and the assertion follows from the claim.

The claim is proved as follows. Let $x, z \in K$ and $q \in \mathbb{N}$. By uniform continuity on the compact set X, there is a number δ with $0 < \delta < \frac{1}{3q}$ such that
$$d(a,b) < \delta \text{ implies } d(f(a), f(b)) < \frac{1}{3q}.$$
Choosing $p \in \mathbb{N}$ with $p > 3q$ and $d_H(K^p, K) < \delta$ one can construct a $\frac{1}{q}$-chain from x to z in K as required. In fact, one finds for $x, z \in K$ points x_k and x_l in K^p with $d(x, x_k) < \delta$ and $d(z, x_l) < \delta$. Since K^p is a periodic chain, there is a $\frac{1}{p}$-chain in K^p from x_{k+1} to x_l which we denote for convenience by
$$\xi_0 = x_k, \xi_1, \ldots, \xi_n := x_l \text{ with } d(f(\xi_i), \xi_{i+1}) < \frac{1}{p} \text{ for all } i.$$
Since $d_H(K^p, K) < \delta$, we find $\eta_i \in K$ with $d(\xi_i, \eta_i) < \delta$ for $i = 1, \ldots, n-1$. Then, with $\eta_0 := x$ and $\eta_n := z$ we have constructed a $\frac{1}{q}$-chain in K from x to z. In fact, $d(x, \xi_0) = d(x, x_k^p) < \delta$ implies
$$d(f(x), \eta_1) \leq d(f(x), f(\xi_0)) + d(f(\xi_0), \xi_1) + d(\xi_1, \eta_1) \leq \frac{1}{3q} + \frac{1}{p} + \delta < \frac{1}{q}.$$
For $i = 1, \ldots, n-2$,
$$d(f(\eta_i), \eta_{i+1}) \leq d(f(\eta_i), f(\xi_i)) + d(f(\xi_i), \xi_{i+1}) + d(\xi_{i+1}, \eta_{i+1})$$
$$\leq \frac{1}{3q} + \frac{1}{p} + \delta < \frac{1}{q}.$$

Finally, for $i = n-1$,
$$d(f(\eta_{n-1}), \eta_n) \leq d(f(\eta_{n-1}), f(\xi_{n-1})) + d(f(\xi_{n-1}), x_l) + d(x_l, z)$$
$$\leq \frac{1}{3q} + \frac{1}{p} + \delta < \frac{1}{q}.$$

This shows that we have constructed a $\frac{1}{q}$-chain in K from x to z. $\square$

Many features of the qualitative behavior of a dynamical system are preserved under conjugacies defined in Definition 2.3.4. This includes fixed points, periodic orbits and limit sets. The next result shows that this also holds for chain transitivity.

Theorem 3.3.7. *Let $\Phi_1, \Phi_2 : \mathbb{Z} \times X \longrightarrow X$ be two dynamical systems generated by $f_1, f_2 : X \to X$, respectively, on a compact metric state space X and let $h : X \to X$ be a topological conjugacy for Φ and Ψ. Then h maps every chain transitive set of Φ_1 onto a chain transitive set of Φ_2.*

Proof. The conjugacy h is a homeomorphism and X is compact by assumption. Hence for all $\varepsilon > 0$ there exists a $\delta > 0$ such that for all $z \in X$ it holds that $N(z, \varepsilon) \subset h^{-1}(N(h(z), \delta))$. For a chain transitive set $N_2 \subset X$ of f_2, we claim that $N_1 := h^{-1}(N_2)$ is a chain transitive set of f_1: Take $p_1, q_1 \in N_1$ and fix $\varepsilon > 0$, $T > 0$. Choose δ as above and let ξ_2 be a δ-chain from $p_2 = h(p_1)$ to $q_2 = h(q_1)$. Then $h^{-1}(\xi_2) =: \xi_1$ is an ε-chain from p_1 to q_1. Now the assertion follows by considering h^{-1}. $\square$

3.4. Exercises

Exercise 3.4.1. Let $C_n, n \in \mathbb{N}$, be a decreasing sequence of (nonvoid) compact connected sets in a metric space X, i.e., $C_{n+1} \subset C_n$ for all $n \in \mathbb{N}$. Prove that $C := \bigcap_{n \in \mathbb{N}} C_n$ is a (nonvoid) compact connected set.
Hint: Suppose that $C = U \cup V$ with $U \cap V = \varnothing$ for open subsets $U, V \subset C$. Show that for $n \in \mathbb{N}$ the sets
$$F_n := C_n \cap (C_1 \setminus U) \cap (C_1 \setminus V)$$
form a decreasing sequence of compact sets and use that any decreasing sequence of nonvoid compact sets has nonvoid intersection.

Exercise 3.4.2. Let Φ be a continuous flow on a metric space X. (i) Show that a fixed point $x_0 = \Phi(t, x_0), t \in \mathbb{R}$, gives rise to the chain transitive set $\{x_0\}$. (ii) Show that for a T-periodic point $x_0 = \Phi(T, x_0)$ the orbit $\{\Phi(t, x_0) \,|\, t \in \mathbb{R}\}$ is a chain transitive set.

Exercise 3.4.3. Let Φ be a continuous flow on a compact connected metric space X and assume that the periodic points are dense in X. Show that X is chain transitive.

Exercise 3.4.4. Determine the chain components in the interval $[0,1] \subset \mathbb{R}$ of the ordinary differential equation
$$\dot{x} = \begin{cases} x^2 \sin(\frac{\pi}{x}) & \text{for } x \in (0,1], \\ 0 & \text{for } x = 0. \end{cases}$$

Exercise 3.4.5. Prove part (i) of Proposition 3.1.15: Let $h : X \to X$ be a C^0 conjugacy of dynamical systems $\Phi, \Psi : \mathbb{R} \times X \longrightarrow X$ on a compact metric state space X. Then for $Y \subset X$ the limit sets satisfy $h(\alpha(Y)) = \alpha(h(Y))$ and $h(\omega(Y)) = \omega(h(Y))$.

Exercise 3.4.6. Prove Proposition 3.1.13: Let Φ be a continuous flow on a compact metric space X. (i) Let $x, y \in X$ and suppose that for all $\varepsilon, T > 0$ there is an (ε, T)-chain from x to y. Then for the time-reversed flow Φ^* has the property that for all $\varepsilon, T > 0$ there is an (ε, T)-chain from y to x. (ii) A chain transitive set K for Φ is also chain transitive for the time-reversed flow.

Exercise 3.4.7. Prove Proposition 3.3.5(i) and (ii): For a dynamical system Φ in discrete time on a compact metric space X the following holds: (i) Every chain transitive set K of Φ is also chain transitive for the time-reversed dynamical system Φ^*. (ii) Let Y be a maximal chain transitive set. If $K \cap Y \neq \varnothing$ for a chain transitive set K, then $K \subset Y$.

Exercise 3.4.8. Let $f : X \to X$ and $g : Y \to Y$ be homeomorphisms on compact metric spaces X and Y, respectively. Show that the chain components for the dynamical system generated by $(f,g) : X \times Y \to X \times Y, (f,g)(x,y) := (f(x), g(y))$ are the products of the chain components of f and g.

Exercise 3.4.9. (i) Give an example of a compact metric space X where for some point $x \in X$ the boundary of the ε-neighborhood $N(x, \varepsilon)$ is different from $\{y \in X \mid d(y,x) = \varepsilon\}$. (ii) Show that for every metric space X and all $\varepsilon > \varepsilon' > 0$ the boundaries of the corresponding neighborhoods around a point $x \in X$ satisfy
$$\partial N(x, \varepsilon) \cap \partial N(x, \varepsilon') = \varnothing.$$

Exercise 3.4.10. Let the unit circle $\mathbb{S}^1$ be parametrized by $x \in [0, 1)$ and consider the dynamical system generated by $f : \mathbb{S}^1 \to \mathbb{S}^1$,
$$f(x) := \begin{cases} 2x & \text{for } x \in [0, \frac{1}{2}), \\ 2x - 1 & \text{for } x \in [\frac{1}{2}, 1). \end{cases}$$

(i) Show that f is continuous if one chooses as a metric on $\mathbb{S}^1$,
$$d(x,y) := \left| e^{2\pi i x} - e^{2\pi i y} \right|.$$

Intuitively and topologically this means that the points 0 and 1 are identified. Topologically equivalent is also $\tilde{d}(x,y) = \text{dist}(x-y, \mathbb{Z})$. (ii) Explain why this dynamical system is also called bit shift (use the binary representation of the real numbers). (iii) Use the characterization from (ii) to determine all periodic points. (iv) Show that $\mathbb{S}^1$ is a chain transitive set for f.

Exercise 3.4.11. For a system in discrete time, give an example of an ω-limit set $\omega(x)$ which is not connected.

3.5. Orientation, Notes and References

Orientation. This chapter has introduced limit sets for time tending to $\pm\infty$ and the notion of chain transitivity for dynamical systems on compact metric spaces: In continuous time, a set K is chain transitive if for all $x, y \in K$ and all $\varepsilon, T > 0$ there is a chain with jump sizes at most ε after time T. The maximal chain transitive sets contain all limit sets and they coincide with the connected components of the chain recurrent set; cf. Theorem 3.2.6. Hence a classification of the maximal chain transitive sets, also called chain components, gives insight into the global behavior of dynamical systems. Similar properties hold in discrete time. In the next chapter, we will use this approach in order to describe the Lyapunov spaces of linear autonomous dynamical systems in $\mathbb{R}^d$ by analyzing the induced systems on projective space.

Notes and references. Our definitions of conjugacies are restricted to dynamical systems defined on the same state space. This is only in order to simplify the notation, the notion of conjugacy extends in a natural way to systems on different state spaces.

We remark that in dimensions $d = 1$ and $d = 2$ limit sets in the continuous-time case simplify: Any limit set $\alpha(x)$ and $\omega(x)$ from a single point x of a differential equation in $\mathbb{R}^1$ consists of a single fixed point. A nonempty, compact limit set of a differential equation in $\mathbb{R}^2$, which contains no fixed points, is a periodic orbit. This is a main result of Poincaré-Bendixson theory. More generally, any nonempty, compact limit set of a differential equation in $\mathbb{R}^2$ consists of fixed points and connecting orbits (i.e., heteroclinic orbits for solutions tending to equilibria for $t \to \pm\infty$), or is a periodic orbit. Poincaré-Bendixson theory is a classical subject of the theory of ordinary differential equation, cf., e.g., Amann [4], Hirsch, Smale, and Devaney [70] or Teschl [133].

The monograph Alongi and Nelson [2] discusses chain transitivity and recurrence with many proofs given in detail. In particular, it is shown in [2, Theorem 2.7.18] that the result of Proposition 3.1.11 holds for arbitrary maximal chain transitive sets, i.e., for chain components $\mathcal{M}$: For all $x, y \in \mathcal{M}$ and all $\varepsilon > 0$ there is a chain with jump sizes less than ε from x

to y with all jump times $T_i = 1$. The proof is similar to the proof of Proposition 3.1.10, but more lengthy. A proof of Blaschke's Theorem, Theorem 3.2.3, is given in [**2**, Proposition C.0.15]. The characterization of compact metric spaces mentioned for the proof of Blaschke's Theorem can also be found in Bruckner, Bruckner, and Thompson [**22**, Theorem 9.58].

For additional details on the concepts and results of this chapter we refer the reader to Alongi and Nelson [**2**], Ayala-Hoffmann et al. [**11**], and Robinson [**117**]. A concise and slightly different treatment of the discrete-time case is given in Easton [**42**, Chapter 2].

An important question concerning the difference between ε-chains and trajectories is the following: Can one find arbitrarily close to an infinitely long chain a trajectory? For diffeomorphisms, the shadowing lemma due to Bowen gives an affirmative answer under hyperbolicity assumptions; cf. Katok and Hasselblatt [**75**, Section 18.1].

It is worth noting that we have dealt only with parts of the theory of flows on metric spaces based on Conley's ideas: One can construct a kind of Lyapunov function which strictly decreases along trajectories outside the chain recurrent set; cf. Robinson [**117**, Section 9.1]. Hence systems which are obtained by identifying the chain components with points are also called gradient-like systems. The notions of attractors, repellers and Morse decompositions will be treated in Chapter 8. Finally, we have not considered the important subject of Conley indices which classify isolated invariant sets; cf. Easton [**42**, Chapter 2] and Mischaikow [**107**].

Chapter 4

Linear Systems in Projective Space

In this chapter we return to matrices $A \in gl(d, \mathbb{R})$ and the dynamical systems defined by them. Geometrically, the invertible linear map e^{At} on $\mathbb{R}^d$ associated with A maps k-dimensional subspaces onto k-dimensional subspaces. In particular, the flow $\Phi_t = e^{At}$ induces a dynamical system on projective space, i.e., the set of all one-dimensional subspaces, and, more generally, on every Grassmannian, i.e., the set of all k-dimensional subspaces, $k = 1, \ldots, d$. As announced at the end of Chapter 2, we will characterize certain properties of A through these associated systems. More precisely, we will show in the present chapter that the Lyapunov spaces uniquely correspond to the chain components of the induced dynamical system on projective space. Chapter 5 will deal with the technically more involved systems on the Grassmannians.

Section 4.1 shows for continuous-time systems that the chain components in projective space characterize the Lyapunov spaces. Section 4.2 proves an analogous result in discrete time.

4.1. Linear Flows Induced in Projective Space

This section shows that the projections of the Lyapunov spaces coincide with the chain components in projective space.

We start with the following motivating observations. Consider the system in $\mathbb{R}^2$ given by

(4.1.1) $$\begin{bmatrix} \dot{x}_1 \\ \dot{x}_2 \end{bmatrix} = \begin{bmatrix} 0 & 1 \\ -1 & 0 \end{bmatrix} \begin{bmatrix} x_1 \\ x_2 \end{bmatrix}.$$

The nontrivial trajectories consist of circles around the origin (this is the linear oscillator $\ddot{x} = -x$.) The slope along a trajectory is $k(t) := \frac{x_2(t)}{x_1(t)}$. Using the quotient rule, one finds that it satisfies the differential equation

$$\frac{d}{dt}k(t) = \dot{k} = \frac{\dot{x}_2 x_1 - x_2 \dot{x}_1}{x_1^2} = -\frac{x_1^2}{x_1^2} - \frac{x_2^2}{x_1^2} = -1 - k^2,$$

as long as $x_1(t) \neq 0$. For $x_1(t) \to 0$ one finds $k(t) \to \infty$. Thus this nonlinear differential equation, a Riccati equation, has solutions with a bounded interval of existence. Naturally, this can also be seen by the solution formula for $k(t)$ with initial condition $k(0) = k_0$,

$$k(t) = \tan(-t + \arctan k_0), \ t \in \left(-\arctan k_0 - \frac{\pi}{2}, -\arctan k_0 + \frac{\pi}{2}\right).$$

Geometrically, this Riccati differential equation describes the evolution of a one-dimensional subspace (determined by the slope) under the flow of the differential equation (4.1.1). Note that for $x_1 \neq 0$ the points (x_1, x_2) and $(1, \frac{x_2}{x_1})$ generate the same subspace. The Riccati equation can describe this evolution only on a bounded time interval, since it uses the parametrization of the subspaces given by the slope, which must be different from $\pm\infty$, i.e., it breaks down on the x_2-axis. The analysis in projective space will avoid the artificial problem resulting from parametrizations.

These considerations are also valid in higher dimensions. Consider for a solution of $\dot{x} = Ax(t)$ with $x_1(t) \neq 0$ the vector $K(t) := \left[\frac{x_2(t)}{x_1(t)}, \ldots, \frac{x_d(t)}{x_1(t)}\right]^\top \in \mathbb{R}^{d-1}$. Partition $A = (a_{ij}) \in gl(d, \mathbb{R})$ in

$$A = \begin{bmatrix} a_{11} & A_{12} \\ A_{21} & A_{22} \end{bmatrix},$$

where $A_{12} = (a_{12}, \ldots, a_{1d})$, $A_{21} = (a_{21}, \ldots, a_{d1})^\top$ and $A_{22} \in gl(d-1, \mathbb{R})$. Then the function $K(\cdot)$ satisfies the Riccati differential equation

(4.1.2) $$\dot{K} = A_{21} + A_{22}K - Ka_{11} - KA_{12}K.$$

In fact, one finds from

$$\dot{x}_1 = a_{11}x_1 + (a_{12}, \ldots, a_{1d})\begin{bmatrix} x_2 \\ \vdots \\ x_d \end{bmatrix} \text{ and } \begin{bmatrix} \dot{x}_2 \\ \vdots \\ \dot{x}_d \end{bmatrix} = \begin{bmatrix} a_{21} \\ \vdots \\ a_{d1} \end{bmatrix}x_1 + A_{22}\begin{bmatrix} x_2 \\ \vdots \\ x_d \end{bmatrix}$$

the expression

$$\dot{K} = \begin{bmatrix} \dot{x}_2 \\ \vdots \\ \dot{x}_d \end{bmatrix}\frac{x_1}{x_1^2} - \begin{bmatrix} x_2 \\ \vdots \\ x_d \end{bmatrix}\frac{\dot{x}_1}{x_1^2} = A_{21} + A_{22}K - Ka_{11} - KA_{12}K.$$

Conversely, the same computations show that for any solution $K(t) = (k_2(t), \ldots, k_d(t))^\top$ of the Riccati equation (4.1.2) (as long as it exists), the solution of $\dot{x} = Ax$ with initial condition

$$x_1(0) = 1, x_j(0) = k_j(0), j = 2, \ldots, d.$$

satisfies $K(t) = \left[\frac{x_2(t)}{x_1(t)}, \ldots, \frac{x_d(t)}{x_1(t)}\right]^\top$. Hence the vectors $K(t), t \in \mathbb{R}$, determine the curve in projective space which describes the evolution of the one-dimensional subspace spanned by $x(0)$, as long as the first coordinate is different from 0.

This discussion shows that the behavior of lines in $\mathbb{R}^d$ under the flow e^{At} is locally described by a certain Riccati equation (as in the linear oscillator case, one may use different parametrizations when $x_1(t)$ approaches 0). If one wants to discuss the limit behavior as time tends to infinity, this local description is not adequate and one should consider a compact state space.

For the diagonal matrix $A = \mathrm{diag}(1, -1)$ in Example 3.1.2 one obtains two one-dimensional Lyapunov spaces, each corresponding to two opposite points on the unit circle. These points are chain components of the flow on the unit circle. Opposite points should be identified in order to get a one-to-one correspondence between Lyapunov spaces and chain components in this simple example. Thus, in fact, the space of lines, i.e., projective space, is better suited for the analysis than the unit sphere.

The projective space $\mathbb{P}^{d-1}$ for $\mathbb{R}^d$ can be constructed in the following way. Introduce an equivalence relation on $\mathbb{R}^d \setminus \{0\}$ by saying that x and y are equivalent, $x \sim y$, if there is $\alpha \neq 0$ with $x = \alpha y$. The quotient space $\mathbb{P}^{d-1} := \mathbb{R}^d \setminus \{0\}/\sim$ is the projective space. Clearly, it suffices to consider only vectors x with Euclidean norm $\|x\| = 1$. Thus, geometrically, projective space is obtained by identifying opposite points on the unit sphere $\mathbb{S}^{d-1}$ or it may be considered as the space of lines through the origin. We write $\mathbb{P}: \mathbb{R}^d \setminus \{0\} \to \mathbb{P}^{d-1}$ for the projection and usually, denote the elements of $\mathbb{P}^{d-1}$ by $p = \mathbb{P}x$, where $0 \neq x \in \mathbb{R}^d$ is any element in the corresponding equivalence class. A metric on $\mathbb{P}^{d-1}$ is given by

$$(4.1.3) \qquad \mathrm{d}(\mathbb{P}x, \mathbb{P}y) := \min\left(\left\|\frac{x}{\|x\|} - \frac{y}{\|y\|}\right\|, \left\|\frac{x}{\|x\|} - \frac{-y}{\|y\|}\right\|\right).$$

Note that for a point x in the unit sphere $\mathbb{S}^{d-1}$ and a subspace W of $\mathbb{R}^d$ one has
(4.1.4)
$$\mathrm{dist}(x, W \cap \mathbb{S}^{d-1}) = \inf_{y \in W \cap \mathbb{S}^{d-1}} \|x - y\| = \min_{y \in W} \mathrm{d}(\mathbb{P}x, \mathbb{P}y) =: \mathrm{dist}(\mathbb{P}x, \mathbb{P}W).$$

Any matrix in $Gl(d, \mathbb{R})$ (in particular, matrices of the form e^{At}) induces an invertible map on the projective space $\mathbb{P}^{d-1}$. The flow properties of

$\Phi_t = e^{At}, t \in \mathbb{R}$, are inherited by the induced maps and we denote by $\mathbb{P}\Phi_t$ the induced dynamical system on projective space. More precisely, the projection $\mathbb{P}$ is a semiconjugacy, i.e., it is a continuous surjective map satisfying for every $t \in \mathbb{R}$ the conjugacy property

$$\begin{array}{ccc} \mathbb{R}^d \setminus \{0\} & \xrightarrow{\Phi_t} & \mathbb{R}^d \setminus \{0\} \\ \mathbb{P} \downarrow & & \downarrow \mathbb{P} \\ \mathbb{P}^{d-1} & \xrightarrow{\mathbb{P}\Phi_t} & \mathbb{P}^{d-1} \end{array}.$$

We will not need that the projective flow $\mathbb{P}\Phi$ is generated by a differential equation on projective space which, in fact, is a $(d-1)$-dimensional differentiable manifold. Instead, we only need that projective space is a compact metric space and that $\mathbb{P}\Phi$ is a continuous flow; in Exercise 4.3.1, the reader is asked to verify this in detail. Nevertheless, the following differential equation in $\mathbb{R}^d$ leaving the unit sphere $\mathbb{S}^{d-1}$ invariant is helpful to understand the properties of the flow in projective space.

Lemma 4.1.1. *For $A \in gl(d, \mathbb{R})$ let $\Phi_t = e^{At}, t \in \mathbb{R}$, be its linear flow in $\mathbb{R}^d$. The flow Φ projects onto a flow on $\mathbb{S}^{d-1}$, given by the differential equation*

$$\dot{s} = h(s, A) = (A - s^\top As\ I)s, \text{ with } s \in \mathbb{S}^{d-1}.$$

Proof. Exercise 4.3.2. □

Naturally, the flow on the unit sphere also projects to the projective flow $\mathbb{P}\Phi$. In order to determine the global behavior of the projective flow we first show that points outside of the Lyapunov spaces $L_j := L(\lambda_j)$ are not chain recurrent; cf. Definition 1.4.1.

Lemma 4.1.2. *Let $\mathbb{P}\Phi_t$ be the projection to $\mathbb{P}^{d-1}$ of a linear flow $\Phi_t = e^{At}$. If $x \notin \bigcup_{j=1}^\ell L(\lambda_j)$, then $\mathbb{P}x$ is not chain recurrent for the induced projective flow.*

Proof. We may suppose that A is given in real Jordan form, since a linear conjugacy in $\mathbb{R}^d$ yields a topological conjugacy in projective space which preserves the chain transitive sets by Proposition 3.1.15. The following construction shows that for $\varepsilon > 0$ small enough there is no (ε, T)-chain from $\mathbb{P}x$ to $\mathbb{P}x$. It may be viewed as a generalization of Example 3.2.1 where a scalar system was considered.

Recall the setting of Theorem 1.4.4. The Lyapunov exponents are ordered such that $\lambda_1 > \ldots > \lambda_\ell$ with associated Lyapunov spaces $L_j = L(\lambda_j)$. Then

$$V_j = L_\ell \oplus \ldots \oplus L_j \text{ and } W_j = L_j \oplus \ldots \oplus L_1$$

4.1. Linear Flows Induced in Projective Space

define flags of subspaces

$$\{0\} = V_{\ell+1} \subset V_\ell \subset \ldots \subset V_1 = \mathbb{R}^d, \{0\} = W_0 \subset W_1 \subset \ldots \subset W_\ell = \mathbb{R}^d.$$

For $x \notin \bigcup_{j=1}^\ell L(\lambda_j)$ there is a minimal j such that $x \in V_j \setminus V_{j+1}$ and hence there are unique $x_i \in L(\lambda_i)$ for $i = j, \ldots, \ell$ with

$$x = x_\ell + \ldots + x_j.$$

Here $x_j \neq 0$ and at least one $x_i \neq 0$ for some $i \geq j+1$. Hence $x \notin W_j = L(\lambda_j) \oplus \ldots \oplus L(\lambda_1)$ and $V_{j+1} \cap W_j = \{0\}$. We may suppose that x is on the unit sphere $\mathbb{S}^{d-1}$ and has positive distance $\delta > 0$ to the intersection $W_j \cap \mathbb{S}^{d-1}$. By (4.1.4) it follows that $\delta > 0$ is the distance of $\mathbb{P}x$ to the projection $\mathbb{P}W_j$.

The solution formulas show that for all $0 \neq y \in \mathbb{R}^d$,

$$\frac{e^{At}y}{\|e^{At}y\|} = \frac{e^{At}y_\ell}{\|e^{At}y\|} + \ldots + \frac{e^{At}y_j}{\|e^{At}y\|} + \ldots + \frac{e^{At}y_1}{\|e^{At}y\|}.$$

If $y \notin V_{j+1}$ one has for $i \geq j+1$ that $\frac{e^{At}y_i}{\|e^{At}y\|} \to 0$ for $t \to \infty$. Also for some $i \leq j$ one has $y_i \neq 0$ and $\frac{e^{At}y_i}{\|e^{At}y\|} \in L(\lambda_i)$. This implies that for $t \to \infty$,

$$\operatorname{dist}(\mathbb{P}\Phi_t(y), \mathbb{P}W_j) = \operatorname{dist}\left(\frac{e^{At}y}{\|e^{At}y\|}, W_j \cap \mathbb{S}^{d-1}\right) \to 0.$$

There is $0 < 2\varepsilon < \delta$ such that the 2ε-neighborhood N of $\mathbb{P}W_j$ has void intersection with $\mathbb{P}V_{j+1}$. We may take $T > 0$ large enough such that for all initial values $\mathbb{P}y$ in the compact set $\operatorname{cl} N$ and all $t \geq T$,

$$\operatorname{dist}(\mathbb{P}\Phi_t(y), \mathbb{P}W_j) < \varepsilon.$$

Now consider an (ε, T) chain starting in $\mathbb{P}x_0 = \mathbb{P}x \notin \bigcup_{j=1}^\ell L(\lambda_j)$ and let $T_0 > T$ such that $\operatorname{dist}(\mathbb{P}\Phi_{T_0}(x_0), \mathbb{P}W_j) < \varepsilon$. The next point $\mathbb{P}x_1$ of the chain has distance less than ε to $\mathbb{P}\Phi_{T_0}(x_0)$, hence

$$\operatorname{dist}(\mathbb{P}x_1, \mathbb{P}W_j) \leq d(\mathbb{P}x_1, \mathbb{P}\Phi_{T_0}(x_0)) + \operatorname{dist}(\mathbb{P}\Phi_{T_0}(x_0), \mathbb{P}W_j) < 2\varepsilon < \delta.$$

Thus $\mathbb{P}x_1 \in N$ and it follows that $\operatorname{dist}(\mathbb{P}\Phi_t(x_1), \mathbb{P}W_j) < \varepsilon$ for all $t \geq T$. Repeating this construction along the (ε, T)-chain, one sees that the final point $\mathbb{P}x_n$ has distance less than δ from $\mathbb{P}W_j$ showing, by definition of δ, that $\mathbb{P}x_n \neq \mathbb{P}x_0 = \mathbb{P}x$. $\square$

The characteristics of the projected flow $\mathbb{P}\Phi$ are summarized in the following result. In particular, it shows that the topological properties of this projected flow determine the decomposition of $\mathbb{R}^d$ into the Lyapunov spaces; cf. Definition 1.4.1.

Theorem 4.1.3. *Let $\mathbb{P}\Phi$ be the projection onto $\mathbb{P}^{d-1}$ of a linear flow $\Phi_t(x) = e^{At}x$. Then the following assertions hold.*

(i) $\mathbb{P}\Phi$ has ℓ chain components $\mathcal{M}_1, \ldots, \mathcal{M}_\ell$, where ℓ is the number of Lyapunov exponents $\lambda_1 > \ldots > \lambda_\ell$.

(ii) One can number the chain components such that $\mathcal{M}_j = \mathbb{P}L(\lambda_j)$, the projection onto $\mathbb{P}^{d-1}$ of the Lyapunov space $L_j = L(\lambda_j)$ corresponding to the Lyapunov exponent λ_j.

(iii) The sets

$$\mathbb{P}^{-1}\mathcal{M}_j := \{x \in \mathbb{R}^d \mid x = 0 \text{ or } \mathbb{P}x \in \mathcal{M}_j\}$$

coincide with the Lyapunov spaces and hence yield a decomposition of $\mathbb{R}^d$ into linear subspaces

$$\mathbb{R}^d = \mathbb{P}^{-1}\mathcal{M}_1 \oplus \ldots \oplus \mathbb{P}^{-1}\mathcal{M}_\ell.$$

Proof. We may assume that A is given in Jordan canonical form $J^{\mathbb{R}}$, since coordinate transformations map the real generalized eigenspaces and the chain transitive sets into each other. Lemma 4.1.2 shows that points outside of a Lyapunov space L_j cannot project to a chain recurrent point. Hence it remains to show that the flow $\mathbb{P}\Phi$ restricted to a projected Lyapunov space $\mathbb{P}L_j$ is chain transitive. Then assertion (iii) is an immediate consequence of the fact that the L_i are linear subspaces. We may assume that the corresponding Lyapunov exponent, i.e., the common real part of the eigenvalues, is zero. First, the proof will show that the projected sum of the corresponding eigenspaces is chain transitive. Then the assertion is proved by analyzing the projected solutions in the corresponding generalized eigenspaces.

Step 1: The projected eigenspace for the eigenvalue 0 is chain transitive, since it is connected and consists of equilibria; see Proposition 3.2.5.

Step 2: For a complex conjugate eigenvalue pair $\mu, \bar{\mu} = \pm i\nu, \nu > 0$, an element $x_0 \in \mathbb{R}^d$ with coordinates $(y_0, z_0)^\top$ in the real eigenspace satisfies

$$y(t, x_0) = y_0 \cos \nu t - z_0 \sin \nu t, \ z(t, x_0) = z_0 \cos \nu t + y_0 \sin \nu t.$$

Thus it defines a $\frac{2\pi}{\nu}$-periodic solution on $\mathbb{R}^d$ and together they form a two-dimensional subspace of periodic solutions. The projection to $\mathbb{P}^{d-1}$ is also periodic and hence chain transitive. The same is true for the whole eigenspace of $\pm i\nu$.

Step 3: Now consider for $k = 1, \ldots, m$ a collection of eigenvalue pairs $\pm i\nu_k, \nu_k > 0$ such that all ν_k are rational, i.e., there are $p_k, q_k \in \mathbb{N}$ with $\nu_k = \frac{p_k}{q_k}$. Then the corresponding eigensolutions have periods $\frac{2\pi}{\nu_k} = 2\pi \frac{q_k}{p_k}$. It follows that these solutions have the common (nonminimal) period $2\pi q_1 \ldots q_m$. Then the projected sum of the eigenspaces consists of periodic solutions and

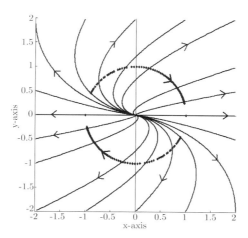

Figure 4.1. The flow for a two-dimensional Jordan block

hence is chain transitive. If the ν_k are arbitrary real numbers, we can approximate them by rational numbers $\tilde{\nu}_k$. This can be used to construct (ε, T)-chains, where, by Proposition 3.1.10, it suffices to construct (ε, T)-chains with jump times $T_i \in (1, 2]$. Replacing in the matrix the ν_k by $\tilde{\nu}_k$, one obtains matrices which are arbitrarily close to the original matrix. By Corollary 1.1.2(ii), for every $\varepsilon > 0$ one may choose the $\tilde{\nu}_k$ such that for every $x \in \mathbb{R}^d$ the corresponding solution $\tilde{\Phi}_t x$ satisfies

$$\left\| \Phi_t x - \tilde{\Phi}_t x \right\| < \varepsilon \text{ for all } t \in [0, 2].$$

This also holds for the distance in projective space showing that the projected sum of all eigenspaces for complex conjugate eigenvalue pairs is chain transitive. Next, we may also add the eigenspace for the eigenvalue 0 and see that the projected sum of all real eigenspaces is chain transitive. This follows, since the component of the solution in the eigenspace for 0 is constant (cf. Proposition 2.1.6(i)).

Step 4: Call the subspaces of $\mathbb{R}^d$ corresponding to the Jordan blocks Jordan subspaces. Consider first initial values in a Jordan subspace corresponding to a real eigenvalue, i.e., by assumption to the eigenvalue zero. The projective eigenvector p (i.e., an eigenvector projected on $\mathbb{P}^{d-1}$) is an equilibrium for $\mathbb{P}\Phi$. For all other initial values the projective solutions tend to p for $t \to \pm\infty$, since for every initial value the component corresponding to the eigenvector has the highest polynomial growth; cf. the solution formula (1.3.2). This shows that the projective Jordan subspace is chain transitive. Figures 4.1 and 4.2 illustrate the situation for a two-dimensional and a three-dimensional Jordan block, respectively. In Figure 4.1 solutions of the linear system in $\mathbb{R}^2$ (with positive real part of the eigenvalues) and their projections to the unit circle are indicated, while Figure 4.2 shows projected solutions on the sphere $\mathbb{S}^2$ in $\mathbb{R}^3$ (note that here the eigenspace is the vertical axis). The analogous statement holds for Jordan subspaces corresponding to a complex-conjugate pair of eigenvalues.

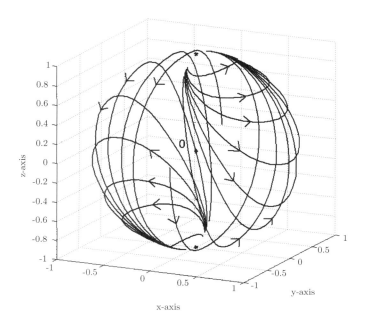

Figure 4.2. The projected flow for a three-dimensional Jordan block

Step 5: It remains to show that the projected sum of all Jordan subspaces is chain transitive. By **Step 4** the components in every Jordan subspace converge for $t \to \pm\infty$ to the corresponding real eigenspace, and hence the sum converges to the sum of the real eigenspaces. By **Step 3** the projected sum of the real eigenspaces is chain transitive. This finally proves that the Lyapunov spaces project to chain transitive sets in projective space. □

Remark 4.1.4. Theorem 4.1.3 shows that the Lyapunov spaces are characterized topologically by the induced projective flow. Naturally, the magnitude of the Lyapunov exponents is not seen in projective space, only their order. The proof of Lemma 4.1.2 also shows that the chain components $\mathcal{M}_j$ corresponding to the Lyapunov exponents λ_j are ordered in the same way by a property of the flow in projective space: Two Lyapunov exponents satisfy $\lambda_i < \lambda_j$, if and only if there exists a point p in projective space with $\alpha(p) \subset \mathcal{M}_i$ and $\omega(p) \subset \mathcal{M}_j$; cf. Exercise 4.3.3.

Surprisingly enough, one can reconstruct the actual values of the Lyapunov exponents from the behavior on the unit sphere based on the differential equation given in Lemma 4.1.1. This is shown in Exercise 4.3.2.

The chain components are preserved under conjugacies of the flows on projective space.

Corollary 4.1.5. *For $A, B \in gl(d, \mathbb{R})$ let $\mathbb{P}\Phi$ and $\mathbb{P}\Psi$ be the associated flows on $\mathbb{P}^{d-1}$ and suppose that there is a topological conjugacy h of $\mathbb{P}\Phi$ and $\mathbb{P}\Psi$. Then the chain components $\mathcal{N}_1, \ldots, \mathcal{N}_\ell$ of $\mathbb{P}\Psi$ are of the form $\mathcal{N}_i = h(\mathcal{M}_i)$, where $\mathcal{M}_i$ is a chain component of $\mathbb{P}\Phi$. In particular, the number of Lyapunov spaces of Φ and Ψ agrees.*

Proof. By Proposition 3.1.15(iii) the maximal chain transitive sets, i.e., the chain components, are preserved by topological conjugacies. The second assertion follows by Theorem 4.1.3. □

4.2. Linear Difference Equations in Projective Space

In this section it is shown that for linear difference equations the projections of the Lyapunov spaces coincide with the chain components in projective space.

Consider a linear difference equation of the form

$$x_{n+1} = Ax_n, n \in \mathbb{Z},$$

where $A \in Gl(d, \mathbb{R})$. According to the discussion in Section 2.3, A generates a continuous dynamical system Φ in discrete time with time-1 map $\varphi_1 = \Phi(1, \cdot) = A$. By linearity, this induces a dynamical system $\mathbb{P}\Phi$ in discrete time on projective space $\mathbb{P}^{d-1}$ with time-1 map $\mathbb{P}\varphi = \mathbb{P}\Phi(1, \cdot)$ given by

$$p \mapsto \mathbb{P}(Ax) \text{ for any } x \text{ with } \mathbb{P}x = p.$$

This can also be obtained by first considering the induced map on the unit sphere $\mathbb{S}^{d-1}$ and then identifying opposite points. The system on the unit sphere projects to the projective flow $\mathbb{P}\Phi$. The characteristics of the projected dynamical system $\mathbb{P}\Phi$ are summarized in the following result. In particular, it shows that the topological properties of this projected flow determine the decomposition of $\mathbb{R}^d$ into Lyapunov spaces (recall Definition 1.5.4.)

Theorem 4.2.1. *Let $\mathbb{P}\Phi$ be the projection onto $\mathbb{P}^{d-1}$ of a linear dynamical system $\Phi(n, x) = A^n x, n \in \mathbb{Z}, x \in \mathbb{R}^d$, associated with $x_{n+1} = Ax_n$. Then the following assertions hold.*

(i) $\mathbb{P}\Phi$ has ℓ chain components $\mathcal{M}_1, \ldots, \mathcal{M}_\ell$, where ℓ is the number of Lyapunov exponents $\lambda_1 > \ldots > \lambda_\ell$.

(ii) One can number the chain components such that $\mathcal{M}_j = \mathbb{P}L(\lambda_j)$, the projection onto $\mathbb{P}^{d-1}$ of the Lyapunov space $L(\lambda_j)$ corresponding to the Lyapunov exponent λ_j.

(iii) The sets
$$\mathbb{P}^{-1}\mathcal{M}_j := \{x \in \mathbb{R}^d \mid x = 0 \text{ or } \mathbb{P}x \in \mathcal{M}_j\}$$
coincide with the Lyapunov spaces and hence yield a decomposition of $\mathbb{R}^d$ into linear subspaces
$$\mathbb{R}^d = \mathbb{P}^{-1}\mathcal{M}_1 \oplus \ldots \oplus \mathbb{P}^{-1}\mathcal{M}_\ell.$$

Proof. We may assume that A is given in Jordan canonical form, since coordinate transformations map the generalized eigenspaces and the chain transitive sets into each other.

Analogously to Lemma 4.1.2 and its proof one sees that points outside of the Lyapunov spaces are not chain recurrent. This follows from Theorem 1.5.8. Hence it remains to show that the system $\mathbb{P}\Phi$ restricted to a projected Lyapunov space $\mathbb{P}L(\lambda_j)$ is chain transitive. Then assertion (iii) is an immediate consequence of the fact that the L_i are linear subspaces. We go through the same steps as for the proof of Theorem 4.1.3. Here we may assume that all eigenvalues have modulus 1.

Step 1: The projected eigenspace for a real eigenvalue μ is chain transitive, since it is connected and consists of equilibria; see Proposition 3.3.5(iii).

Step 2: Consider a complex conjugate eigenvalue pair $\mu, \bar{\mu} = \alpha \pm i\beta, \beta > 0$, with $|\mu| = |\bar{\mu}| = 1$. Then an element $x_0 \in \mathbb{R}^d$ with coordinates $(y_0, z_0)^\top$ in the real eigenspace satisfies

$$\varphi(n, x_0) = A^n x_0 = \begin{bmatrix} \cos\beta & -\sin\beta \\ \sin\beta & \cos\beta \end{bmatrix}^n \begin{bmatrix} y_0 \\ z_0 \end{bmatrix}.$$

This means that we apply n times a rotation by the angle β, i.e., a single rotation by the angle $n\beta$. If $\frac{2\pi}{\beta}$ is rational, there are $p, q \in \mathbb{N}$ with $\frac{2\pi}{\beta} = \frac{p}{q}$, and hence $p\beta = 2\pi q$. Then $\varphi(p, x_0) = x_0$ and hence x_0 generates a p-periodic solution in $\mathbb{R}^d$. These solutions form a two-dimensional subspace of periodic solutions. The projections are also periodic and hence, by Proposition 3.3.5(iii), one obtains a chain transitive set. The same is true for the whole real eigenspace of μ.

Now consider for $k = 1, \ldots, m$ a collection of eigenvalue pairs $\mu_k, \bar{\mu}_k = \alpha_k \pm i\beta_k, \beta_k > 0$ such that all $\frac{2\pi}{\beta_k}$ are rational, i.e., there are $p_k, q_k \in \mathbb{N}$ with $\frac{2\pi}{\beta_k} = \frac{p_k}{q_k}$. Then the corresponding eigensolutions have periods p_k. It follows that these solutions have the common (not necessarily minimal) period $2\pi p_1 \ldots p_m$. Hence the projected sum of the real eigenspaces is chain transitive.

4.2. Linear Difference Equations in Projective Space

If the β_k are arbitrary real numbers, we can approximate them by rational numbers $\tilde\beta_k$. This can be used to construct ε-chains: Replacing in the matrix the β_k by $\tilde\beta_k$, one obtains matrices $\tilde A$ which are close to the original matrix. The matrices $\tilde A$ may be chosen such that $\|Ax - \tilde A x\| < \varepsilon$ for every $x \in \mathbb{R}^d$ with $\|x\| = 1$. This also holds for the distance in projective space showing that the projected sum of all real eigenspaces for complex conjugate eigenvalue pairs is chain transitive.

Step 3: By Steps 1 and 2 and using similar arguments one shows that the projected sum of all real eigenspaces is chain transitive.

Step 4: Call the subspaces of $\mathbb{R}^d$ corresponding to the Jordan blocks Jordan subspaces. Consider first initial values in a Jordan subspace corresponding to a real eigenvalue. The projective eigenvector p (i.e., an eigenvector projected on $\mathbb{P}^{d-1}$) is an equilibrium for $\mathbb{P}\Phi$. For all other initial values the projective solutions tend to p for $n \to \pm\infty$, since they induce the highest polynomial growth in the component corresponding to the eigenvector. This shows that the projective Jordan subspace is chain transitive. The analogous statement holds for Jordan subspaces corresponding to a complex-conjugate pair of eigenvalues.

Step 5: It remains to show that the projected sum of all Jordan subspaces is chain transitive. This follows, since for $n \to \pm\infty$ the components in every Jordan subspace converge to the corresponding eigenspace, and hence the sum converges to the sum of the eigenspaces. The same is true for the projected sum of all generalized eigenspaces. This, finally, shows that the Lyapunov spaces project to chain transitive sets in projective space. □

Theorem 4.2.1 shows that the Lyapunov spaces are characterized topologically by the induced projective system. Naturally, the magnitudes of the Lyapunov exponents are not seen in projective space, only their order. Furthermore the chain components $\mathcal{M}_j$ corresponding to the Lyapunov exponents λ_j are ordered in the same way by a property of the flow in projective space: Two Lyapunov exponents satisfy $\lambda_i < \lambda_j$, if and only if there exists a point p in projective space with $\alpha(p) \subset \mathcal{M}_i$ and $\omega(p) \subset \mathcal{M}_j$.

How do the chain components behave under conjugacy of the flows on $\mathbb{P}^{d-1}$?

Corollary 4.2.2. *For $A, B \in Gl(d, \mathbb{R})$ let $\mathbb{P}\Phi$ and $\mathbb{P}\Psi$ be the associated dynamical systems on $\mathbb{P}^{d-1}$ and suppose that there is a topological conjugacy h of $\mathbb{P}\Phi$ and $\mathbb{P}\Psi$. Then the chain components $\mathcal{N}_1, \ldots, \mathcal{N}_\ell$ of $\mathbb{P}\Psi$ are of the form $\mathcal{N}_i = h(\mathcal{M}_i)$, where $\mathcal{M}_i$ is a chain component of $\mathbb{P}\Phi$. In particular, the number of chain components of $\mathbb{P}\Phi$ and $\mathbb{P}\Psi$ agree.*

Proof. This is a consequence of Theorem 3.3.7. □

4.3. Exercises

Exercise 4.3.1. (i) Prove that the metric (4.1.3) is well defined and turns the projective space $\mathbb{P}^{d-1}$ into a compact metric space. (ii) Show that the linear flow $\Phi_t(x) = e^{At}x, x \in \mathbb{R}^d, t \in \mathbb{R}$, induces a continuous flow $\mathbb{P}\Phi$ on projective space.

Exercise 4.3.2. Let $x(t, x_0)$ be a solution of $\dot{x} = Ax$ with $A \in gl(d, \mathbb{R})$. Write $s(t) = \frac{x(t,x_0)}{\|x(t,x_0)\|}, t \in \mathbb{R}$, for the projection to the unit sphere in the Euclidean norm. (i) Show that $s(t)$ is a solution of the differential equation

$$\dot{s}(t) = [A - s(t)^\top As(t) \cdot I]s(t).$$

Observe that this is a differential equation in $\mathbb{R}^d$ which leaves the unit sphere invariant. Give a geometric interpretation! Use this equation to show that eigenvectors corresponding to real eigenvalues give rise to fixed points on the unit sphere. (ii) Prove the following formula for the Lyapunov exponents:

$$\lambda(x_0) = \lim_{t\to\infty} \frac{1}{t} \int_0^t s(\tau)^\top As(\tau) d\tau$$

by considering the 'polar decomposition' $\mathbb{S}^{d-1} \times (0, \infty)$.

Exercise 4.3.3. Consider the chain components given in Theorem 4.1.3. Show that there is $p \in \mathbb{P}^{d-1}$ with $\alpha(p) \subset \mathcal{M}_i$ and $\omega(p) \subset \mathcal{M}_j$ if and only if $\lambda_i < \lambda_j$.

Exercise 4.3.4. Consider the linear difference equation in $\mathbb{R}^2$ given by

$$\begin{bmatrix} x_{k+1} \\ y_{k+1} \end{bmatrix} = \begin{bmatrix} 1 & 1 \\ 1 & 0 \end{bmatrix} \begin{bmatrix} x_k \\ y_k \end{bmatrix}$$

and determine the eigenvalues and the eigenspaces. Show that the line through the initial point $x_0 = 0, y_0 = 1$ converges under the flow to the line with slope $(1 + \sqrt{5})/2$, the golden mean. Explain the relation to the Fibonacci numbers given by the recursion $f_{k+1} = f_k + f_{k-1}$ with initial values $f_0 = 0, f_1 = 1$.

Exercise 4.3.5. Consider the method for calculating $\sqrt{2}$ which was proposed by Theon of Smyrna in the second century B.C.: Starting from $(1, 1)$, iterate the transformation $x \mapsto x + 2y, y \mapsto x + y$. Explain why this gives a method to compute $\sqrt{2}$.
Hint: Argue similarly as in Exercise 4.3.4.

4.4. Orientation, Notes and References

Orientation. This chapter has characterized the Lyapunov spaces of linear dynamical systems by a topological analysis of the induced systems on projective space. Theorems 4.1.3 and 4.2.1 show that the projections of the

Lyapunov spaces $L(\lambda_j)$ to projective space coincide with the chain components of the projected flow. It is remarkable that these topological objects in fact have a 'linear structure'. The proofs are based on the explicit solution formulas and the structure in $\mathbb{R}^d$ provided by the Lyapunov exponents and the Lyapunov spaces. The insight gained in this chapter will be used in the second part of this book in order to derive decompositions of the state space into generalized Lyapunov spaces related to generalized Lyapunov exponents. More precisely, in Chapter 9 we will analyze a general class of linear dynamical systems (in continuous time) and construct a decomposition into generalized Lyapunov spaces. Here the line of proof will be reversed, since no explicit solution formulas are available: first the chain components yielding a linear decomposition are constructed and then associated exponential growth rates are determined.

In the next chapter, a generalization to flows induced on the space of k-dimensional subspaces, the k-Grassmannian, will be given. This requires some notions and facts from mulitilinear algebra, which are collected in Section 5.1. An understanding of the results in this chapter is not needed for the rest of this book, with the exception of some facts from multilinear algebra. They can also be picked up later, when they are needed (in Chapter 11 in the analysis of random dynamical systems).

Notes and references. The characterization of the Lyapunov spaces as the chain components in projective space is folklore (meaning that it is well known to the experts in the field, but it is difficult to find explicit statements and proofs). The differential equation on the unit sphere given in Lemma 4.1.1 is also known as Oja's flow (Oja [**108**]) and plays an important role in principal component analysis in neural networks where dominant eigenvalues are to be extracted. But the idea of using the $\mathbb{S}^{d-1} \times (0, \infty)$ coordinates (together with explicit formulas in Lemma 4.1.1 and Exercise 4.3.2) to study linear systems goes back at least to Khasminskii [**78, 79**].

Theorems 2.2.5 and 2.3.7 characterize the equivalence classes of linear differential and difference equations in $\mathbb{R}^d$ up to topological conjugacy. Thus it is natural to ask for a characterization of the topological conjugacy classes in projective space. Corollaries 4.1.5 and 4.2.2 already used such topological conjugacies of the projected linear dynamical systems in continuous and discrete time. However, the characterization of the corresponding equivalence classes is surprisingly difficult and has generated a number of papers. A partial result in the general discrete-time case has been given by Kuiper [**88**]; Ayala and Kawan [**14**] give a complete solution for continuous-time systems (and a correction to Kuiper's proof) and discuss the literature.

Exercises 4.3.4 and 4.3.5 are taken from Chatelin [**25**, Examples 3.1.1 and 3.1.2].

Chapter 5

Linear Systems on Grassmannians

Every linear system in $\mathbb{R}^d$ induces dynamical systems on the set of k-dimensional subspaces of $\mathbb{R}^d$, called the Grassmannians. This chapter discusses the relation between the dynamical behavior of these systems and exponential growth rates of volumes. The analysis is restricted to continuous-time systems.

Chapter 4 has discussed how the length of vectors grows with time; this only depends on the line through the initial value, i.e., the corresponding element in projective space, and the corresponding projections of the Lyapunov spaces have been characterized by the corresponding induced flow in projective space. The present chapter will discuss how the volume of parallelepipeds spanned by k vectors grows with time and analyze the corresponding induced flow on the set of k-dimensional subspaces. Section 5.2 performs this analysis for continuous-time systems on Grassmannians and the associated volume growth rates. The required notions and results from multilinear algebra are provided in Section 5.1.

It should be noted that the results in the chapter are not needed below, with the exception of some notions from multilinear algebra which are collected in Section 5.1. They will be used in Chapter 11 in the proof of Oseledets' Multiplicative Ergodic Theorem for random dynamical systems.

We begin with a new look at the analysis in Chapter 4, where the induced flow in projective space is considered. The motivation given in Section 4.1 via Riccati differential equations is not restricted to one-dimensional subspaces.

The invertible linear map e^{At} maps any k-dimensional subspace onto a k-dimensional subspace, hence one can analyze the evolution of k-dimensional subspaces under the flow $e^{At}, t \in \mathbb{R}$. For solutions $x^{(1)}(t), \ldots, x^{(k)}(t)$ of $\dot{x} = Ax$ write

$$\left[x^{(1)}(t), \ldots, x^{(k)}(t) \right] = \left[\begin{array}{c} X_1(t) \\ X_2(t) \end{array} \right] \text{ with } X_1(t) \in \mathbb{R}^{k \times k}, X_2(t) \in \mathbb{R}^{(d-k) \times k}.$$

For invertible $X_1(t)$, define $K(t) = X_2(t) X_1^{-1}(t) \in \mathbb{R}^{(d-k) \times k}$. The vectors $x^{(i)}(t)$ generate the same subspace as the columns of $\left[\begin{array}{c} X_1(t) \\ X_2(t) \end{array} \right] X_1^{-1}(t)$. This matrix has the $k \times k$ identity matrix I_k in the upper k rows. For $1 \leq k \leq d$ partition $A \in \mathbb{R}^{d \times d}$ as

$$A = \left[\begin{array}{cc} A_{11} & A_{12} \\ A_{21} & A_{22} \end{array} \right],$$

where $A_{11} \in gl(k, \mathbb{R}), A_{22} \in gl(d-k, \mathbb{R})$ and A_{12} and A_{21} are $k \times (d-k)$ and $(d-k) \times k$ matrices, respectively. Then $K(t)$ is a solution of a matrix Riccati differential equation on $\mathbb{R}^{(d-k) \times k}$ which has the same form as (4.1.2):

$$\dot{K} = A_{21} + A_{22} K - K A_{11} - K A_{12} K.$$

Conversely, every $(d-k) \times k$ matrix solution $K(t)$ of this equation defines a curve of subspaces determined by the linear span of the columns in

$$\left[\begin{array}{c} I_k \\ K(t) \end{array} \right] =: [x^{(1)}(t), \ldots, x^{(k)}(t)]$$

and $\mathrm{span}\{x^{(1)}(t), \ldots, x^{(k)}(t)\} = \mathrm{span}\{e^{At} x^{(1)}(0), \ldots, e^{At} x^{(k)}(0)\}$. Again, one sees that the solutions of the Riccati equation describe the evolution of k-dimensional subspaces under the flow e^{At}. Instead of looking at Riccati differential equations, in this chapter we will analyze the corresponding flow on the set of k-dimensional subspaces, thus avoiding the problem that the solutions of the Riccati equation may have a bounded interval of existence and hence changes of the local coordinate charts might be necessary.

We will determine the volume growth rates and characterize the long time behavior of the linear subspaces in a coordinate free form which in local coordinates are described by Riccati differential equations. As for projective space, we will not need that the underlying spaces form differentiable manifolds. Instead we will only need that the Grassmannians are compact metric spaces. Nevertheless, this is a somewhat technical story and the reader may skip it—the time-varying theory presented in the next chapters does not depend on the ideas discussed here.

5.1. Some Notions and Results from Multilinear Algebra

Every course on linear algebra also includes some multilinear algebra in the form of determinants which determine the volume of full-dimensional

5.1. Some Notions and Results from Multilinear Algebra

parallelepipeds. One may also associate volumes to lower dimensional parallelepipeds. The interplay between geometric intuition and formal manipulations is striking in this area.

The kth Grassmannian $\mathbb{G}_k$ of $\mathbb{R}^d$ can be defined via the following construction: Let $F(k,d)$ be the set of all ordered sets of k linearly independent vectors in $\mathbb{R}^d$. Two elements $X = (x_1, \ldots, x_k)$ and $Y = (y_1, \ldots, y_k)$ in $F(k,d)$ are said to be equivalent, $X \sim Y$, if there exists $T \in Gl(k, \mathbb{R})$ with $X^\top = TY^\top$, where X and Y are interpreted as $d \times k$ matrices. In other words, they are equivalent if and only if they generate the same subspace. The quotient space $\mathbb{G}_k := F(k,d)/\sim$ is the kth Grassmannian (it is a compact, $k(d-k)$-dimensional differentiable manifold.) For $k = 1$ we obtain the projective space $\mathbb{P}^{d-1} = \mathbb{G}_1$ in $\mathbb{R}^d$. Observe that a matrix $A \in Gl(d, \mathbb{R})$ (in particular, $e^{At}, t \in \mathbb{R}$) induces maps on every Grassmannian $\mathbb{G}_k$, since the dimension of a subspace is invariant under A. For $V = \text{span}\{x_1, \ldots, x_k\}$ one has $AV = \text{span}\{Ax_1, \ldots, Ax_k\}$. Furthermore, the flow properties (see Definition 2.1.1) of $\Phi_t = e^{At}, t \in \mathbb{R}$, are inherited by the induced maps.

Let H be a Euclidean vector space of dimension d with scalar product denoted by $\langle \cdot, \cdot \rangle$. A parallelepiped is spanned by k linearly independent vectors $x_1, \ldots, x_k \in H$, i.e., it is of the form

$$(5.1.1) \qquad \{x = \alpha_1 x_1 + \ldots + \alpha_k x_k \mid 0 \leq \alpha_i \leq 1 \text{ for } i = 1, \ldots, k\}.$$

Suppose that $e_1, \ldots, e_k$ is an orthonormal basis of the subspace spanned by $x_1, \ldots, x_k$. Then, with $x_i = \sum_{j=1}^k b_{ij} e_j, i = 1, \ldots, k$, the volume of the parallelepiped is defined as $|\det(b_{ij})_{i,j}|$. Using $b_{ij} = \langle x_i, e_j \rangle$ one computes

$$\langle x_i, x_\ell \rangle = \left\langle x_i, \sum_{j=1}^k \langle x_\ell, e_j \rangle e_j \right\rangle = \sum_{j=1}^k \langle x_\ell, e_j \rangle \langle x_i, e_j \rangle.$$

It follows that

$$\det \begin{bmatrix} \langle x_1, e_1 \rangle & \cdots & \langle x_1, e_k \rangle \\ \vdots & & \vdots \\ \langle x_k, e_1 \rangle & \cdots & \langle x_k, e_k \rangle \end{bmatrix}^2 = \det \begin{bmatrix} \langle x_1, x_1 \rangle & \cdots & \langle x_1, x_k \rangle \\ \vdots & & \vdots \\ \langle x_k, x_1 \rangle & \cdots & \langle x_k, x_k \rangle \end{bmatrix}.$$

Hence the term on the right-hand side is the square of the volume and the definition of the volume is independent of the choice of the orthonormal basis.

A somewhat more abstract framework is the following. Let $\omega : H^k \to \mathbb{R}$ be an alternating k-linear map, thus ω is linear in each of its arguments and for all $i \neq j$,

$$\omega(y_1, \ldots, y_i, \ldots, y_j, \ldots, y_k) = -\omega(y_1, \ldots, y_j, \ldots, y_i, \ldots, y_k).$$

Let $(x_1,\ldots,x_d) \in H^d$ be a basis of H and consider $(y_1,\ldots,y_k) \in H^k$ with $y_i = \sum_{j=1}^{d} b_{ij} x_j$ for all i. Then one computes

$$\omega(y_1,\ldots,y_k) = \sum_{j_1<\ldots<j_k} \det B_{j_1\ldots j_k} \omega(x_{j_1},\ldots,x_{j_k}), \tag{5.1.2}$$

where $B_{j_1\ldots j_k}$ is the $k \times k$-matrix obtained from the columns $j_1,\ldots,j_k$ of $B = (b_{ij})_{i,j}$ and summation is over all ordered k-combinations of $\{1,\ldots,d\}$. An immediate consequence of this formula is that ω is already determined by its values on basis elements of H. In particular, if $e_1,\ldots,e_d$ is an orthonormal basis, then $\omega(e_{j_1},\ldots,e_{j_k}) := 1, 1 \leq j_1 < \ldots < j_k \leq d$, defines via (5.1.2) an alternating k-linear map.

For $1 \leq k \leq d$ the k-fold alternating product $\bigwedge^k H = H \wedge \ldots \wedge H$ is defined as the vector space with an alternating k-linear map $\omega_\wedge : H^k \to \bigwedge^k H : (x_1,\ldots,x_k) \mapsto x_1 \wedge \ldots \wedge x_k$ with:

(i) if $(x_1,\ldots,x_d)$ is a basis of H, then

$$\{x_{j_1} \wedge \ldots \wedge x_{j_k} \mid 1 \leq j_1 < \ldots < j_k \leq d\}$$

is a basis of $\bigwedge^k H$, and

(ii) if $\varphi : H^k \to \mathbb{R}$ is an alternating k-linear map, then there is a unique linear map $\theta : \bigwedge^k H \to \mathbb{R}$ with $\varphi = \theta \circ \omega_\wedge$.

The elements of the form $x_1 \wedge \ldots \wedge x_k$ are called simple vectors; we may think of them as parallelepipeds. From an orthonormal basis $e_1,\ldots,e_d$ of H one obtains a basis of $\bigwedge^k H$ by

$$\{e_{j_1} \wedge \ldots \wedge e_{j_k} \mid 1 \leq j_1 < \ldots < j_k \leq d\}. \tag{5.1.3}$$

By linear extension, this also induces an inner product $\langle \cdot, \cdot \rangle$ on $\bigwedge^k H$ by making the basis above orthonormal:

$$\langle e_{i_1} \wedge \ldots \wedge e_{i_k}, e_{j_1} \wedge \ldots \wedge e_{j_k} \rangle = \det(\langle e_{j_r}, e_{j_s}\rangle)_{j_r,j_s} = \begin{cases} 1 & \text{if all } j_r = j_s, \\ 0 & \text{else.} \end{cases}$$

Hence, for $x_1,\ldots,x_k,y_1,\ldots,y_k \in H$ the inner product is

$$\langle x_1 \wedge \ldots \wedge x_k, y_1 \wedge \ldots \wedge y_k \rangle = \det(\langle x_i, y_j\rangle)_{i,j}. \tag{5.1.4}$$

Now the volume of a parallelepiped spanned by k linearly independent vectors $x_1,\ldots,x_k \in H$ is given by the norm obtained from this inner product, i.e., it equals $\|x_1 \wedge \ldots \wedge x_k\|$ and hence the square of the volume is again given by

$$\det(\langle x_i, x_j\rangle)_{i,j} = \det \begin{bmatrix} \langle x_1, x_1\rangle & \cdots & \langle x_1, x_k\rangle \\ \vdots & & \vdots \\ \langle x_k, x_1\rangle & \cdots & \langle x_k, x_k\rangle \end{bmatrix}.$$

5.1. Some Notions and Results from Multilinear Algebra

Proposition 5.1.1. *Let $(x_1, \ldots, x_k), (y_1, \ldots, y_k)$ be linearly independent k-tuples of vectors in H. Then*

(5.1.5) $$x_1 \wedge \ldots \wedge x_k = \pm y_1 \wedge \ldots \wedge y_k$$

if and only if $(x_1, \ldots, x_k), (y_1, \ldots, y_k)$ span the same subspace and the parallelepipeds spanned by them have the same volume.

Proof. Assumption (5.1.5) implies for $i = 1, \ldots, k$,

$$y_i \wedge (x_1 \wedge \ldots \wedge x_k) = y_i \wedge (y_1 \wedge \ldots \wedge y_k) = 0.$$

Since the x_j are linearly independent, this shows that y_i is linearly dependent on the x_j. Similarly, all x_i are linearly dependent on the y_j. Hence both tuples span the same subspace and, clearly, the volumes are equal. For the converse one finds, with $y_j = \sum_{j=1}^{k} b_{ij} x_j$, $1 \leq i \leq k$, by (5.1.2)

(5.1.6) $$y_1 \wedge \ldots \wedge y_k = \det(b_{ij})_{i,j} \, x_1 \wedge \ldots \wedge x_k.$$

Since the volumes are equal, it follows that $|\det(b_{ij})| = 1$ and hence (5.1.5) follows. $\square$

A consequence of Proposition 5.1.1 is that we can identify the set $\mathbb{G}_k H$ of k-dimensional subspaces in H with the equivalence classes of simple vectors $x_1 \wedge \ldots \wedge x_k$ in $\bigwedge^k H$ differing only by a nonzero factor. In other words, we can identify $\mathbb{G}_k H$ with a subset of projective space $\mathbb{P}(\bigwedge^k H)$ (this is called the Plücker embedding.) It is not hard to see that under the metric (4.1.3) this subset is compact, and in the following we use this metric d_k for $\mathbb{G}_k H$.

The exterior product is the bilinear form $\bigwedge^j H \times \bigwedge^k H \to \bigwedge^{j+k} H$ determined by

$$(y_1 \wedge \ldots \wedge y_j, z_1 \wedge \ldots \wedge z_k) \mapsto y_1 \wedge \ldots \wedge y_k \wedge z_1 \wedge \ldots \wedge z_k.$$

We note the following lemma.

Lemma 5.1.2. *Let $x, x' \in \bigwedge^j H$ and $y, y' \in \bigwedge^k H$ with $\|x - x'\| < \varepsilon$ and $\|y - y'\| < \varepsilon$. Then in $\bigwedge^{j+k} H$ one has*

$$\|x \wedge y - x' \wedge y'\| < \varepsilon \max(\|x\|, \|y'\|).$$

Proof. The triangle inequality and the Hadamard inequality (used in the third line, cf. Exercise 5.3.2) show

$$\begin{aligned}
\|x \wedge y - x' \wedge y'\| &\leq \|x \wedge y - x \wedge y'\| + \|x \wedge y' - x' \wedge y'\| \\
&= \|x \wedge (y - y')\| + \|(x - x') \wedge y'\| \\
&\leq \|x\| \, \|y - y'\| + \|x - x'\| \, \|y'\| \\
&< \varepsilon \max(\|x\|, \|y'\|). \qquad \square
\end{aligned}$$

The following lemma estimates the distance of sums of subspaces.

Lemma 5.1.3. Let $H = X \oplus X^\perp$ be an orthogonal decomposition. Then for all k-dimensional subspaces $V, W \subset X$ and all j-dimensional subspaces $V', W' \subset X^\perp$,

$$d_k(V, W) < \varepsilon \text{ and } d_j(V', W') < \varepsilon \text{ implies } d_{k+j}(V \oplus V', W \oplus W') < \varepsilon.$$

Proof. Consider bases $(v_1, \ldots, v_k)$ of V, $(w_1, \ldots, w_k)$ of W, $(v'_1, \ldots, v'_j)$ of V', and $(w'_1, \ldots, w'_j)$ of W' with

$$\|v_1 \wedge \ldots \wedge v_k\| = \|w_1 \wedge \ldots \wedge w_k\| = \|v'_1 \wedge \ldots \wedge v'_j\| = \|w'_1 \wedge \ldots \wedge w'_j\| = 1.$$

Using the orthogonality assumption one finds

$$\left\|(v_1 \wedge \ldots \wedge v_k) \wedge (v'_1 \wedge \ldots \wedge v'_j)\right\|^2 = \det(\langle v_r, v_s\rangle)_{r,s} \cdot \det(\langle v'_r, v'_s\rangle)_{r,s}$$
$$= \|v_1 \wedge \ldots \wedge v_k\| \|v'_1 \wedge \ldots \wedge v'_k\| = 1,$$

and, analogously, $\|(w_1 \wedge \ldots \wedge w_k) \wedge (w'_1 \wedge \ldots \wedge w'_j)\| = 1$. Clearly, a basis of $V \oplus V'$ is given by $v_1, \ldots, v_k, v'_1, \ldots, v'_j$ and a basis of $W \oplus W'$ is given by $w_1, \ldots, w_k, w'_1, \ldots, w'_j$. Hence (recall that $G_k H$ is identified with a subset of $\mathbb{P}(\bigwedge^k H)$ endowed with the metric in (4.1.3)) it follows that $d_{k+j}(V \oplus V', W \oplus W')$ equals the minimum of

$$\|v_1 \wedge \ldots \wedge v_k \wedge v'_1 \wedge \ldots \wedge v'_j \pm w_1 \wedge \ldots \wedge w_k \wedge w'_1 \wedge \ldots \wedge w'_j\|.$$

Now Lemma 5.1.2 implies

$$d_{k+j}(V \oplus V', W \oplus W') < \varepsilon \max\left(\|v_1 \wedge \ldots \wedge v_k\|, \|w'_1 \wedge \ldots \wedge w'_j\|\right) = \varepsilon. \quad \square$$

5.2. Linear Systems on Grassmannians and Volume Growth

Linear flows $\Phi_t = e^{At}$ map k-dimensional subspaces to k-dimensional subspaces, hence they induce dynamical systems $\mathbb{G}_k \Phi$ on the sets of k-dimensional subspaces, the Grassmannians endowed with appropriate metrics. In this section the chain components and associated exponential growth rates of k-dimensional volumes will be determined generalizing the discussion for projective space, i.e., for the special case $k = 1$. The discussion is based on the notions and facts from multilinear algebra presented in Section 5.1. In particular, we will use that the Grassmannians are compact metric spaces and that the induced flows are continuous.

We endow $H = \mathbb{R}^d$ with the following inner product which is adapted to the decomposition into the Lyapunov spaces $L_j, 1 \leq j \leq \ell$. Take a basis corresponding to the Jordan normal form of $A \in gl(d, \mathbb{R})$, hence for each j, one has a basis $e_1^j, \ldots, e_{d_j}^j$ of L_j which is orthonormal with respect to the Euclidean inner product. Define

(5.2.1) $$\left\langle e_{i_1}^{j_1}, e_{i_2}^{j_2} \right\rangle := \begin{cases} 0 & \text{for } j_1 \neq j_2 \text{ or } i_1 \neq i_2, \\ 1 & \text{for } j_1 = j_2 \text{ and } i_1 = i_2. \end{cases}$$

5.2. Linear Systems on Grassmannians and Volume Growth 87

These vectors form an orthonormal basis for an inner product on $\mathbb{R}^d$ and hence all the constructions from Section 5.2 can be applied to them. Recall that we identify $\mathbb{G}_k H$ with a subset of projective space $\mathbb{P}(\bigwedge^k H)$ and use the associated metric d_k from (4.1.3). Then $\mathbb{G}_k H$ is compact and the following lemma shows that the induced flow is continuous.

Lemma 5.2.1. *For a linear flow $\Phi_t = e^{At}, t \in \mathbb{R}$, the induced flow $\mathbb{G}_k \Phi_t, t \in \mathbb{R}$, on the Grassmannian $\mathbb{G}_k$ mapping a k-dimensional subspace V to $\Phi(t)V$ is a continuous flow on a compact metric space.*

Proof. Continuity follows from the definition of the metric d_k. □

Next, we determine the exponential growth rate of volumes under linear flows. Consider a linear map A on the Hilbert space $\mathbb{R}^d$ endowed with the inner product defined by (5.2.1). For simplicity, we restrict the discussion to parallelepipeds (5.1.1) with volume $\|x_1 \wedge \ldots \wedge x_k\|$. Under the flow e^{At} this parallelepiped is mapped to the k-dimensional parallelepiped spanned by $e^{At}x_1, \ldots, e^{At}x_k$. The exponential growth rate for $t \to \infty$ of the volume is defined by

$$\lambda(x_1 \wedge \ldots \wedge x_k) := \limsup_{t \to \infty} \frac{1}{t} \log \left\| e^{At}x_1 \wedge \ldots \wedge e^{At}x_k \right\|.$$

This generalizes the notion of Lyapunov exponents (see Definition 1.4.1), since $\|e^{At}x_1\|$ may be considered as the length of the one-dimensional parallelepiped spanned by $e^{At}x_1$. We will show that these exponential growth rates for $t \to \pm\infty$ are determined by the chain components of the flow on the Grassmannians.

Definition 5.2.2. Let $A \in gl(d, \mathbb{R})$ be a matrix with flow $\Phi_t = e^{At}$ on $\mathbb{R}^d$ and denote the Lyapunov spaces by $L_j = L(\lambda_j), j = 1, \ldots, \ell$. For $k = 1, \ldots, d$ define the index set

(5.2.2) $I(k) := \{(k_1, \ldots, k_\ell) \mid k_1 + \ldots + k_\ell = k \text{ and } 0 \leq k_i \leq d_i = \dim L_i\}$

and consider the following subsets of the Grassmannian $\mathbb{G}_k$:

(5.2.3) $\qquad \mathcal{M}^k_{k_1, \ldots, k_\ell} = \mathbb{G}_{k_1} L_1 \oplus \ldots \oplus \mathbb{G}_{k_\ell} L_\ell, \ (k_1, \ldots, k_\ell) \in I(k).$

Here the sum on the right-hand side denotes the set of all k-dimensional subspaces V^k with $\dim(V^k \cap L_i) = k_i, \ i = 1, \ldots, \ell$.

Note that the k-dimensional subspaces V^k in (5.2.3) are the direct sums of the subspaces $V^k \cap L_i, i = 1, \ldots, \ell$. Furthermore, for $k = d$ the only index set is $I(d) = (d_1, \ldots, d_\ell)$ and $\mathcal{M}^d_{d_1, \ldots, d_\ell} = \mathbb{R}^d$. We will show in Theorem 5.2.8 that the sets $\mathcal{M}^k_{k_1, \ldots, k_\ell}, (k_1, \ldots, k_\ell) \in I(k)$, are the chain components in $\mathbb{G}_k$.

The following example illustrates Definition 5.2.2.

Example 5.2.3. Consider the matrix

$$A = \begin{bmatrix} 1 & 0 & 0 \\ 0 & 1 & 0 \\ 0 & 0 & -1 \end{bmatrix}.$$

Let $\mathbf{e}_i$ denote the ith standard basis vector. There are the two Lyapunov spaces $L_1 = L(1) = \text{span}(\mathbf{e}_1, \mathbf{e}_2)$ and $L_2 = L(-1) = \text{span}(\mathbf{e}_3)$ in $\mathbb{R}^3$ with dimensions $d_1 = 2$ and $d_2 = 1$. They project to projective space $\mathbb{P}^2$ as $\mathcal{M}_1 = \{\mathbb{P}x \mid 0 \neq x \in \text{span}(\mathbf{e}_1, \mathbf{e}_2)\}$ (identified with a set of one-dimensional subspaces in $\mathbb{R}^3$) and $\mathcal{M}_2 = \{\mathbb{P}\mathbf{e}_3\}$. Thus, in the notation of Definition 5.2.2, one obtains the following sets of the flows $\mathbb{G}_k\Phi$ on the Grassmannians:

$\mathbb{G}_1$: the index set is $I(1) = \{(1,0), (0,1)\}$ and

$$\mathcal{M}_{1,0}^1 = \{\text{span}(x) \mid 0 \neq x \in \text{span}(\mathbf{e}_1, \mathbf{e}_2)\} \text{ and } \mathcal{M}_{0,1}^1 = \{\text{span}(\mathbf{e}_3)\};$$

$\mathbb{G}_2$: the index set is $I(2) = \{(2,0), (1,1)\}$ and

$$\mathcal{M}_{2,0}^2 = \{\text{span}(\mathbf{e}_1, \mathbf{e}_2)\} \text{ and } \mathcal{M}_{1,1}^2 = \{\text{span}(x, \mathbf{e}_3) \mid 0 \neq x \in \text{span}(\mathbf{e}_1, \mathbf{e}_2)\};$$

$\mathbb{G}_3$: the index set is $I(3) = \{(2,1)\}$ and $\mathcal{M}_{2,1}^3 = \{\text{span}(\mathbf{e}_1, \mathbf{e}_2, \mathbf{e}_3)\}$.

By Theorem 4.1.3 the sets $\mathcal{M}_{1,0}^1$ and $\mathcal{M}_{0,1}^1$ are the chain components in $\mathbb{G}_1 = \mathbb{P}^1$. It will follow from Theorem 5.2.8 that the sets $\mathcal{M}_{2,0}^2$ and $\mathcal{M}_{1,1}^2$ are the chain components of the flow in $\mathbb{G}_2$. In fact, one verifies the assumption that for $k = 2$ and $\lambda_1 = 1, \lambda_2 = -1$ the numbers

$$k_1\lambda_1 + k_2\lambda_2 \text{ with } k_1 + k_2 = k$$

are pairwise different: For $(k_1, k_2) = (2, 0)$ one has $k_1\lambda_1 + k_2\lambda_2 = 2$ and for $(k_1, k_2) = (1, 1)$ one has $k_1\lambda_1 + k_2\lambda_2 = 0$. Figure 5.1 shows the chain components $\mathcal{M}_{1,0}^1$ and $\mathcal{M}_{0,1}^1$ and Figure 5.2 shows $\mathcal{M}_{2,0}^2$ and $\mathcal{M}_{1,1}^2$.

We consider the volume growth for k-dimensional parallelepipeds beginning with elements in the sets specified above.

Proposition 5.2.4. *Consider a set $\mathcal{M}_{k_1,\ldots,k_\ell}^k$ in the Grassmannian $\mathbb{G}_k$ as in Definition 5.2.2. Then for every k-dimensional parallelepiped given by $x_1, \ldots, x_k$ contained in a subspace $V \in \mathcal{M}_{k_1,\ldots,k_\ell}^k$ the exponential growth rate of the volume is*

$$\lim_{t \to \pm\infty} \frac{1}{t} \log \|\mathrm{e}^{At}x_1 \wedge \ldots \wedge \mathrm{e}^{At}x_k\| = \sum_{i=1}^{\ell} k_i\lambda_i.$$

In particular, for $k = d$ the exponential growth rate of the volume of full dimensional parallelepipeds equals $\sum_{i=1}^{\ell} d_i\lambda_i$, where d_i is the dimension of the Lyapunov space $L_i, i = 1, \ldots, \ell$.

5.2. Linear Systems on Grassmannians and Volume Growth 89

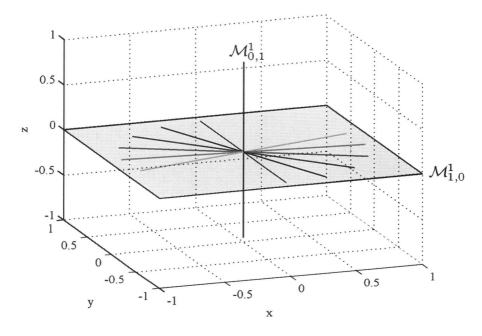

Figure 5.1. The chain components $\mathcal{M}^1_{1,0}$ and $\mathcal{M}^1_{0,1}$ in $\mathbb{G}_1$ for Example 5.2.3

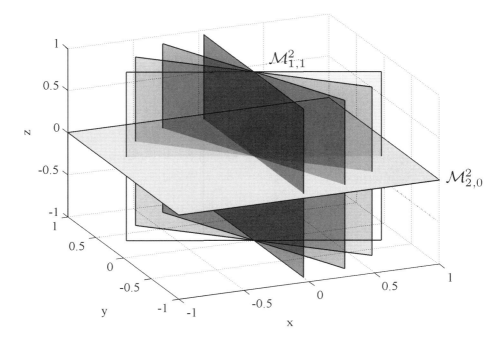

Figure 5.2. The chain components $\mathcal{M}^2_{2,0}$ and $\mathcal{M}^2_{1,1}$ in $\mathbb{G}_2$ for Example 5.2.3

Proof. The linear subspace V generated by $x_1, \ldots, x_k$ is an element of $\mathcal{M}^k_{k_1, \ldots, k_\ell}$, hence $\dim(V \cap L_i) = k_i$ with $\sum_{i=1}^{\ell} k_i = k$. Thus there is a basis of V given by $z_i^1, \ldots, z_i^{k_i} \in L_i, i = 1, \ldots, \ell$, and one finds $a_i^{js} \in \mathbb{R}$ such that

$$x_s = \sum_{i=1}^{\ell} \sum_{j=1}^{k_i} a_i^{js} z_i^j \text{ for } s = 1, \ldots, k.$$

It follows that for $t \in \mathbb{R}$ and $s = 1, \ldots, k$

$$e^{At} x_s = \sum_{i=1}^{\ell} \sum_{j=1}^{k_i} a_i^{js} e^{At} z_i^j = \sum_{i=1}^{\ell} \sum_{j=1}^{k_i} a_i^{js} e^{\lambda_i t} P_i^{js}(t) z_i^j,$$

where the functions $P_i^{js}(t)$ have polynomial growth for $t \to \pm\infty$. Hence the nonsingular $k \times k$-transformation matrix $B(t)$ given by

$$\begin{bmatrix} e^{\lambda_1 t} a_1^{11} P_1^{11}(t) \ldots & e^{\lambda_1 t} a_1^{k_1 1} P_1^{k_1 1}(t) \ldots & e^{\lambda_\ell t} a_\ell^{11} P_\ell^{11}(t) \ldots & e^{\lambda_\ell t} a_\ell^{k_\ell 1} P_\ell^{k_\ell 1}(t) \\ e^{\lambda_1 t} a_1^{12} P_1^{12}(t) \ldots & e^{\lambda_1 t} a_1^{k_1 2} P_1^{k_1 2}(t) \ldots & e^{\lambda_\ell t} a_\ell^{12} P_\ell^{12}(t) \ldots & e^{\lambda_\ell t} a_\ell^{k_\ell 2} P_\ell^{k_\ell 2}(t) \\ \vdots & \vdots & \vdots & \vdots \\ e^{\lambda_1 t} a_1^{1k} P_1^{1k}(t) \ldots & e^{\lambda_1 t} a_1^{k_1 k} P_1^{k_1 k}(t) \ldots & e^{\lambda_\ell t} a_\ell^{1k} P_\ell^{1k}(t) \ldots & e^{\lambda_\ell t} a_\ell^{k_\ell k} P_\ell^{k_\ell k}(t) \end{bmatrix}$$

yields

$$\begin{bmatrix} e^{At} x_1 \\ \vdots \\ e^{At} x_k \end{bmatrix} = B(t) \begin{bmatrix} z_1^1, \ldots, z_1^{k_1}, \ldots, z_\ell^1, \ldots, z_\ell^{k_\ell} \end{bmatrix}^\top.$$

As in (5.1.6) one computes

$$e^{At} x_1 \wedge \ldots \wedge e^{At} x_k = \det B(t) \cdot \left[z_1^1 \wedge \ldots \wedge z_1^{k_1} \wedge \ldots \wedge z_\ell^1 \wedge \ldots \wedge z_\ell^{k_\ell} \right].$$

Now we observe that by multilinearity we can take out of $\det B(t)$ the factor

$$e^{k_1 \lambda_1 t} \ldots e^{k_\ell \lambda_\ell t}.$$

All remaining terms in the determinant have polynomial growth. Now taking the norm, dividing the logarithm by t and letting $t \to \pm\infty$ one finds that

$$\lim_{t \to \pm\infty} \frac{1}{t} \log \left\| e^{At} x_1 \wedge \ldots \wedge e^{At} x_k \right\| = \sum_{i=1}^{\ell} k_i \lambda_i. \qquad \square$$

Next we show that the volume growth rates for arbitrary parallelepipeds are also determined by the growth rates on the sets $\mathcal{M}^k_{k_1, \ldots, k_\ell}$.

Theorem 5.2.5. *For every k-dimensional parallelepiped spanned by vectors $x_1, \ldots, x_k$ in $\mathbb{R}^d$ the exponential growth rate of the volume is given by*

$$\lim_{t \to \infty} \frac{1}{t} \log \left\| e^{At} x_1 \wedge \ldots \wedge e^{At} x_k \right\| = \sum_{i=1}^{\ell} k_i \lambda_i,$$

where $(k_1, \ldots, k_\ell)$ is an element of the index set $I(k)$ from (5.2.2).

5.2. Linear Systems on Grassmannians and Volume Growth

Proof. We argue similarly as in the proof of Proposition 5.2.4, but now it is not sufficient to use for $i = 1, \ldots, \ell$ a basis of the intersection $V \cap L_i$. Instead we have to work with a basis of L_i. Here we may take the basis $e_i^j, j = 1, \ldots, k_i, i = 1, \ldots, \ell$, introduced in the beginning of this section. Then one finds numbers $a_i^{js} \in \mathbb{R}$ such that

$$x_s = \sum_{i=1}^{\ell} \sum_{j=1}^{d_i} a_i^{js} e_i^j \text{ for } s = 1, \ldots, k.$$

It follows that for $t \in \mathbb{R}$ and $s = 1, \ldots, k$,

(5.2.4) $$e^{At} x_s = \sum_{i=1}^{\ell} \sum_{j=1}^{d_i} a_i^{js} e^{At} e_i^j = \sum_{i=1}^{\ell} \sum_{j=1}^{d_i} a_i^{js} e^{\lambda_i t} P_i^{js}(t) e_i^j,$$

where the functions $P_i^{js}(t)$ have polynomial growth for $t \to \pm \infty$. By formula (5.1.2) it follows that

(5.2.5) $$e^{At} x_1 \wedge \ldots \wedge e^{At} x_k = \sum \det B_{i_1 \ldots i_k}^{j_1 \ldots j_k} \left[e_{i_1}^{j_1} \wedge \ldots \wedge e_{i_k}^{j_k} \right].$$

Here $B_{i_1 \ldots i_k}^{j_1 \ldots j_k}$ is the $k \times k$-submatrix of the $k \times d$-matrix $B = B(t)$ formed by the coefficients of the e_i^j in (5.2.4) and comprising the columns determined by $i_1, \ldots, i_k$ and $j_1, \ldots, j_k$; summation is over all ordered k sets of column indices determined by $(i_1, \ldots, i_k)$ and $(j_1, \ldots, j_k)$.

Fix a column of B determined by i_r, j_r. Then every entry of this column contains the factor $e^{\lambda_{i_r} t}$, and hence this is also true for the columns of the matrices $B_{i_1 \ldots i_k}^{j_1 \ldots j_k}$. Thus by multilinearity one can take out of $\det B_{i_1 \ldots i_k}^{j_1 \ldots j_k}$ the factor

$$e^{\lambda_{i_1} t} \ldots e^{\lambda_{i_k} t} = e^{(\lambda_{i_1} + \ldots + \lambda_{i_k}) t}.$$

All remaining terms in the determinant have polynomial growth. Now take the factor $e^{\Lambda t}$ out of the sum in (5.2.5) where

(5.2.6) $$\Lambda := \max (\lambda_{i_1} + \ldots + \lambda_{i_k}) = \sum_{i=1}^{\ell} k_i \lambda_i \text{ with } \sum_{i=1}^{\ell} k_i = k.$$

Here the maximum is taken over all tuples $(i_1, \ldots, i_k)$ and $(j_1, \ldots, j_k)$ with $\det B_{i_1 \ldots i_k}^{j_1 \ldots j_k} \neq 0$ and k_i is determined by the number of λ_{i_r} which coincide. Taking the norm, dividing the logarithm by t and letting $t \to \infty$ one finds that

$$\lim_{t \to \infty} \frac{1}{t} \log \left\| e^{At} x_1 \wedge \ldots \wedge e^{At} x_k \right\| = \Lambda = \sum_{i=1}^{\ell} k_i \lambda_i. \qquad \square$$

Remark 5.2.6. Similar arguments can be applied to determine the limit for $t \to -\infty$: The exponential growth rate of the volume is obtained by using the minimum, instead of the maximum, of $\lambda_{i_1} + \ldots + \lambda_{i_k}$.

To complete this section, we show that the chain components on $\mathbb{G}_k$ are given by the sets specified in Definition 5.2.2. Our proof needs an additional hypothesis (but cf. Section 5.4). The procedure is very similar to the one in Chapter 4, and we only sketch the arguments. First we show that no point outside of these sets is chain recurrent.

Lemma 5.2.7. *Assume that the numbers $\sum_{i=1}^{\ell} k_i \lambda_i$ with $(k_1, \ldots, k_\ell) \in I(k)$ from (5.2.2) are pairwise different. Let $V \in \mathbb{G}_k$ and suppose that $V \notin \bigcup \mathcal{M}_{k_1,\ldots,k_\ell}^k$, where the union is taken over all $(k_1, \ldots, k_\ell) \in I(k)$. Then V is not chain recurrent for the flow on $\mathbb{G}_k$.*

Proof. We may suppose that A is given in real Jordan form, since a linear conjugacy in $\mathbb{R}^d$ yields a topological conjugacy in $\mathbb{G}_k$ which preserves the chain transitive sets by Proposition 3.1.15. We use the identification of $\mathbb{G}_k$ with a compact subset of $\mathbb{P}\left(\bigwedge^k H\right)$. The following construction shows that for $\varepsilon > 0$ small enough there is no (ε, T)-chain from V to V.

Due to our assumption, the multi-index $(k_1, \ldots, k_\ell)$ determining the maximum in (5.2.6) is unique. Similarly, as in the proof of Lemma 4.1.2, one can argue that
$$\mathrm{dist}(\mathrm{e}^{At} y_1 \wedge \ldots \wedge \mathrm{e}^{At} y_k, \mathcal{M}_{k_1,\ldots,k_\ell}^k) \to 0.$$
By assumption, V has positive distance to every set from Definition 5.2.2. Consider all these sets and define
$$W_{(k_1,\ldots,k_\ell)} := \{z_1 \wedge \ldots \wedge z_k \in \bigoplus \mathbb{P}^{-1} \mathcal{M}_{j_1,\ldots,j_\ell}^k\} \subset \mathbb{P}(\bigwedge^k \mathbb{R}^d),$$
where the sum is taken over all multi-indices $(j_1, \ldots, j_\ell) \in I(k)$ with $\sum_{i=1}^{\ell} j_i \lambda_i \geq \sum_{i=1}^{\ell} k_i \lambda_i$. Then one can again argue similarly as in the proof of Lemma 4.1.2. $\square$

The following theorem describes the chain components in the Grassmannians. Recall that the Lyapunov spaces projected to projective space are the chain components of the associated flow; see Theorem 4.1.3.

Theorem 5.2.8. *Let $A \in gl(d, \mathbb{R})$ be a matrix with flow $\Phi_t = \mathrm{e}^{At}$ in $\mathbb{R}^d$. Then the sets $\mathcal{M}_{k_1,\ldots,k_\ell}^k$, $(k_1, \ldots, k_\ell) \in I(k)$ from Definition 5.2.2 are chain transitive. Hence, if the numbers $\sum_{i=1}^{\ell} k_i \lambda_i$ with $(k_1, \ldots, k_\ell) \in I(k)$ from (5.2.2) are pairwise different, these are the chain components.*

Proof. Due to Lemma 5.2.7, we only have to prove that the flow restricted to each set $\mathcal{M}_{k_1,\ldots,k_\ell}^k$ is chain transitive.

5.2. Linear Systems on Grassmannians and Volume Growth

(i) As a first step, we show that for every Lyapunov space L_j and every $k \leq d_j = \dim L_j$ the flow restricted to the set

$$\mathbb{G}_k L_j := \{V \in \mathbb{G}_k \mid V \subset L_j\}$$

is chain transitive. This follows similarly as the determination of chain transitive sets in projective space $\mathbb{P}^{d-1} = \mathbb{G}_1$. Note that by Remark 1.3.7 the set $\mathbb{G}_k L_j$ contains an equilibrium, if $k \geq 2$, and if $k = 1$ existence of an equilibrium (for a real eigenvalue) or of a periodic trajectory (in the real eigenspace for a complex conjugate pair of eigenvalues) is guaranteed.

Consider a Jordan block of dimension at least k. Then any subspace $V \in \mathbb{G}_k L_j$ corresponding to this Jordan block is attracted for $t \to \pm\infty$ to the subspace spanned by the first k elements of a corresponding basis. This yields a chain transitive set in $\mathbb{G}_k L_j$ and the Jordan blocks of dimension at least k give rise to a continuum of equilibria in $\mathbb{G}_k L_j$. Furthermore, also the subspaces generated by k elements corresponding to upper parts of different Jordan blocks are invariant and subspaces in the direct sum corresponding to these Jordan blocks are attracted for $t \to \pm\infty$ by this set of subspaces. Proceeding in this way, one sees that the flow restricted to $\mathbb{G}_k L_j$ is chain transitive.

(ii) Next we prove the assertion of the theorem by induction over k. For $k = 1$, the set $I(1)$ of multi-indices consists of the k-tuples with all $k_i = 0$ except for one $k_j = 1$. Thus the sets

$$\mathcal{M}^1_{k_1,\ldots,k_\ell} = \mathbb{G}_{k_1} L_1 \oplus \ldots \oplus \mathbb{G}_{k_\ell} L_\ell = \{V \in \mathbb{G}_1 \mid V \subset L_j\}$$

coincide with the Lyapunov spaces projected to $\mathbb{P}^{d-1}$. Theorem 4.1.3 shows that these are the chain components and that the flow restricted to any of them is chain transitive. Next suppose that the assertion holds for $1, \ldots, k-1 \geq 1$. We show that for $(k_1, \ldots, k_\ell) \in I(k)$ the flow restricted to

$$\mathcal{M}^k_{k_1,\ldots,k_\ell} = \{V^k \in \mathbb{G}_k \mid \dim(V^k \cap L_i) = k_i \text{ for all } i = 1, \ldots, \ell\}$$

is chain transitive. Take $V^k, \tilde{V}^k \in \mathcal{M}^k_{k_1,\ldots,k_\ell}$. Then they satisfy

$$V^k = \bigoplus_{i=1}^{\ell} \left[V^k \cap L_i\right] \text{ and } \tilde{V}^k = \bigoplus_{i=1}^{\ell} \left[\tilde{V}^k \cap L_i\right].$$

Since $k \geq 2$, either all $k_j = 0$ except for one which is equal to k and we are in the situation of part (i) and chain transitivity follows; or there is a subindex r with $0 < k_r < k$. In the latter case, define

$$V^{k_r} := V^k \cap L_r \text{ and } \tilde{V}^{k_r} := \tilde{V}^k \cap L_r,$$

$$V^{k-k_r} := \bigoplus_{i=1, i \neq r}^{\ell} \left[V^k \cap L_i\right] \text{ and } \tilde{V}^{k-k_r} := \bigoplus_{i=1, i \neq r}^{\ell} \left[\tilde{V}^k \cap L_i\right].$$

Since V^{k_r} and $\tilde{V}^{k_r}$ are in $\mathbb{G}_{k_r}L_r$, there are by part (i) for every $\varepsilon > 0$ chains in $\mathbb{G}_{k_r}L_r$ from V^{k_r} to $\tilde{V}^{k_r}$. By the induction hypothesis and $k - k_r < k$, there are for every $\varepsilon > 0$ chains from V^{k-k_r} to $\tilde{V}^{k-k_r}$ in the chain transitive set
$$\{V \in \mathbb{G}_{k-k_r} \mid \dim(V \cap L_i) = k_i \text{ for all } i = 1, \ldots, \ell, \ i \neq r\}.$$
As in part (i), one finds that this set contains an equilibrium or a periodic solution. Hence, using Proposition 3.1.11, we may assume that all jump times are equal to 1 and that the numbers of jumps coincide. Write the chains as
$$V_0 = V^{k_r}, V_1, \ldots, V_n = \tilde{V}^{k_r} \text{ and } W_0 = V^{k-k_r}, W_1, \ldots, W_n = \tilde{V}^{k-k_r},$$
with
$$d(\Phi(1, V_j), V_{j+1}) < \varepsilon \text{ and } d(\Phi(1, W_j), W_{j+1}) < \varepsilon.$$
Note that $V_j, \Phi(1, V_j) \subset L_r$ and $W_j, \Phi(1, W_j) \subset \bigoplus_{i=1, i\neq r}^{\ell} L_i$, hence these subspaces are contained in orthogonal subspaces. Then Lemma 5.1.3 shows that
$$d(\Phi(1, V_j \oplus W_j), V_{j+1} \oplus W_{j+1}) = d(\Phi(1, V_j) \oplus \Phi(1, W_j), V_{j+1} \oplus W_{j+1}) < \varepsilon$$
for all j. It follows that $V^k = V^{k_r} \oplus V^{k-k_r} = V_0 \oplus W_0$ and $\tilde{V}^k = \tilde{V}^{k_r} \oplus \tilde{V}^{k-k_r} = V_n \oplus W_n$ are connected by chains within $\mathcal{M}^k_{k_1,\ldots,k_\ell}$. □

5.3. Exercises

Exercise 5.3.1. Consider the differential equation $\dot{x} = Ax$ with
$$A = \begin{bmatrix} -1 & 0 & 0 & 0 \\ 0 & 0 & 1 & 0 \\ 0 & -1 & 0 & 0 \\ 0 & 0 & 0 & 1 \end{bmatrix}.$$
Determine the chain components in the Grassmannians $\mathbb{G}_k, 1 \leq k \leq 4$.

Exercise 5.3.2. Prove the Hadamard inequality in the alternating product $\bigwedge^2 H = H \wedge H$ of a Euclidean vector space H: For $x, y \in H$,
$$\|x \wedge y\| \leq \|x\| \ \|y\| \text{ for } x, y \in H.$$
The generalized Hadamard inequality states that for $x_1, \ldots, x_k \in \bigwedge^k H$ and $1 < i < k$,
$$\|x_1 \wedge \ldots \wedge x_k\| \leq \|x_1 \wedge \ldots \wedge x_i\| \ \|x_{i+1} \wedge \ldots \wedge x_k\|.$$
(You may also look it up in a book on multilinear algebra!) See also Wach [136] and Kulczycki [90] for a discussion of this type of inequalities.

5.4. Orientation, Notes and References

Orientation. This chapter has generalized the characterization of the induced flows on projective space to the Grassmannians $\mathbb{G}_k$. The chain components and the exponential growth rates of volumes have been characterized.

One point is left open in this chapter: Is the assumption on the exponential growth rates of volumes in Theorem 5.2.8 necessary? The difficulty lies in the fact that the exponential growth rates of volumes and the chain components may not be linearly ordered. Hence, in general, one cannot deduce from the volume growth rates to which chain component the elements in $\mathbb{G}_k$ converge. However, one can show that Theorem 5.2.8 holds without this assumption, since it specifies the finest Morse decompositions (cf. Chapter 8) whose elements coincide with the chain components; cf. Colonius and Kliemann [**30**, Theorem 6].

Notes and references. For background material on the projective space and Grassmannian and flag manifolds, Boothby [**19**] is a good resource. The discussion of Riccati equations is taken from Helmke and Moore [**65**, pp. 10–13]. Using local charts, the Riccati differential equation can be extended to the Grassmann manifold. Thus Theorem 5.2.8 describes the chain components of this differential equation on the compact manifold $\mathbb{G}_k$. A more detailed description of the phase portrait of various classes of matrix Riccati differential equations has been given by Shayman [**126**]. It should be noted that the Riccati differential equations of the calculus of variations and optimal control theory are associated with a symplectic structure. They are also analyzed in [**126**], now on the Lagrange-Grassmann manifold.

Theorem 5.2.8 can be generalized to flag manifolds: For natural numbers $d_1 < \ldots < d_k \leq d$ define a flag of subspaces $\mathbb{R}^d$ by

$$\mathbb{F}(d_1, \ldots, d_k) = \{(V_1, \ldots, V_k) \mid V_i \subset V_{i+1} \text{ and } \dim V_i = d_i \text{ for all } i\}.$$

For $k = d$ this is the complete flag $\mathbb{F}$. Braga Barros and San Martin [**20**] and San Martin and Seco [**122**] develop far reaching generalizations, including as very special cases the analysis of the chain components on the Grassmann and flag manifolds in $\mathbb{R}^d$.

For the notions and facts from multilinear algebra in Section 5.1 we refer, e.g., to Federer [**46**, Chapter 1], Greub [**55**], Merris [**103**], Marcus [**99**], Heil [**64**], Lamprecht [**93**].

Part 2

Time-Varying Matrices and Linear Skew Product Systems

Chapter 6

Lyapunov Exponents and Linear Skew Product Systems

In the second part of this book we allow that the considered matrices depend on time and consider the associated nonautonomous (or time-varying) linear differential and difference equations. The goal is to develop an associated linear algebra which extends the theory valid for the autonomous case. More specifically, we will consider families of nonautonomous linear systems and we will be concerned with spectral theory and associated subspace decompositions of the state space. Thus we generalize the decomposition of the state space into Lyapunov spaces as presented in Theorem 1.4.3 for differential equations and Theorem 1.5.6 for difference equations, where the state space is decomposed into invariant subspaces which are characterized by the property that solutions starting in a subspace realize the same exponential growth rate or Lyapunov exponent for time tending to $\pm\infty$. These results were derived using the characterization of Lyapunov exponents by eigenvalues which allowed us to use the Jordan normal form for matrices.

It will turn out that also for nonautonomous linear systems Lyapunov exponents (also called characteristic numbers) play a central role. However, additional assumptions and constructions are necessary in order to develop results analogous to the autonomous case. In particular, the relation to eigenvalues breaks down and hence the Jordan form cannot be used. In other words, developing a linear algebra for nonautonomous or time-varying systems $\dot{x} = A(t)x$ and $x_n = A_n x_n$ means defining appropriate concepts to generalize eigenvalues, linear eigenspaces and their dimensions that characterize the behavior of the solutions of a time-varying system and that reduce to the constant matrix case if $A(t) \equiv A$. As mentioned above (and already observed by Lyapunov at the end of the nineteenth century), the eigenvalues and eigenspaces of the families $\{A(t), t \in \mathbb{R}\}$ and $\{A_n, n \in \mathbb{Z}\}$ do not provide the appropriate concept: Even if the real parts of all eigenvalues of all matrices $A(t)$, $t \in \mathbb{R}$, are negative, the origin $0 \in \mathbb{R}^d$ need not be a stable fixed point of $\dot{x} = A(t)x$. Instead one has to look at the exponential growth rates of solutions, the Lyapunov exponents.

This chapter contains several examples elucidating the situation for time-varying systems. The framework of linear skew product systems is introduced and a number of classes of linear skew product systems are presented. In the following chapters several of them will be analyzed in detail.

Section 6.1 establishes existence of unique solutions for time-varying linear differential equations and continuous dependence on initial values and parameters. Section 6.2 discusses basic properties of Lyapunov exponents for these differential equations. In particular, counterexamples are presented, which show that, without additional assumptions, nonautonomous linear differential equations miss the most important properties of autonomous linear differential equations. Consequently, Section 6.3 proposes several approaches to embed nonautonomous linear differential equations into dynamical systems having the form of a linear skew product flow over a base flow. Finally, Section 6.4 discusses the discrete time case and introduces skew product dynamical systems for nonautonomous linear difference equations.

6.1. Existence of Solutions and Continuous Dependence

This section derives basic properties of nonautonomous linear differential equations. First, existence and uniqueness of solutions is proved. Then it is shown that the solutions depend continuously on initial values and parameters, i.e., entries in the system matrix.

In Chapter 9 on topological linear systems and Chapter 11 on random linear systems it will be relevant to allow that the system matrix $A(t)$ is discontinuous with respect to t. A corresponding solution may not be differentiable and hence will not satisfy the differential equation for all t. This difficulty can be solved by replacing the differential equation $\dot{x}(t) = A(t)x(t)$ by its integrated version.

More explicitly, let $A : \mathbb{R} \to gl(d, \mathbb{R})$ be a locally integrable matrix function, i.e., it is Lebesgue integrable on every bounded interval. We will consider solutions, also called Carathéodory solutions, which are defined in

6.1. Existence of Solutions and Continuous Dependence

the following way. A solution of the initial value problem

(6.1.1) $$\dot{x} = A(t)x, \ x(t_0) = x_0 \in \mathbb{R}^d,$$

is a continuous function satisfying

(6.1.2) $$x(t) = x_0 + \int_{t_0}^{t} A(s)x(s)ds, t \in \mathbb{R}.$$

Thus a solution $x(\cdot)$ is an absolutely continuous function and hence is differentiable outside of a set of Lebesgue measure zero and here satisfies equation (6.1.1) (see, e.g., Craven [38, Proposition 5.4.5]). Naturally, one may also consider solutions which are only defined on subintervals. If the function $A(\cdot)$ is continuous, the fundamental theorem of analysis shows that a function $x(\cdot)$ as in (6.1.2) is continuously differentiable and satisfies the differential equation for every t.

The following simple example illustrates that a solution for a discontinuous matrix function $A(t)$ may not be differentiable. Let $a(t) = 1$ for $t \geq 0$ and $a(t) = -1$ for $t < 0$. Then the solution of

$$\dot{x} = a(t)x, x(0) = 1$$

is $x(t) = e^t$ for $t > 0$ and $x(t) = e^{-t}$ for $t < 0$. Thus the derivatives in $t = 0$ from the left and from the right do not coincide.

In the scalar case, one can write down the solution of (6.1.1). Suppose that $a : \mathbb{R} \to \mathbb{R}$ is locally integrable. Then the solution of

$$\dot{x} = a(t)x, \ x(t_0) = x_0 \in \mathbb{R},$$

is (Exercise 6.5.5) given by

$$x(t) = e^{\int_{t_0}^{t} a(s)ds} x_0, t \in \mathbb{R}.$$

For higher dimensional equations such an explicit formula is not available; see Exercise 6.5.6. The next theorem on existence and uniqueness of solutions presents a generalization of Theorem 1.1.1 for time-varying matrices.

Theorem 6.1.1. *Let $A : \mathbb{R} \to gl(d, \mathbb{R})$ be a locally integrable matrix function.*

(i) The solutions of $\dot{x} = A(t)x$ form a d-dimensional vector space $sol(A) \subset C(\mathbb{R}, \mathbb{R}^d)$.

(ii) The matrix initial value problems $\dot{X}(t) = A(t)X(t)$ with $X(t_0) = I_d \in Gl(d, \mathbb{R})$ have unique solutions $X(t, t_0)$, called principal fundamental solutions.

(iii) For each initial condition $x(t_0) = x_0 \in \mathbb{R}^d$, the solution of (6.1.1) is unique and given by

(6.1.3) $$\varphi(t, t_0, x_0) = X(t, t_0)x_0.$$

Proof. For $t_1 > t_0$ the formula

$$(Fy)(t) := x_0 + \int_{t_0}^{t} A(s)y(s)\, ds,\ t \in [t_0, t_1],$$

defines a continuous linear transformation on the Banach space of continuous functions $C([t_0, t_1], \mathbb{R}^d)$ with the maximum norm such that for all $y_1, y_2 \in C([t_0, t_1], \mathbb{R}^d)$,

$$\max_{t \in [t_0, t_1]} \|F(y_1)(t) - F(y_2)(t)\| \leq \left(\int_{t_0}^{t_1} \|A(s)\|\, ds\right) \max_{t \in [t_0, t_1]} \|y_1(t) - y_2(t)\|.$$

If $\int_{t_0}^{t_1} \|A(s)\|\, ds < 1$, then F is a contraction on the complete metric space $C([t_0, t_1], \mathbb{R}^d)$. Hence Banach's fixed point theorem implies that the equation $y = F(y)$ has a unique solution which is the solution of the integral equations. The case $\int_{t_0}^{t_1} \|A(s)\|\, ds \geq 1$ can be treated by considering appropriate subintervals. The same arguments also show unique solvability for $t_1 < t_0$ and unique solvability of initial value problems for the matrix equation $\dot{X}(t) = A(t)X(t)$. Then the solution formula (6.1.3) follows, since the right-hand side, too, satisfies (6.1.2). Assertion (i) is a consequence. □

The following results are a consequence of unique solvability.

Proposition 6.1.2. *Consider for a locally integrable function $A : \mathbb{R} \to gl(d, \mathbb{R})$ the equation $\dot{x} = A(t)x$ and let $t, t_1, t_0 \in \mathbb{R}$.*

(i) The solutions satisfy the 2-parameter cocycle property

(6.1.4) $$\varphi(t, t_1, \varphi(t_1, t_0, x_0)) = \varphi(t, t_0, x_0).$$

(ii) The principal fundamental solutions satisfy

$$X(t_0, t_0) = I,\ X(t, t_1)X(t_1, t_0) = X(t, t_0),\ X(t, t_0)^{-1} = X(t_0, t)$$

and X is absolutely continuous in each argument.

(iii) With $X(t) := X(t, 0), t \in \mathbb{R}$ one has

(6.1.5) $$X(t, t_0) = X(t)X(t_0)^{-1} \text{ and hence } \varphi(t, t_0, x_0) = X(t)X(t_0)^{-1}x_0.$$

Proof. Assertion (i) follows from unique solvability of initial value problems, since the left- and the right-hand side are, by definition, solutions of the differential equation that satisfy (6.1.2). The equations in (ii) are a consequence of (i), (iii) follows from (ii), and absolute continuity of $X(t, t_0)$ in the first argument is clear by definition. Concerning the second argument, let $Y(t), t \in I$ be the unique (absolutely continuous) matrix solution of

$$\dot{Y}(t) = -Y(t)A(t) \text{ with } Y(t_0) = I.$$

6.1. Existence of Solutions and Continuous Dependence

Then the product rule shows

$$\frac{d}{dt}[Y(t)X(t,t_0)] = \dot{Y}(t)X(t,t_0) + Y(t)\frac{d}{dt}X(t,t_0)$$
$$= -Y(t)A(t)X(t,t_0) + Y(t)A(t)X(t,t_0) = 0,$$

hence

$$Y(t)X(t,t_0) \equiv Y(t_0)X(t_0,t_0) = I_d.$$

It follows that $X(t_0,t) = X(t,t_0)^{-1} = Y(t)$ is absolutely continuous with respect to t. □

The formula in (6.1.5) is the variation-of-constants formula for homogeneous linear differential equations. In order to show continuity properties with respect to parameters, we need the following useful result called Gronwall's lemma.

Lemma 6.1.3. *Let $I := [t_0, t_1] \subset \mathbb{R}$ be an interval and consider a continuous function $u : I \to [0, \infty)$ such that for a constant $\alpha \geq 0$ and a locally integrable function $\beta : I \to [0, \infty)$,*

$$(6.1.6) \qquad u(t) \leq \alpha + \int_{t_0}^{t} \beta(s)u(s)ds \text{ for all } t \in I.$$

Then

$$(6.1.7) \qquad u(t) \leq \alpha e^{\int_{t_0}^{t} \beta(s)ds} \text{ for all } t \in I.$$

Proof. Suppose first $\alpha > 0$. Define the auxiliary function $v(t) = \alpha + \int_{t_0}^{t} \beta(s)u(s)ds, t \in I$, which satisfies $u(t) \leq v(t)$ and $v(t) \geq \alpha > 0$. With $\dot{v}(t) = \beta(t)u(t)$ it follows for $u(t) \neq 0$ that

$$\frac{d}{dt}\log v(t) = \frac{\dot{v}(t)}{v(t)} \leq \frac{\beta(t)u(t)}{u(t)} = \beta(t).$$

If $u(t) = 0$ the inequality $0 = \frac{d}{dt}\log v(t) \leq \beta(t)$ holds trivially. Integrating this inequality one finds

$$\log v(t) - \log v(t_0) \leq \int_{t_0}^{t} \beta(s)ds.$$

With $v(t_0) = \alpha$ application of the exponential function yields

$$u(t) \leq v(t) = \alpha + \int_{t_0}^{t} \beta(s)u(s)ds \leq \alpha e^{\int_{t_0}^{t} \beta(s)ds}$$

and (6.1.7) follows. If $\alpha = 0$ one has for every $\varepsilon > 0$,

$$u(t) \leq \varepsilon + \int_{t_0}^{t} \beta(s)u(s)ds.$$

Hence the result above implies the assertion for $\alpha = 0$, since
$$u(t) \leq \varepsilon e^{\int_{t_0}^{t} \beta(s)ds} \to 0 \text{ for } \varepsilon \to 0. \qquad \square$$

The next theorem shows that solutions depend continuously on initial values and on parameters occurring in the system matrix $A(\cdot)$.

Theorem 6.1.4. *Let Γ be a metric space and consider a matrix-valued function $A : \mathbb{R} \times \Gamma \to gl(d, \mathbb{R})$, such that for every $\gamma \in \Gamma$ the function $A(\cdot, \gamma) : \mathbb{R} \to gl(d, \mathbb{R})$ is locally integrable and that for almost every $t \in \mathbb{R}$ the function $A(t, \cdot) : \Gamma \to gl(d, \mathbb{R})$ is continuous and, for a locally integrable function $\beta : \mathbb{R} \to [0, \infty)$,*
$$\sup_{\gamma \in \Gamma} \|A(t, \gamma)\| \leq \beta(t) \text{ for almost all } t \in \mathbb{R}.$$
Then the solution $\varphi(t, t_0, x_0, \gamma), t \in \mathbb{R}$, of $\dot{x}(t) = A(t, \gamma)x(t), x(t_0) = x_0$, is continuous with respect to $(t, x_0, \gamma) \in \mathbb{R} \times \mathbb{R}^d \times \Gamma$.

Proof. Consider for a sequence $(x_n, \gamma_n) \to (x_0, \gamma_0)$ in $\mathbb{R}^d \times \Gamma$ the corresponding solutions $\varphi(t, t_0, x_n, \gamma_n), t \in \mathbb{R}$. Then one estimates for $n \in \mathbb{N}$ and $t > t_0$,

$$\|\varphi(t, t_0, x_n, \gamma_n) - \varphi(t, t_0, x_0, \gamma_0)\|$$
$$= \left\| x_n + \int_{t_0}^{t} A(s, \gamma_n)\varphi(s, t_0, x_n, \gamma_n)ds - x_0 - \int_{t_0}^{t} A(s, \gamma_0)\varphi(s, t_0, x_0, \gamma_0)ds \right\|$$
$$\leq \|x_n - x_0\| + \int_{t_0}^{t} \|A(s, \gamma_n) - A(s, \gamma_0)\| \|\varphi(s, t_0, x_0, \gamma_0)\| \, ds$$
$$+ \int_{t_0}^{t} \|A(s, \gamma_n)\| \|\varphi(s, t_0, x_n, \gamma_n) - \varphi(s, t_0, x_0, \gamma_0)\| \, ds.$$

Now apply Gronwall's lemma, Lemma 6.1.3, on $I := [t_0, t_1]$ with $\beta(t) := \|A(t, \gamma_n)\|$ and
$$\alpha := \|x_n - x_0\| + \int_{t_0}^{t_1} \|A(s, \gamma_n) - A(s, \gamma_0)\| \|\varphi(s, t_0, x_0, \gamma_0)\| \, ds.$$

One gets for $t \in [t_0, t_1]$,
$$\|\varphi(t, t_0, x_n, \gamma_n) - \varphi(t, t_0, x_0, \gamma_0)\|$$
$$\leq [\|x_n - x_0\| + \int_{t_0}^{t} \|A(s, \gamma_n) - A(s, \gamma_0)\| \|\varphi(s, t_0, x_0, \gamma_0)\| \, ds] e^{\int_{t_0}^{t} \beta(s)ds}.$$

Then the convergence $\varphi(t, t_0, x_n, \gamma_n) \to \varphi(t, t_0, x_0, \gamma_0)$ for $n \to \infty$ follows by Lebesgue's Theorem on dominated convergence. Furthermore, if $t_n \to t \in [t_0, t_1]$ then $\varphi(t_n, t_0, x_n, \gamma_n)$, too, converges to $\varphi(t, t_0, x_0, \gamma_0)$. Since $t_1 > t_0$ is arbitrary, the assertion follows for all $t > t_0$. For $t_1 < t_0$ one argues analogously. $\square$

6.1. Existence of Solutions and Continuous Dependence

Consider for $\dot{x} = A(t)x$ a matrix function $W(t) \in gl(d,\mathbb{R}), t \in \mathbb{R}$, where the columns are solutions of a homogeneous differential equation. This is called Wronskian (or Wronski matrix) and

$$w(t) := \det W(t)$$

is called a Wronski determinant. Recall that the trace $\mathrm{tr}A$ of a quadratic matrix is the sum of the diagonal elements. The following is Liouville's formula.

Theorem 6.1.5. *Every Wronski determinant of $\dot{x} = A(t)x$ satisfies the scalar homogeneous linear differential equation*

$$\dot{w}(t) = [\mathrm{tr}A(t)]w(t),$$

hence

$$w(t) = w(t_0)e^{\int_{t_0}^{t} \mathrm{tr}A(s)ds} \text{ for } t, t_0 \in \mathbb{R}.$$

Proof. Let $W(t) = (w_{ij}(t))_{i,j=1,\ldots,d}$ and denote the ith row of $W(t)$ by $w_i(t) = (w_{i1}(t), \ldots, w_{id}(t))$. Then one obtains for the rows

(6.1.8) $$\dot{w}_i(t) = a_{i1}(t)w_1(t) + \ldots + a_{id}(t)w_d(t).$$

The Laplace expansion for the determinant has the form

(6.1.9) $$\det W(t) = \sum_{\sigma \in \Sigma_d} (\mathrm{sgn}\,\sigma)w_{1\sigma(1)}(t)\ldots w_{d\sigma(d)}(t),$$

where Σ_d is the permutation group and $\mathrm{sgn}\,\sigma$ is the signature of a permutation σ, hence equal to $+1$, if the number of transpositions which make up σ is even and equal to -1 otherwise. Differentiation and, again, use of the Laplace expansion yields

$$\frac{d}{dt}\det W(t)$$
$$= \sum_{\sigma \in \Sigma_d}(\mathrm{sgn}\,\sigma)\dot{w}_{1\sigma(1)}(t)\ldots w_{d\sigma(d)}(t) + \cdots + \sum_{\sigma \in \Sigma_d}(\mathrm{sgn}\,\sigma)w_{1\sigma(1)}(t)\ldots \dot{w}_{d\sigma(d)}(t)$$

$$= \det \begin{bmatrix} \dot{w}_1(t) \\ w_2(t) \\ \vdots \\ w_d(t) \end{bmatrix} + \ldots + \det \begin{bmatrix} w_1(t) \\ w_2(t) \\ \vdots \\ \dot{w}_d(t) \end{bmatrix}$$

$$= a_{11}(t)\det W(t) + \ldots + a_{dd}(t)\det W(t).$$

The last equality follows by inserting (6.1.8), using multilinearity of the determinant and observing that determinants with identical rows vanish.

For example, one obtains for the first summand

$$\det \begin{bmatrix} \dot{w}_1(t) \\ w_2(t) \\ \vdots \\ w_d(t) \end{bmatrix} = \det \begin{bmatrix} a_{11}(t)w_1(t) + \ldots + a_{1d}(t)w_d(t) \\ w_2(t) \\ \vdots \\ w_d(t) \end{bmatrix} = a_{11}(t) \det W(t).$$

$\square$

Remark 6.1.6. A geometric interpretation of Theorem 6.1.5 is as follows: Suppose that $X(t)$ is a fundamental matrix, i.e., its columns are d linear independent solutions $x_1(t), \ldots, x_d(t)$ of $\dot{x} = A(t)x$. Then

$$\det X(t) = \det X(t_0) e^{\int_{t_0}^{t} \operatorname{tr} A(s) ds}$$

relates to the volume $\det X(t_0)$ of the parallelepiped spanned by $x_1(t_0), \ldots, x_d(t_0)$ at time t_0 to the volume $\det X(t)$ of its image under the solution map at time t; cf. also the discussion of volume growth rates in Chapter 5.

6.2. Lyapunov Exponents

This section discusses exponential growth rates of solutions by introducing Lyapunov exponents and deriving some of their basic properties. Then examples are constructed that show that, in general, they do not lead to subspace decompositions of the state space and that they are not characterized by eigenvalues of the system matrices.

We consider nonautonomous differential equations $\dot{x} = A(t)x$ as in (6.1.1). Recall that (absolutely continuous) solutions with $x(t_0) = x_0 \in \mathbb{R}^d$ are denoted by $\varphi(t, t_0, x_0), t \in \mathbb{R}$. We omit the argument t_0 if $t_0 = 0$. In analogy to the definition for autonomous linear differential equations, Definition 1.4.1, the exponential growth rate of a solution is given by

$$\limsup_{t \to \infty} \frac{1}{t} \log \|\varphi(t, t_0, x_0)\|.$$

Instead of lim sup one may also consider lim inf or the corresponding limits for $t \to -\infty$. Using the 2-parameter cocycle property (6.1.4), one sees that it is sufficient to consider solutions $\varphi(t, x_0)$ with initial time $t_0 = 0$.

Definition 6.2.1. Let $\varphi(\cdot, x_0)$ be the solution of the initial value problem $\dot{x} = A(t)x, 0 \neq x(0) = x_0 \in \mathbb{R}^d$. Its Lyapunov exponent or exponential growth rate is defined as

$$\lambda(x_0) = \limsup_{t \to \infty} \frac{1}{t} \log \|\varphi(t, x_0)\|,$$

where log denotes the natural logarithm and $\|\cdot\|$ is any norm in $\mathbb{R}^d$.

6.2. Lyapunov Exponents

If $\lambda(x_0) < 0$, the function $\varphi(t, x_0)$ converges to $0 \in \mathbb{R}^d$ for $t \to \infty$. If $A(t) \equiv A$ is constant, the results in Section 1.4 show that the Lyapunov exponents are the real parts of the eigenvalues $\mu_i = \lambda_i + \imath \omega_i$ of A and the limit superior is a limit; furthermore, the state space can be decomposed into the Lyapunov spaces, $\mathbb{R}^d = \bigoplus_{j=1}^{\ell} L(\lambda_j)$, where $L(\lambda_j)$ is the subspace of all initial points which have the exponential growth rate λ_j for $t \to \pm\infty$.

The general nonautonomous case is much more complicated. It is immediately clear that for an arbitrary matrix function $A(t), t \in \mathbb{R}$, the behaviors for positive and negative times are, in general, different. Hence one cannot expect that the exponential growth rates for positive and negative times are related. The following scalar example shows that the limit superior in the definition of Lyapunov exponents may not be a limit. Consider

$$\dot{x} = (\cos t - t \sin t - 2)x, \tag{6.2.1}$$

which has the solutions

$$x(t) = e^{t \cos t - 2t} x(0), t \in \mathbb{R}.$$

The exponential growth rates are

$$\limsup_{t \to \infty} \frac{1}{t} \log \|x(t)\| = \limsup_{t \to \infty} \frac{1}{t} (t \cos t - 2t) = -1,$$

$$\liminf_{t \to \infty} \frac{1}{t} \log \|x(t)\| = \liminf_{t \to \infty} \frac{1}{t} (t \cos t - 2t) = -3.$$

Furthermore, a boundedness condition on the matrices is needed in order to get a finite Lyapunov exponent, since the exponential growth rates may be infinite as shown by the scalar equation

$$\dot{x} = 2tx$$

with solutions $x(t) = e^{t^2} x(0)$. Here $\limsup_{t \to \infty} \frac{1}{t} \log \|x(t)\| = \lim_{t \to \infty} t = \infty$.

Throughout the rest of this section, we suppose the boundedness assumption

$$\limsup_{t \to \infty} \frac{1}{t} \int_0^t \|A(s)\| \, ds < \infty. \tag{6.2.2}$$

In order to show that the number of Lyapunov exponents is bounded by the dimension of the state space, we note the following lemma.

Lemma 6.2.2. *Let $f, g : [0, \infty) \to (0, \infty)$ be locally integrable functions and denote $\lambda(f) := \limsup_{t \to \infty} \frac{1}{t} \int_0^t \log f(s) ds \leq \infty$, and analogously $\lambda(g)$. Then $\lambda(fg) \leq \lambda(f) + \lambda(g)$, and $\lambda(f+g) \leq \max(\lambda(f), \lambda(g))$ with $\lambda(f+g) = \max(\lambda(f), \lambda(g))$ if $\lambda(f) \neq \lambda(g)$.*

Proof. Exercise 6.5.1 or Exercise 6.5.7. □

In particular, it follows that for $0 \neq x, y \in \mathbb{R}^d$ the Lyapunov exponents satisfy $\lambda(x+y) \leq \max\{\lambda(x), \lambda(y)\}$. The next proposition states basic properties of Lyapunov exponents.

Proposition 6.2.3. *(i) There are $1 \leq \ell \leq d$ different Lyapunov exponents, which we order according to $\lambda_\ell < \ldots < \lambda_1 < \infty$.*

(ii) On every basis of $\mathbb{R}^d$, the maximal Lyapunov exponent is attained.

(iii) There exists a basis of $\mathbb{R}^d$, called a normal basis, such that each Lyapunov exponent is attained on some basis element.

Proof. (i) For $0 \neq x_0 \in \mathbb{R}^d$ the solution satisfies for $t \geq 0$,

$$\|\varphi(t, x_0)\| = \left\| x_0 + \int_0^t A(s)\varphi(s, x_0)ds \right\| \leq \|x_0\| + \int_0^t \|A(s)\| \, \|\varphi(s, x_0)\| \, ds.$$

Hence Gronwall's lemma, Lemma 6.1.3, implies

$$\|\varphi(t, x_0)\| \leq \|x_0\| \, e^{\int_0^t \|A(s)\| ds}$$

and hence

$$\frac{1}{t} \log \|\varphi(t, x_0)\| \leq \frac{1}{t} \log \|x_0\| + \frac{1}{t} \int_0^t \|A(s)\| \, ds.$$

Now assumption (6.2.2) shows that the limit superior for $t \to \infty$ is finite.

In order to show that there are at most d different Lyapunov exponents, suppose that we have p Lyapunov exponents ordered such that $\lambda_p < \ldots < \lambda_1$. We have to show that $p \leq d$. The sets

$$W_\lambda := \{x \in \mathbb{R}^d \mid \lambda(x) \leq \lambda \text{ or } x = 0\}$$

are linear subspaces, since for $\alpha \neq 0$ and $0 \neq x, y \in \mathbb{R}^d$,

$$\lambda(\alpha x) = \lambda(x) \text{ and } \lambda(x+y) \leq \max\{\lambda(x), \lambda(y)\}.$$

Abbreviate $W_i := W_{\lambda_i}$. These subspaces form the flag

$$\{0\} \subset W_p \subset W_{p-1} \subset \ldots \subset W_1 \subset \mathbb{R}^d$$

and the inclusions of the W_i are proper. Thus $p \leq d$ follows.

(ii) Suppose that $\lambda(x_0)$ is maximal and let $x_1, \ldots, x_d$ be a basis of $\mathbb{R}^d$, hence $x_0 = \alpha_1 x_1 + \ldots + \alpha_d x_d$ with $\alpha_1, \ldots, \alpha_d \in \mathbb{R}$. This implies

$$\varphi(t, x_0) = \varphi(t, \alpha_1 x_1 + \ldots + \alpha_d x_d) = \alpha_1 \varphi(t, x_1) + \ldots + \alpha_d \varphi(t, x_d).$$

Thus by Lemma 6.2.2,

$$\lambda(x_0) = \limsup_{t \to \infty} \frac{1}{t} \log \|\varphi(t, x_0)\| \leq \max_{i=1,\ldots,d} \limsup_{t \to \infty} \frac{1}{t} \log \|\varphi(t, x_i)\|$$

and hence there is x_i with $\lambda(x_i) = \lambda(x_0)$.

6.2. Lyapunov Exponents

(iii) Consider the flag
$$\{0\} \subset W_\ell \subset W_{\ell-1} \subset \ldots \subset W_1 \subset \mathbb{R}^d$$
corresponding to the Lyapunov exponents $\lambda_\ell < \ldots < \lambda_1$. Extend a basis of W_ℓ to a basis of $W_{\ell-1}$, etc. By definition, the Lyapunov exponent for a basis element of W_ℓ equals λ_ℓ, and the basis elements of $W_{\ell-1}$, which are not in W_ℓ, have Lyapunov exponent $\lambda_{\ell-1}$, etc. □

Next we discuss the role of the eigenvalues of $A(t)$ for fixed t. Consider a linear nonautonomous equation of the form

(6.2.3) $$\dot{x} = A(t)x \text{ in } \mathbb{R}^2.$$

We will show that the eigenvalues of the matrices $A(t)$ do not determine the stability properties of the equation. Thus the stability behavior of the autonomous differential equations with 'frozen' coefficients $A(t_0), t_0 \in \mathbb{R}$,

(6.2.4) $$\dot{x} = A(t_0)x$$

can be different from the stability behavior of (6.2.3). How can this be the case? In order to construct such examples, observe that (6.2.3) can only be unstable if there is a time t such that the Euclidean norm of a solution $x(t)$ increases, i.e., there is $t \in \mathbb{R}$ such that

(6.2.5) $$\frac{d}{dt}\|x(t)\|^2 = 2x(t)^\top \dot{x}(t) = 2x(t)^\top A(t)x(t) > 0.$$

Thus if all eigenvalues of $A(t)$ are in the open left half-plane $\mathbb{C}_-$ and (6.2.5) holds, it follows that $A(t)$ is in

$$\mathcal{B} := \{B \in gl(2,\mathbb{R}) \mid \text{spec}(B) \subset \mathbb{C}_- \text{ and } x^\top Bx > 0 \text{ for some } x \in \mathbb{R}^2\}.$$

For $\omega > 0$ let

$$G(\omega) := \begin{bmatrix} 0 & -\omega \\ \omega & 0 \end{bmatrix} \text{ with } e^{tG(\omega)} = \begin{bmatrix} \cos\omega t & -\sin\omega t \\ \sin\omega t & \cos\omega t \end{bmatrix},$$

which is a rotation in the plane by the angle ωt. Now pick $B \in \mathcal{B}$ and define

(6.2.6) $$A(t) := e^{tG(\omega)} B e^{-tG(\omega)}, t \in \mathbb{R}.$$

Note that the eigenvalues of $A(t)$ coincide with the eigenvalues of B and hence are in $\mathbb{C}_-$ since $B \in \mathcal{B}$. In order to determine the behavior of the solutions $x(t)$ of the corresponding equation (6.2.3) consider $y(t) := e^{-tG(\omega)}x(t), t \in \mathbb{R}$. Then the product rule shows that $y(t)$ is a solution of an autonomous differential equation given by

$$\begin{aligned}\dot{y}(t) &= -G(\omega)e^{-tG(\omega)}x(t) + e^{-tG(\omega)}\dot{x}(t) \\ &= -G(\omega)e^{-tG(\omega)}x(t) + e^{-tG(\omega)}e^{tG(\omega)}Be^{-tG(\omega)}x(t) \\ &= [B - G(\omega)]y(t).\end{aligned}$$

Since $x(t) = e^{tG(\omega)}y(t)$ is obtained by rotating $y(t)$ by the angle ωt, one sees that $x(t) \to \infty$ for $t \to \infty$ if and only if $y(t) \to \infty$ for $t \to \infty$. Now the stability behavior of $y(t)$ is determined by the eigenvalues of $B - G(\omega)$. In particular, if we can find a matrix $B \in \mathcal{B}$ and $\omega > 0$ such that $\text{spec}(B - G(\omega)) \not\subset \mathbb{C}_-$, the matrix function (6.2.6) yields a desired counterexample where (6.2.3) is unstable, while the eigenvalues of every matrix $A(t)$ are in $\mathbb{C}_-$.

Hinrichsen and Pritchard [68, Example 3.3.7] give the example

$$B = \begin{bmatrix} -1 & -5 \\ 0 & -1 \end{bmatrix} \text{ and } \omega = 1$$

with the unbounded solution for the initial value $[1, 0]^\top$:

$$x(t, x_0) = \begin{bmatrix} e^t(\cos t + \frac{1}{2}\sin t) \\ e^t(\sin t - \frac{1}{2}\cos t) \end{bmatrix}, t \in \mathbb{R}.$$

Here B has the double eigenvalue -1 and the vector $x = [-1, 1]^\top$ satisfies $x^\top B x > 0$. Another example is

$$B = \begin{bmatrix} -10 & 12 \\ 0 & -1 \end{bmatrix} \text{ and } \omega = -6,$$

which yields a classical system due to Vinograd [135] with $A(t) \in gl(2, \mathbb{R})$ given by

$$\begin{bmatrix} -1 - 9\cos^2(6t) + 12\sin(6t)\cos(6t) & 12\cos^2(6t) + 9\sin(6t)\cos(6t) \\ 12\sin^2(6t) + 9\sin t \cos(6t) & -1 - 9\sin^2(6t) - 12\sin(6t)\cos(6t) \end{bmatrix}.$$

The eigenvalues of $A(t)$ are -1 and -10. Here $B - G(-6)$ has the eigenvalues 2 and -13, hence the system is unstable and, in fact, $\dot{x} = A(t)x$ has the exponentially growing solution

$$x(t) = \begin{bmatrix} e^{2t}(\cos 6t + 2\sin 6t) + 2e^{-13t}(2\cos 6t - \sin 6t) \\ e^{2t}(2\cos 6t - \sin 6t) - 2e^{-13t}(\cos 6t + 2\sin 6t) \end{bmatrix}.$$

The vector $x = [1, 1]^\top$ satisfies $x^\top B x > 0$. Another classical example, due to Markus and Yamabe [100], is constructed with

$$B = \begin{bmatrix} \frac{1}{2} & 1 \\ -1 & -1 \end{bmatrix} \text{ and } \omega = -1;$$

here B has the eigenvalues $\frac{1}{4}(-1 \pm \sqrt{7})$ and $B - G(-1)$ has the eigenvalues -1 and $\frac{1}{2}$. The following example is due to Nemytskii and Vinograd [23]:

$$A(t) = \begin{bmatrix} 1 - 4(\cos 2t)^2 & 2 + 2\sin 4t \\ -2 + 2\sin 4t & 1 - 4(\sin 2t)^2 \end{bmatrix}.$$

6.2. Lyapunov Exponents

Then one computes that the unbounded function

$$(6.2.7) \qquad x(t) = \begin{bmatrix} e^t \sin 2t \\ e^t \cos 2t \end{bmatrix}$$

is a solution. The eigenvalues $\lambda_1(t)$ and $\lambda_2(t)$ of the matrix $A(t)$ are equal to -1, since

$$\det[A(t) - \lambda I] = \lambda^2 - 2\lambda + 1.$$

Note that all examples constructed in this way have periodic coefficients with period $\frac{\omega}{2\pi}$.

Remark 6.2.4. A decomposition into Lyapunov spaces which is similar to the autonomous case can be approached as follows: There are $\ell \leq d$ real numbers $\lambda_1 > \ldots > \lambda_\ell$ and $m_1, \ldots, m_\ell \in \mathbb{N}$ with $\sum_{j=1}^{\ell} m_j = d$, and subspaces $L_1, \ldots, L_\ell$ of dimension $m_1, \ldots, m_\ell$ (called the multiplicities of the λ_j) such that

$$\mathbb{R}^d = L_1 \oplus \ldots \oplus L_\ell$$

and for all $x \neq 0$,

$$(6.2.8) \qquad \lambda(x) = \lim_{t \to \infty} \frac{1}{t} \log \|\varphi(t, x)\| \in \{\lambda_1, \ldots, \lambda_\ell\}$$

and

$$(6.2.9) \qquad \lambda_j = \lim_{t \to \pm\infty} \frac{1}{t} \log \|\varphi(t, x)\| \quad \text{if and only if } x \in L_j.$$

But equation (6.2.1) shows that this property can only hold under additional assumptions.

Lyapunov exponents measure the exponential growth rates of the lengths of vectors under the solution map of $\dot{x} = A(t)x$. Similarly, one may analyze the exponential growth rates of the volumes of parallelepipeds, as discussed for the autonomous case in detail in Chapter 5. For full-dimensional parallelepipeds, Liouville's formula from Theorem 6.1.5 determines the volume growth, as indicated in Remark 6.1.6. This immediately gives the exponential growth rate of volumes as

$$\limsup_{t \to \infty} \frac{1}{t} \int_0^t \operatorname{tr} A(s) ds.$$

The trace of a matrix is equal to the sum of the eigenvalues. Hence in the autonomous case $\dot{x} = Ax$, the exponential growth rate of volumes is given by the sum of the Lyapunov exponents $\sum_{j=1}^{\ell} d_j \lambda_j$ where $d_j = \dim L(\lambda_j)$. In the nonautonomous case, this is only an upper bound.

Proposition 6.2.5. *For a normal basis $x_1, \ldots, x_d$ of $\mathbb{R}^d$ as in Proposition 6.2.3(iii) one has*

(6.2.10) $$\limsup_{t \to \infty} \frac{1}{t} \int_0^t \operatorname{tr} A(s) ds \leq \sum_{j=1}^d \lambda(x_j).$$

Proof. By Liouville's formula (cf. Remark 6.1.6), the fundamental solution with $X(0) = [x_1(0), \ldots, x_d(0)]$ satisfies

$$e^{\int_{t_0}^t \operatorname{tr} A(s) ds} = \det X(t) \det X(0)^{-1}.$$

In the Laplace expansion (6.1.9) for the determinant every summand is the product of component functions of the solutions $x_j(t)$. Now repeated application of Lemma 6.2.2 implies the assertion. □

The following example shows that inequality (6.2.10) may be strict.

Example 6.2.6. Consider the differential equation for the matrix function defined for $t > 0$ by

$$A(t) := \begin{bmatrix} -\sin \log t - \cos \log t & 0 \\ 0 & \sin \log t + \cos \log t \end{bmatrix}.$$

The solution is given by

$$\begin{bmatrix} x_1(t) \\ x_2(t) \end{bmatrix} = \begin{bmatrix} \exp(-t \sin \log t) x_1(0) \\ \exp(t \sin \log t) x_2(0) \end{bmatrix}.$$

The only Lyapunov exponent is $\lambda = 1$. Since $\operatorname{tr} A(t) = 0$, one obtains for the volume growth rate

$$\limsup_{t \to \infty} \frac{1}{t} \int_0^t \operatorname{tr} A(s) ds = 0.$$

Thus in the nonautonomous case the volume growth rate may not be determined by the Lyapunov exponents. Furthermore, in contrast to eigenvalues, Lyapunov exponents may change discontinuously under perturbations as the following example, due to Vinograd, shows.

Example 6.2.7. Consider the matrix functions

$$A(t) := \begin{bmatrix} 0 & 0 \\ 0 & f(t) \end{bmatrix}, \quad A_\varepsilon(t) := \begin{bmatrix} 0 & \varepsilon \\ \varepsilon & f(t) \end{bmatrix}$$

with $\varepsilon > 0$ and for $n \in \mathbb{N}_0$,

$$f(t) := \begin{cases} -1 & \text{for} \quad (2n)^2 \leq t < (2n+1)^2, \\ 1 & \text{for} \quad (2n+1)^2 \leq t < (2n+2)^2. \end{cases}$$

The solutions of $\dot{x} = A(t)x$ are $x_1(t) = x_1(0), x_2(t) = x_2(0) + e^{\int_0^t f(s)ds}$. One computes

$$\int_{(2n)^2}^{(2n+1)^2} f(s)ds = -4n - 1 \text{ and } \int_{(2n+1)^2}^{(2n+2)^2} f(s)ds = 4n - 3.$$

Hence the Lyapunov exponent for $\varepsilon = 0$ is $\lambda = 0$, while the equation $\dot{x} = A_\varepsilon(t)x$ has the Lyapunov exponents $\lambda_1 \leq -\frac{1}{2} + \varepsilon$ and $\lambda_2 \geq \frac{1}{2}$.

6.3. Linear Skew Product Flows

In spite of the counterexamples collected above, for many time-varying systems it turns out that the Lyapunov exponents are appropriate generalizations of eigenvalues, since they capture the key properties of (real parts of) eigenvalues and of the associated subspace decomposition of $\mathbb{R}^d$. These systems are linear skew product flows for which the base is a dynamical system that describes the changes in the coefficient matrix. The basic idea is to exploit certain recurrence properties of the coefficient functions. Examples for this type of systems include periodic and almost periodic differential equations, random differential equations, systems over ergodic or chain recurrent bases, linear robust systems, and bilinear control systems. This section introduces periodic linear differential equations, robust (or controlled) linear systems, and random linear dynamical systems, without going into technical details. These are the main classes of systems for which the linear algebra in Chapter 1 will be generalized. This will be worked out by Floquet theory in Chapter 7, the Morse spectrum and Selgrade's theorem in Chapter 9, and the multiplicative ergodic theorem in Chapter 11. Some other relevant classes of linear skew product systems are also mentioned in the present section.

We begin with the following motivation. Recall that a continuous-time dynamical system or flow with state space X is a map $\Phi : \mathbb{R} \times X \to X$ satisfying

(6.3.1) $\qquad \Phi(0, x) = x$ and $\Phi(t + s, x) = \Phi(t, \Phi(s, x))$

for all $x \in X$ and all $s, t \in \mathbb{R}$. Let $A : \mathbb{R} \longrightarrow gl(d, \mathbb{R})$ be a locally integrable function and consider the linear differential equation $\dot{x} = A(t)x$. If we add the new component

$$\dot{t} = 1,$$

we obtain an autonomous differential equation in the augmented state space $\mathbb{R} \times \mathbb{R}^d$. More formally, denote the solutions of $\dot{x} = A(t)x$ at time t with initial condition $x(\tau) = x_0$ by $\psi(t, \tau, x)$. Then define a flow $\Phi = (\theta, \varphi) :$

$\mathbb{R} \times \mathbb{R} \times \mathbb{R}^d \longrightarrow \mathbb{R} \times \mathbb{R}^d$ by

$$\theta : \mathbb{R} \times \mathbb{R} \longrightarrow \mathbb{R}, \ \theta_t \tau := t + \tau,$$

$$\varphi : \mathbb{R} \times \mathbb{R} \times \mathbb{R}^d \longrightarrow \mathbb{R}^d, \ \varphi(t, \tau, x_0) := \psi(t + \tau, \tau, x_0).$$

Then obviously

$$\Phi(0, \tau, x_0) = (\theta_0 \tau, \varphi(0, \tau, x_0)) = (\theta_0 \tau, \psi(\tau, \tau, x_0)) = (\tau, x_0)$$

and Φ satisfies

$$\Phi(t + s, \tau, x_0) = (\theta_{t+s} \tau, \varphi(t+s, \tau, x_0)) = (\theta_t(\theta_s \tau), \varphi(t, \theta_s \tau, \varphi(s, \tau, x_0)))$$
$$= \Phi(t, \Phi(s, \tau, x_0)).$$

In fact, the flow property for the 'base component' θ is clear and for the map ψ we use the cocycle property of nonautonomous differential equations from Proposition 6.1.2(i) showing, as claimed, that

$$\varphi(t + s, \tau, x_0) = \psi(t + s + \tau, \tau, x_0) = \psi(t + s + \tau, s + \tau, \psi(s + \tau, \tau, x_0))$$
$$= \varphi(t, \theta_s \tau, \varphi(s, \tau, x_0)).$$

We emphasize that the solutions of $\dot{x} = A(t)x$ themselves do not define a flow. The additional component θ 'keeps track of time'. This construction is, however, not useful for the study of stability properties. The component θ on $\mathbb{R}$, which describes a time shift, has void ω-limit sets and hence no recurrence properties of the coefficients can be taken into account. More elaborate constructions of the base space are necessary. A classical way is to assume that the differential equation is embedded into a family of differential equations and then to add topological or measure theoretic assumptions on the way the time dependence is generated. We follow this approach and define linear skew product flows and their Lyapunov exponents.

Definition 6.3.1. A linear skew product flow is a dynamical system Φ with state space $X = B \times \mathbb{R}^d$ of the form $\Phi = (\theta, \varphi) : \mathbb{R} \times B \times \mathbb{R}^d \longrightarrow B \times \mathbb{R}^d$ where $\theta : \mathbb{R} \times B \longrightarrow B$, and $\varphi : \mathbb{R} \times B \times \mathbb{R}^d \longrightarrow \mathbb{R}^d$ is linear in its $\mathbb{R}^d$-component, i.e., for each $(t, b) \in \mathbb{R} \times B$ the map $\Phi(t, b) := \varphi(t, b, \cdot) : \mathbb{R}^d \longrightarrow \mathbb{R}^d$ is linear. A skew product flow is called measurable if B is a measurable space and Φ is measurable; it is called continuous, if B is a metric space and Φ is continuous.

The construction above is a special case with base space $B = \mathbb{R}$. By the dynamical system properties (6.3.1) of Φ the first component θ, which is independent of the second argument, defines a dynamical system on B. It is called the base component or base flow on the base space B. The second component φ is called a cocycle over θ and satisfies the cocycle property

$$\varphi(0, b, x) = x \text{ and } \varphi(t + s, b, x) = \varphi(t, \theta_s b, \varphi(s, b, x)).$$

6.3. Linear Skew Product Flows

Note that the cocycle or skew-component φ is not a dynamical system by itself.

We may consider exponential growth rates for the linear part of solutions.

Definition 6.3.2. Let $\Phi : \mathbb{R} \times B \times \mathbb{R}^d \longrightarrow B \times \mathbb{R}^d$ be a linear skew product flow. For $0 \neq x_0 \in \mathbb{R}^d$ and $b \in B$ the Lyapunov exponent is defined as

$$\lambda(x_0, b) = \limsup_{t \to \infty} \frac{1}{t} \log \|\varphi(t, b, x_0)\|,$$

where log denotes the natural logarithm and $\|\cdot\|$ is any norm in $\mathbb{R}^d$.

If $\lambda(x_0, b) < 0$, the skew-component $\varphi(t, b, x_0)$ converges to $0 \in \mathbb{R}^d$ for $t \to \infty$.

Next we introduce several examples of linear skew product flows.

Example 6.3.3. Consider a linear differential equation of the form $\dot{x} = A^0(t)x$ where A^0 is an element of the space $C_b(\mathbb{R}, gl(d, \mathbb{R}))$ of all bounded uniformly continuous functions on $\mathbb{R}$ with values in $gl(d, \mathbb{R})$ with norm given by

$$(6.3.2) \qquad \|A(\cdot)\| = \sum_{k=1}^{\infty} \frac{1}{2^k} \max_{t \in [-k,k]} \|A(t)\|, \, A(\cdot) \in C_b(\mathbb{R}, gl(d, \mathbb{R})).$$

The topology generated by the associated metric is called the topology of uniform convergence on compact subsets of $\mathbb{R}$.

Define the time shift or translation

$\theta : \mathbb{R} \times C_b(\mathbb{R}, gl(d, \mathbb{R})) \to C_b(\mathbb{R}, gl(d, \mathbb{R})), [\theta(t, A(\cdot))](\tau) := A(\tau + t), \tau \in \mathbb{R},$

and let the base space B be the closure of the set of time shifts of the coefficient matrix $A^0(\cdot)$ in $C_b(\mathbb{R}, gl(d, \mathbb{R}))$,

$$B := \text{cl}\{\theta(t, A^0(\cdot)) \mid t \in \mathbb{R}\}.$$

This construction allows one to take into account recurrence properties of the matrix function A^0. The set B is invariant under the map θ. We claim that the base space B is a compact metric space. First we show that for any sequence (t_n) in $\mathbb{R}$ the sequence $\theta(t_n, A^0(\cdot)) = A^0(\cdot + t_n), n \in \mathbb{N}$ in B contains a convergent subsequence. For $k \in \mathbb{N}$ the restrictions to $[-k, k]$ are clearly bounded. By uniform continuity of A^0 they are equicontinuous, i.e., for $\varepsilon > 0$ there is $\delta > 0$ such that for $t, s \in [-k, k]$ with $|t - s| < \delta$ we have

$$\|A^0(t + t_n) - A^0(s + t_n)\| < \varepsilon \text{ for all } n.$$

Hence the Arzelà-Ascoli theorem implies that there is a subsequence of (t_n) such that the corresponding functions $A^0(\cdot + t_n)$ converge uniformly on $[-k, k]$. We label this subsequence by $(t_j^k)_{j \in \mathbb{N}}$ and then set $s_k := t_k^k$. Clearly, the functions $A^0(\cdot + s_k)$ converge in B.

For a general sequence $A_n, n \in \mathbb{N}$ in B we approximate A_n by $A^0(\cdot + t_n)$ for an appropriate $t_n \in \mathbb{R}$, then extract a convergent subsequence of A_n by applying again the Arzelà-Ascoli theorem and a Cantor diagonal argument to the sequence $A^0(\cdot + t_n)$. (The reader is asked to write down the details in Exercise 6.5.9).

Similar arguments show that the map

$$\theta : \mathbb{R} \times B \to B, \theta(t, A(\cdot))(\tau) = A(\tau + t), \tau \in \mathbb{R},$$

is continuous: Let $t_n \to t$ and $A_n(\cdot) = A^0(\cdot + s_n) \to A(\cdot) \in B$ with $s_n \in \mathbb{R}$. Then for all $k \in \mathbb{N}$,

$$\max_{\tau \in [-k,k]} \|\theta(t_n, A_n)(\tau) - A(\tau)\|$$

$$= \max_{\tau \in [-k,k]} \|A^0(\tau + s_n + t_n) - A(\tau)\|$$

$$\leq \max_{\tau \in [-k,k]} \|A^0(\tau + s_n + t_n) - A(\tau + t_n)\| + \max_{\tau \in [-k,k]} \|A(\tau + t_n) - A(\tau)\|.$$

The second summand converges to 0, uniformly in k, since $A(\cdot)$ is uniformly continuous. It follows that $d(\theta(t_n, A_n), A) \to 0$ in the metric (6.3.2), since also $d(A^0(\cdot + s_n), A(\cdot)) \to 0$. For general $A_n(\cdot)$ one again uses approximations by elements of the form $A^0(\cdot + s_n)$. Now, continuity of θ implies that the set B is invariant under the flow θ.

Consider the evaluation map $ev\,(A(\cdot)) : B \to gl(d, \mathbb{R}), A(\cdot) := A(0)$, which is continuous, and note that $\theta(t, A)$ is mapped to $A(t)$. Then one verifies that $\Phi := (\theta, \varphi)$ is a linear skew product flow on $X = B \times \mathbb{R}^d$ where the cocycle $\varphi(t, A(\cdot), x)$ denotes the solution at time t of

$$\dot{x}(\tau) = ev\,(\theta_\tau A(\cdot))\, x(\tau) = A(\tau)x(\tau), \tau \in \mathbb{R},$$

with $x(0) = x$. Continuity of φ, and hence of Φ, follows from Theorem 6.1.4 applied with the metric space $X := B$, and using continuity of θ.

The discussion up to now is motivated by the argument that the analysis of a single nonautonomous differential equation may be simplified by embedding it into a family of differential equations. Here the base space, constructed topologically, should be as small as possible. On the other hand, the formalism of skew product flows also allows one to analyze quite large families of differential equations. The next two examples concern such situations. The first one uses another topological construction.

Example 6.3.4. Consider a family of linear differential equations of the form

$$\dot{x} = A(u(t))x := A_0 x + \sum_{i=1}^{m} u_i(t) A_i x, u \in \mathcal{U},$$

6.3. Linear Skew Product Flows

where $A_0, \ldots, A_m \in gl(d, \mathbb{R})$, $\mathcal{U} = \{u : \mathbb{R} \longrightarrow U, \text{locally integrable}\}$ and $U \subset \mathbb{R}^m$. This defines a linear skew product flow via the following construction: The base component is defined as the shift $\theta : \mathbb{R} \times \mathcal{U} \longrightarrow \mathcal{U}$, $\theta(t, u) = u(t+\cdot)$, and the skew-component consists of the solutions $\varphi(t, u, x)$, $t \in \mathbb{R}$ of the differential equation corresponding to the function u. Then $\Phi : \mathbb{R} \times \mathcal{U} \times \mathbb{R}^d \longrightarrow \mathcal{U} \times \mathbb{R}^d$, $\Phi(t, u, x) = (\theta(t, u), \varphi(t, u, x))$ defines a linear skew product flow. The functions u can be interpreted as time-varying perturbations, then we speak of a robust linear system. Alternatively, the functions u may be interpreted as (time dependent) controls which may be chosen in order to achieve a desired system behavior. This example class is the subject of Chapter 9. In particular, in Section 9.5 we will show that for compact and convex range U the base space $\mathcal{U}$, which is a huge set of functions, will become a compact metrizable space, and we will establish continuity properties of this flow.

The next example uses measure theoretical constructions (or stochastic processes), instead of topology.

Example 6.3.5. Let $(\Omega, \mathcal{F}, P)$ be a probability space, i.e., a set Ω with σ-algebra $\mathcal{F}$ and probability measure P and let $\theta : \mathbb{R} \times \Omega \longrightarrow \Omega$ be a measurable flow such that the probability measure P is invariant under θ. This means that for all $t \in \mathbb{R}$ the probability measures $\theta_t P$ on Ω defined by $(\theta_t P)(X) := P(\theta_t^{-1}(X))$, $X \in \mathcal{F}$, coincide with P. Flows of this form are often called metric dynamical systems (not to be confused with a system on a metric space). A random linear dynamical system is a measurable skew product flow $\Phi = (\theta, \varphi) : \mathbb{R} \times \Omega \times \mathbb{R}^d \longrightarrow \Omega \times \mathbb{R}^d$, where θ is a metric dynamical system on $(\Omega, \mathcal{F}, P)$ and each $\varphi : \mathbb{R} \times \Omega \times \mathbb{R}^d \longrightarrow \mathbb{R}^d$ is linear in its $\mathbb{R}^d$-component. This example class will be analyzed in detail in Chapter 11.

A further example class is provided by differential equations linearized over an invariant set.

Example 6.3.6. Consider a differential equation

$$\dot{y} = f(y), \ y(0) = y_0 \in \mathbb{R}^d,$$

where $f : \mathbb{R}^d \to \mathbb{R}^d$ is C^1 and suppose that $B \subset \mathbb{R}^d$ is an invariant set, i.e., the solutions $\theta(t, y_0), t \in \mathbb{R}$, exist for all $y_0 \in B$ and remain in B, i.e., $\theta(t, x_0) \in B$ for all $t \in \mathbb{R}$. Denote the Jacobian of f along a trajectory $\theta(t, y_0)$ by $f'(\theta(t, y_0))$ and consider the coupled system

(6.3.3) $\qquad \dot{y} = f(y), \ y(0) = y_0 \in \mathbb{R}^d,$

(6.3.4) $\qquad \dot{x} = f'(\theta(t, y_0))x, \ x(0) = x_0 \in \mathbb{R}^d.$

Then $\Phi : \mathbb{R} \times B \times \mathbb{R}^d \longrightarrow B \times \mathbb{R}^d$ is a skew product flow, where $\Phi = (\theta, \varphi)$ is defined as follows: The base flow is θ and $\varphi(t, y_0, x_0)$ is the solution of the linear differential equation (6.3.4). A special case is $B = \{e\}$ where e is an equilibrium of $\dot{y} = f(y)$. Here the linearized system is given by the autonomous linear differential equation $\dot{x} = f'(e)x$. In this case it is of paramount interest to deduce properties of the nonlinear differential equation $\dot{y} = f(y)$ near the equilibrium e from properties of the autonomous linear differential equation $\dot{x} = f'(e)x$, which is completely understood. The idea leads to stable and unstable manifolds as well as the Hartman-Grobman theorem which gives a (local) topological conjugacy result.

6.4. The Discrete-Time Case

The panorama opened up in the previous sections has a complete analogue in the discrete-time situation. In this section, some basic facts on nonautonomous linear difference equations and the linear skew product systems generated by them are presented.

Let $A : \mathbb{Z} \to Gl(d, \mathbb{R})$ be a matrix function and consider the associated nonautonomous linear difference equation

(6.4.1) $$x_{n+1} = A(n)x_n, n \in \mathbb{Z}.$$

Basic existence and uniqueness results are given by the following theorem.

Theorem 6.4.1. *Let $A : \mathbb{Z} \to Gl(d, \mathbb{R})$ be given.*

(i) The solutions of $x_{n+1} = A(n)x_n, n \in \mathbb{Z}$, form a d-dimensional vector space.

(ii) The initial value problems $X_{n+1} = A(n)X_n$ in $Gl(d, \mathbb{R})$ with $X(n_0) = I_d$ have unique solutions $X(n, n_0), n \in \mathbb{Z}$, called principal fundamental solution. The matrices $X(n, n_0)$ are invertible for all $n, n_0 \in \mathbb{Z}$.

(iii) For each initial condition $x_{n_0} = x \in \mathbb{R}^d$, the solution of (6.4.1) is unique and given by

(6.4.2) $$\varphi(n, n_0, x) = X(n, n_0)x.$$

(iv) The solutions satisfy the 2-parameter cocycle property

$$\varphi(n, n_0, x) = \varphi(n, m, \varphi(m, n_0, x)) \text{ for } n, m, n_0 \in \mathbb{Z}.$$

Proof. For two solutions $x_n, y_n, n \in \mathbb{Z}$, the (pointwise) sum and the multiplication by a scalar $\alpha \in \mathbb{R}$ satisfy for all $n \in \mathbb{Z}$,

$$x_n + y_n = A(n)(x_n + y_n) \text{ and } \alpha x_n = A(n)(\alpha x_n),$$

hence they are again solutions. This shows that the solutions form a vector space. By induction one sees that the unique solution of $X_{n+1} = A(n)X_n$

6.4. The Discrete-Time Case

with $X(n_0) = I$ is given by

$$X(n, n_0) = A(n-1)\ldots A(n_0) = \prod_{j=n_0}^{n-1} A(j) \text{ for } n > n_0,$$

$$X(n, n_0) = A(n)^{-1}\ldots A(n_0-1)^{-1} = \prod_{j=n}^{n_0-1} A(j)^{-1} \text{ for } n < n_0.$$

We also let $\prod_{j=n_0}^{n_0-1} A(j) := I_d$. In particular, the matrices $X(n, n_0)$ are nonsingular, since $A(j) \in Gl(d, \mathbb{R})$ for all j. Analogously, for any given initial value x at time n_0 the unique solution of (6.4.1) is given by

$$\varphi(n, n_0, x) = \prod_{j=n_0}^{n-1} A(j)x \text{ for } n \geq n_0, \varphi(n, n_0, x) = \prod_{j=n}^{n_0-1} A(j)^{-1}x \text{ for } n < n_0.$$

It follows that

$$\varphi(n, n_0, x) = X(n, n_0)x, n \in \mathbb{Z},$$

and the maximal number of linearly independent solutions equals d. The cocycle property follows from unique solvability. $\square$

The next theorem shows that solutions depend continuously on initial values and on parameters occurring in the system matrix $A(\cdot)$.

Theorem 6.4.2. *Let Γ be a metric space and consider a function $A : \mathbb{Z} \times \Gamma \to Gl(d, \mathbb{R})$, such that for every $n \in \mathbb{Z}$ the function $A(n, \cdot) : \Gamma \to Gl(d, \mathbb{R})$ is continuous. Then for every $n \in \mathbb{Z}$ the solution $\varphi(n, n_0, x, \gamma)$ of $x_{n+1} = A(n, \gamma)x_n, x_{n_0} = x$, is continuous with respect to $(x, \gamma) \in \mathbb{R}^d \times \Gamma$.*

Proof. Let $(x^k, \gamma^k) \to (x, \gamma)$ in $\mathbb{R}^d \times \Gamma$ and consider the corresponding solutions $\varphi(n, n_0, x^k, \gamma^k), n \in \mathbb{Z}$. Then one can estimate for $k \in \mathbb{N}$ and $n \geq n_0$,

$$\left\|\varphi(n, n_0, x^k, \gamma^k) - \varphi(n, n_0, x, \gamma)\right\| = \left\|\prod_{j=n_0}^{n-1} A(j, \gamma^k)x^k - \prod_{j=n_0}^{n-1} A(j, \gamma)x\right\|$$

$$\leq \prod_{j=n_0}^{n-1} \left\|A(j, \gamma^k)\right\| \left\|x^k - x\right\| + \prod_{j=n_0}^{n-1} \left\|A(j, \gamma^k) - A(j, \gamma)\right\| \|x\|.$$

The assertion follows for $k \to \infty$. For $n \leq n_0$ one argues analogously. $\square$

For nonautonomous linear difference equations $x_{n+1} = A(n)x_n$ the exponential growth rate, or Lyapunov exponent, of a solution $\varphi(n, n_0, x)$ is given by

$$\limsup_{n \to \infty} \frac{1}{n} \log \|\varphi(n, n_0, x)\|.$$

By the cocycle property it is sufficient to consider solutions with initial time $n_0 = 0$. If $n_0 = 0$, we write the solution as $\varphi(n, x)$ and the principal fundamental solution as $X(n)$, and we define

$$\lambda(x) = \limsup_{n \to \infty} \frac{1}{n} \log \|\varphi(n, x)\| = \limsup_{n \to \infty} \frac{1}{n} \log \|X(n)x\|, \ 0 \neq x \in \mathbb{R}^d.$$

If $A(n) \equiv A$ is constant, Theorem 1.5.6 shows that the Lyapunov exponents are given by $\log |\mu_i|$ where μ_i are the eigenvalues of A, and the theorem provides a corresponding decomposition of the state space into the Lyapunov spaces.

For nonautonomous systems, the same arguments as for continuous time show that $\lambda(x + y) \leq \max\{\lambda(x), \lambda(y)\}$, the number of different Lyapunov exponents is at most d, and the maximal Lyapunov exponent is attained on every base of $\mathbb{R}^d$; see Section 6.2.

Finally, we define linear skew product dynamical systems in discrete time in the following way. Recall that a dynamical system in discrete time with state space X is given by a map $\Phi : \mathbb{Z} \times X \to X$ with $\Phi(0, x) = x$ and $\Phi(n + m, x) = \Phi(n, \Phi(m, x))$ for all $x \in X$ and $m, n \in \mathbb{Z}$.

Definition 6.4.3. A linear skew product dynamical system in discrete time is a dynamical system Φ with state space $X = B \times \mathbb{R}^d$ of the form $\Phi = (\theta, \varphi) : \mathbb{Z} \times B \times \mathbb{R}^d \longrightarrow B \times \mathbb{R}^d$, where $\theta : \mathbb{Z} \times B \longrightarrow B$ and $\varphi : \mathbb{Z} \times B \times \mathbb{R}^d \longrightarrow \mathbb{R}^d$ is linear in its $\mathbb{R}^d$-component, i.e., for each $(n, b) \in \mathbb{Z} \times B$ the map $\Phi(n, b) := \varphi(n, b, \cdot) : \mathbb{R}^d \longrightarrow \mathbb{R}^d$ is linear. A skew product system Φ is called measurable if B is a measurable space and Φ is measurable; it is called continuous, if B is a metric space and Φ is continuous.

Hence $\theta : \mathbb{Z} \times B \longrightarrow B$ is a map with the dynamical system properties $\theta(0, \cdot) = \mathrm{id}_B$ and $\theta(n + m, b) = \theta(n, \theta(m, b))$ for all $n, m \in \mathbb{Z}$ and $b \in B$. For conciseness, we also write $\theta := \theta(1, \cdot)$ for the time-one map and hence

$$\theta^n b = \theta(n, b) \text{ for all } n \in \mathbb{Z} \text{ and } b \in B.$$

The map $\Phi = (\theta, \varphi) : \mathbb{Z} \times B \times \mathbb{R}^d \longrightarrow B \times \mathbb{R}^d$ has the form

$$\Phi(n, b, x_0) = (\theta^n b, \varphi(n, b, x))$$

and the dynamical system property of Φ means for the second component that it satisfies the cocycle property

$$\varphi(0, b, x) = x, \ \varphi(n + m, b, x) = \varphi(n, \theta^m b, \varphi(m, b, x))$$

for all $b \in B, x \in \mathbb{R}^d$ and $n, m \in \mathbb{Z}$.

Equivalently, consider for a given map θ on B a map A defined on B with values in the group $Gl(d, \mathbb{R})$ of invertible $d \times d$ matrices. Then the solutions $\varphi(n, b, x), n \in \mathbb{Z}$ of the nonautonomous linear difference equations

(6.4.3) $$x_{n+1} = A(\theta^n b)x_n$$

define a linear skew product dynamical system $\Phi = (\theta, \varphi)$ in discrete time: The dynamical system property of the base component θ on the base space B is clear and the solution formula (or direct inspection) shows the cocycle property for $n, m \in \mathbb{N}_0$,

$$\varphi(n+m, b, x) = \prod_{j=0}^{n+m-1} A(\theta^j b)x = \prod_{j=0}^{n-1} A(\theta^j(\theta^m b)) \prod_{j=0}^{m-1} A(\theta^j b)x$$
$$= \varphi(n, \theta^m b, \varphi(m, b, x)),$$

and analogously for all $n, m \in \mathbb{Z}$. We say that the map A is the generator of Φ. Since the map $\varphi : \mathbb{Z} \times B \times \mathbb{R}^d \longrightarrow \mathbb{R}^d$ is linear in the $\mathbb{R}^d$-argument, the maps $\Phi(n, b) = \varphi(n, b, \cdot)$ are linear on $\mathbb{R}^d$ and the cocycle property can be written as

(6.4.4) $\quad \Phi(0, b) = I_d, \ \Phi(n+m, b) = \Phi(n, \theta^m b) \Phi(m, b)$

for all $b \in B$ and $n, m \in \mathbb{Z}$. One also sees that for all $b \in B$,

(6.4.5) $\quad \Phi(n, b) = A(\theta^{n-1} b) \cdots A(b)$ for $n \in \mathbb{N}_0$.

Conversely, a linear skew product dynamical system in discrete time defines a nonautonomous linear difference equation of the form (6.4.3) with $A(b) := \Phi(1, b) = \varphi(1, b, \cdot), b \in B$.

Examples of linear skew product systems in discrete time can be constructed similarly to the case of continuous time. In particular, the matrix function may be periodic (cf. Section 7.1) or one has a random system in the following sense.

Example 6.4.4. Let $(\Omega, \mathcal{F}, P)$ be a probability space, i.e., a set Ω with σ-algebra $\mathcal{F}$ and probability measure P and let $\theta : \mathbb{Z} \times \Omega \longrightarrow \Omega$ be a measurable dynamical system such that the probability measure P is invariant under θ. This means that for all $n \in \mathbb{Z}$ the probability measures $\theta_n P$ on Ω defined by $(\theta_n P)(X) := P\{\theta_n^{-1}(X)\}, X \in \mathcal{F}$, coincide with P. Systems of this form are called (discrete-time) metric dynamical systems. A random linear dynamical system in discrete time is a measurable skew product dynamical system $\Phi = (\theta, \varphi) : \mathbb{R} \times \Omega \times \mathbb{R}^d \longrightarrow \Omega \times \mathbb{R}^d$, where θ is a metric dynamical system on $(\Omega, \mathcal{F}, P)$ and each $\varphi : \mathbb{Z} \times \Omega \times \mathbb{R}^d \longrightarrow \mathbb{R}^d$ is linear in its $\mathbb{R}^d$-component. This class will be analyzed in detail in Chapter 11.

6.5. Exercises

Exercise 6.5.1. Prove Lemma 6.2.2: Let $f, g : [0, \infty) \to (0, \infty)$ be locally integrable functions and denote $\lambda(f) := \limsup_{t \to \infty} \frac{1}{t} \int_0^t \log f(s) ds \leq \infty$, and

analogously $\lambda(g)$. Then $\lambda(fg) \leq \lambda(f)+\lambda(g)$, and $\lambda(f+g) \leq \max(\lambda(f), \lambda(g))$ with $\lambda(f+g) = \max(\lambda(f), \lambda(g))$ if $\lambda(f) \neq \lambda(g)$.
Hint: For the second assertion consider the product $f \cdot (1 + \frac{g}{f})$.

Exercise 6.5.2. Consider the differential equations $\dot{x} = A(t)x$ and denote the solution with initial condition $x(t_0) = x_0 \in \mathbb{R}^d$ by $\varphi(t, t_0, x_0), t \in \mathbb{R}$. Show that the following set of exponential growth rates is independent of t_0:

$$\left\{ \limsup_{t \to \infty} \frac{1}{t} \log \|\varphi(t, t_0, x_0)\| \mid 0 \neq x_0 \in \mathbb{R}^d \right\}.$$

Exercise 6.5.3. Consider the differential equation $\dot{x} = A(t)x$ and $\dot{y} = B(t)y$. For initial values x_0 and y_0 at time 0, denote the corresponding solutions by $x(t, x_0)$ and $y(t, y_0)$, respectively. Suppose that there is a continuous matrix function $Z(t), t \in \mathbb{R}$ which together with its inverse $Z(t)^{-1}$ is bounded on $\mathbb{R}$, such that

$$Z(t)x(t, x_0) = y(t, Z(0)x_0) \text{ for all } t \in \mathbb{R} \text{ and all } x_0 \in \mathbb{R}.$$

Show that the Lyapunov exponents for x_0 and $y_0 = Z(0)x_0$ of these differential equations coincide. Formulate and prove the analogous statement for linear difference equations. $Z(t)$ is a (time-varying) conjugacy, often called a Lyapunov transformation, e.g., Hahn [**61**, Definition 61.2].

Exercise 6.5.4. For $\dot{x} = A(t)x$ show that the Lyapunov exponent of $0 \neq x_0 \in \mathbb{R}^d$ is

$$\lambda(x_0) = \limsup_{t \to \infty} \frac{1}{t} \log \|x(t, x_0)\| = \limsup_{t \to \infty} \frac{1}{t} \int_0^t s(\tau)^\top A(\tau) s(\tau) d\tau,$$

where $s(t)$ is the projection to the unit sphere

$$s(t) = \frac{x(t, x_0)}{\|x(t, x_0)\|}, t \in \mathbb{R}.$$

Exercise 6.5.5. Prove the following assertion from Section 6.1: Suppose that $a : \mathbb{R} \to \mathbb{R}$ is locally integrable. Then the solution of

$$\dot{x} = a(t)x, \ x(t_0) = x_0 \in \mathbb{R},$$

is given by

$$x(t) = e^{\int_{t_0}^t a(s)ds} x_0, t \in \mathbb{R}.$$

Exercise 6.5.6. The following example shows that for higher dimensional systems $\dot{x} = A(t)x$ the formula from Exercise 6.5.5,

$$x(t) = e^{\int_{t_0}^t A(s)ds} x(t_0),$$

is false. Let

$$A(t) := \begin{bmatrix} 0 & 1 \\ 0 & 0 \end{bmatrix} \text{ for } t \in [0, 1] \text{ and } A(t) := \begin{bmatrix} 0 & 0 \\ -1 & 0 \end{bmatrix} \text{ for } t > 1.$$

Note that the two matrices above do not commute. Compute the solution $x(2)$ of $\dot{x} = A(t)x$ for the initial value $x(0) = [1,0]^\top$ and compare with $e^{\int_0^2 A(s)ds}[1,0]^\top$.

Exercise 6.5.7. Let $f : [0,\infty) \to \mathbb{R}$ be a continuous function. If $a \in \mathbb{R}$ is such that $f(t)e^{at}$ is bounded for $t \geq 0$, then $f(t)e^{a't}$ with $a' < a$ is also bounded for $t \geq 0$. If $b \in \mathbb{R}$ is such that $f(t)e^{bt}$ is unbounded for $t \geq 0$, then $f(t)e^{b't}$ with $b' > b$ is also unbounded for $t \geq 0$. Thus one finds a unique number separating the set A of all real numbers a with $f(t)e^{at}$ bounded and the set B of all numbers b with $f(t)e^{bt}$ unbounded. Let $-\lambda$ be this number and call λ the type number of f. If one of the sets A or B is void, the type number is defined as ∞ and $-\infty$, respectively. Show the following assertions: (i) The type number is $\lambda = \limsup_{t \to \infty} \frac{1}{t} \log |f(t)|$. (ii) If $\lambda_1 \geq \lambda_2 \geq \ldots \geq \lambda_n$ are the type numbers of the functions $f_i, i = 1, \ldots, n$, and λ, λ' those of $f_1 + \ldots + f_n$ and $f_1 \cdot \ldots \cdot f_n$ respectively, then $\lambda \leq \lambda_1, \lambda' \leq \lambda_1 + \lambda_2 + \ldots + \lambda_n$. In addition, $\lambda = \lambda_1$ if $\lambda_1 > \lambda_2$, and $\lambda \leq \lambda_1$ if $\lambda_1 = \lambda_2$. (iii) If $\lambda_1 > \lambda_2 > \ldots > \lambda_n > -\infty$ are the type numbers of the functions $f_i, i = 1, \ldots, n$, then these functions are linearly independent on $[0, \infty)$.

The last assertion yields another proof that a linear differential equation $\dot{x} = A(t)x$ in $\mathbb{R}^d$ has at most d different Lyapunov exponents (Proposition 6.2.3).

Exercise 6.5.8. Consider the following example:

$$A(t) = \begin{bmatrix} -1 - 2\cos 4t & 2 + 2\sin 4t \\ -2 + 2\sin 4t & -1 + 2\sin 4t \end{bmatrix}.$$

Show that the eigenvalues of $A(t)$ for each $t \in \mathbb{R}$ are in the left half-plane and that the solution for the initial value $x_1(0) = 0, x_2(0) = 1$ is the unbounded function

$$x(t) = \begin{bmatrix} e^t \sin 2t \\ e^t \cos 2t \end{bmatrix}.$$

Exercise 6.5.9. Consider the linear skew product system in Example 6.3.3 and write down the details for the arguments.

6.6. Orientation, Notes and References

Orientation. This chapter has set the stage for the rest of this book: As seen in Section 6.2, the stability theory for nonautonomous linear differential or difference equations is significantly more complicated than in the autonomous case, since the analysis of eigenvalues of the right-hand sides for fixed time does not determine the exponential growth rates of solutions. Instead of following up the theory for single nonautonomous linear differential or difference equations, we take this as a motivation to analyze instead families of such equations forming linear skew product flows where, in addition to the linear equation, a flow in the base space is present which determines the behavior of the linear part. Sections 6.3 and 6.4 for continuous time and discrete time, respectively, have sketched several ways on how to construct linear skew product flows and have given first insight into the scope of systems that may be considered as linear skew product flows. The rest of this book will develop the corresponding theory for three classes of systems: periodic systems, topological systems, and measurable systems. Here the Lyapunov exponents, corresponding decompositions into Lyapunov spaces, and the stability properties will be analyzed. The periodic case presented in Chapter 7 is relatively simple since periodic equations can be transformed into autonomous equations and hence their properties can be inferred. Our main reason to include this chapter is that it provides intuition for the topological theory of linear skew product systems in Chapter 9 and the measurable theory in Chapter 11. These chapters, however, will need completely different techniques which will be prepared for the topological case in Chapter 8 and for the measurable theory in Chapter 10.

Notes and references. Basic references for Lyapunov exponents are Cesari [**24**], Hahn [**61**], Bylov et al. [**23**]. Here conditions are formulated which lead to nonautonomous differential equations with nicer properties; in particular, regular equations [**61**, § 64] have properties related to the discussion in Proposition 6.2.5.

The introductory discussion of linear skew product flows is based on Arnold [**6**], Bronstein and Kopanskii [**21**], Colonius and Kliemann [**29**], Cong [**32**], and Robinson [**117**]. A theory of Lyapunov exponents for nonautonomous differential equations is also given in Barreira and Pesin [**15**, Chapter 1] with the aim to analyze nonlinear differential equations based on linearization; cf. Example 6.3.6. Kloeden and Rasmussen [**81**] present an approach to nonautonomous systems based on pullback constructions. The construction in Example 6.3.3, in particular, for the special case of almost periodic coefficient functions, has been a starting point for the theory of linear skew product flows, classical references are Sacker and Sell [**119**] and Miller [**106**]. An early reference for dynamical systems obtained by time shifts is Bebutov [**18**].

Carathéodory solutions of the differential equation (6.1.1) are by definition integrals of Lebesgue integrable functions. As indicated, such functions, called absolutely continuous, are differentiable for almost all t and satisfy the differential equation for these t. It is worth mentioning that a continuous function, which satisfies the differential equation for almost all t, is not necessarily absolutely continuous and hence is not a Carathéodory solution.

In Section 6.2 we follow Josić and Rosenbaum [**73**] in the construction of unstable nonautonomous differential equations with eigenvalues in the left half-plane for every t. They even show that for every matrix $B \in \mathcal{B}$ there is $\omega > 0$ such that the matrix $B - G(\omega)$ is unstable and hence the differential equation $\dot{x} = A(t)$ with matrix $A(t)$ defined by (6.2.6) is unstable. This paper also contains further construction ideas.

Linear skew product flows obtained by linearization, as in Example 6.3.6, are of tremendous relevance in theory and applications. We do not treat them here for two reasons: There are many excellent presentations in the literature (examples are Amann [**4**], Robinson [**117**] and Barreira and Pesin [**15**]). Furthermore, the interesting questions in linearization theory concern the behavior of the nonlinear part $\dot{y} = f(y)$ outside of the base component. This is obvious for the case of an equilibrium e, where $B = \{e\}$, but also holds for more complicated invariant sets; cf. also Wiggins [**139**]. For these analyses additional concepts and techniques are relevant that go beyond the scope of this book.

The type number introduced in Exercise 6.5.7 is Lyapunov's original definition in [**97**] (with the opposite sign). We learned the example in Exercise 6.5.8 from Ludwig Arnold.

Chapter 7

Periodic Linear Differential and Difference Equations

We have seen in the preceding chapter that the general theory of time-varying (or nonautonomous) linear differential equations $\dot x = A(t)x$ and difference equations $x_{n+1} = A(n)x_n$ presents great difficulties, since many properties of the autonomous equations are not valid. Hence we restrict our further analysis to certain classes of matrix functions $A(\cdot)$. Historically the first complete theory for a class of time-varying linear systems was initiated by Floquet [50] in 1883 for linear differential equations with periodic coefficients. The basic idea of Floquet theory is to transform a periodic equation into an autonomous equation, without changing the Lyapunov exponents (called here Floquet exponents). This works for linear difference and differential equations.

The counterexamples in Section 6.2 have shown that the eigenvalues of $A(t)$ do not give the desired information on stability. Instead we will construct a theory for Lyapunov exponents and associated (time-varying) decompositions of the state space. In Section 7.1 a Floquet theory for periodic linear difference equations is developed. Analogously, Section 7.2 presents the main results of Floquet theory for periodic linear differential equations. Here the role of Lyapunov exponents and Lyapunov spaces are emphasized which have been introduced in Chapter 1 for the autonomous case. Section 7.3 discusses in detail a prominent example, the Mathieu equation including a stability diagram which indicates for which parameter values stability holds.

7.1. Floquet Theory for Linear Difference Equations

We begin the discussion of periodic linear skew product systems by considering a linear difference equation with periodic coefficients. This is the first class of linear skew product systems for which we will describe the Lyapunov exponents and the corresponding Lyapunov spaces.

Definition 7.1.1. A periodic linear difference equation

$$(7.1.1) \qquad x_{n+1} = A(n)x_n, n \in \mathbb{Z},$$

is given by a matrix function $A : \mathbb{Z} \to Gl(d, \mathbb{R})$ satisfying the periodicity condition $A(n+p) = A(n), n \in \mathbb{Z}$, for some period $p \in \mathbb{N}$.

An immediate observation is that equation (7.1.1) may be reduced to an autonomous equation by taking p steps at once. Thus define for a solution $x_n, n \in \mathbb{Z}$, a new sequence $y_k := x_{kp}, k \in \mathbb{Z}$. Then periodicity of $A(\cdot)$ shows that

$$x_{kp+1} = A(kp)x_{kp} = A(0)x_{kp} = A(0)y_k,$$
$$x_{kp+2} = A(kp+1)x_{kp+1} = A(1)A(0)x_{kp} = A(1)A(0)y_k,$$
$$\vdots$$
$$x_{kp+p} = \prod_{j=1}^{p} A(p-j) x_{kp} = \prod_{j=1}^{p} A(p-j) y_k.$$

Hence y_k is a solution of the autonomous linear difference equation in $\mathbb{R}^d$ given by

$$(7.1.2) \qquad y_{k+1} = X(p)y_k \text{ with } X(p) := \prod_{j=1}^{p} A(p-j).$$

Conversely, a solution y_k of this equation determines a solution of $x_{n+1} = A(n)x_n$ via the equations above. The results on exponential growth rates and eigenvalues from Section 1.5 immediately apply to equation (7.1.2). This motivates the following definition.

Definition 7.1.2. For the p-periodic equation (7.1.1) the eigenvalues $\alpha_j \in \mathbb{C}$ of $X(p)$ are called the Floquet multipliers, and the Floquet exponents are defined by

$$\lambda_j := \frac{1}{p} \log |\alpha_j|.$$

The following analysis of an associated skew product system will reveal more insight into the relation of the Floquet exponents and the behavior of equation (7.1.1). We will need the following lemma concerning roots of invertible matrices. Its proof is motivated by the standard definition of powers $a^b := e^{b \cdot \log a}$ for $0 \neq a, b \in \mathbb{C}$ (here log denotes the principal value of the complex logarithm.) The scalar example $(-1)^{\frac{1}{2}}$ shows that for real a one has to distinguish between roots with complex and real entries.

Lemma 7.1.3. *Let $S \in Gl(d, \mathbb{C})$ be an invertible matrix and $n \in \mathbb{N}$. Then there exists $R \in Gl(d, \mathbb{C})$ such that $R^n = S$. If $S \in Gl(d, \mathbb{R})$, then there is $Q \in Gl(d, \mathbb{R})$ with $Q^n = S^2$. The eigenvalues ζ of S are given by $\zeta = \mu^n, \mu \in \operatorname{spec}(R)$, in particular, $|\mu| = |\zeta|^{1/n}$, and the algebraic multiplicity of an eigenvalue ζ of S equals the sum of the algebraic multiplicities of the eigenvalues μ of R with $\mu^n = \zeta$. Analogous assertions hold for the eigenvalues of S^2 and Q.*

Proof. One may suppose that S is given in Jordan normal form. Then it suffices to consider the Jordan blocks separately. So we may suppose that for an eigenvalue $\mu \in \mathbb{C}$ ($\mu \neq 0$ since S is invertible)

$$S = \mu I + N = \mu \left(I + \frac{1}{\mu} N\right),$$

where N is the associated nilpotent matrix with $N^m = 0$ for some m (given by the size of the block).

Recall the series expansion $\log(1+z) = \sum_{j=1}^{\infty} \frac{(-1)^{j+1}}{j} z^j$, $|z| < 1$, and observe that $\left[e^{\frac{1}{n} \log(1+z)}\right]^n = 1+z$. We claim that $S = R^n$ with

(7.1.3) $$R := \exp\left\{\frac{1}{n}\left[(\log \mu) I + \sum_{j=1}^{m} \frac{(-1)^{j+1}}{j}\left(\frac{1}{\mu} N\right)^j\right]\right\}.$$

In fact, the matrices $(\log \mu) I$ and $\sum_{j=1}^{m} \frac{(-1)^{j+1}}{j \mu^j} N^j$ commute and one finds

$$R^n = \exp(\log \mu) \cdot \exp\left[\sum_{j=1}^{m} \frac{(-1)^{j+1}}{j}\left(\frac{1}{\mu} N\right)^j\right] = \mu\left(I + \frac{1}{\mu} N\right) = S.$$

The proof is analogous to the arguments showing that the series expansions of the exponential function and the logarithm are inverse to each other (based on the Cauchy product of power series, here, however, only m summands occur in (7.1.3)).

Now suppose that $S \in Gl(d, \mathbb{R})$. There is a matrix $R \in Gl(d, \mathbb{C})$ given by (7.1.3) with $R^n = S$. Then the matrix $\bar{R}$ with complex conjugate entries commutes with R, since this is true for all summands in R and $\bar{R}$. In particular, one finds

$$\overline{(R\bar{R})} = \overline{(\bar{R}R)} = R\bar{R},$$

showing that $Q := R\bar{R} \in Gl(d, \mathbb{R})$. Since S has real entries it follows that $\bar{R}^n = S$ implying
$$Q^n = (R\bar{R})^n = R^n \bar{R}^n = S^2.$$
The eigenvalues of S are of the form $\zeta = \mu^n$, where μ are the eigenvalues of R, with the indicated algebraic multiplicities, since this holds within each Jordan block of S; note that each Jordan block subspace is invariant for R by construction. Similarly, one argues for the eigenvalues of S^2 and Q. $\square$

The fact that the moduli of the eigenvalues of R are determined by S is remarkable, since the matrix R itself with $R^n = S$ is not unique: For any n-th root $\zeta \in \mathbb{C}$ of unity one obtains that also $(R \cdot \zeta I)^n = R^n \zeta^n = R^n = S$.

Recall that the principal fundamental solution $X(n), n \in \mathbb{Z}$, is defined as the unique solution of the matrix difference equation
$$X(n+1) = A(n)X(n) \text{ with initial value } X(0) = I_d,$$
and the solutions of $x_{n+1} = A(n)x_n, x_0 = x$, are given by $x_n = X(n)x$. The following lemma shows consequences of the periodicity assumption for $A(n)$ for the principal fundamental solution.

Lemma 7.1.4. *The principal fundamental solution $X(n)$ of $x_{n+1} = A(n)x_n$ with p-periodic $A(\cdot)$ satisfies for all $n, m \in \mathbb{Z}$,*
$$X(n) = \prod_{j=1}^{n} A(n-j) \text{ and } X(n+mp) = X(n)X(mp) = X(n)X(p)^m.$$

Proof. The first assertion is immediate from the definitions. Then the second assertion follows for $n, m \geq 0$ from p-periodicity by

$X(n + mp)$
$= A(n + mp - 1) \ldots A(mp) \ldots A(2p - 1) \ldots A(p)A(p - 1) \ldots A(1)A(0)$
$= A(n - 1) \ldots A(0) \ldots A(p - 1) \ldots A(0)A(p - 1) \ldots A(1)A(0)$
$= X(n)X(p)^m.$

Analogously one argues for $n < 0$ and for $m < 0$. $\square$

We use Lemma 7.1.3 to analyze the Floquet exponents.

Proposition 7.1.5. *There is a matrix $Q \in gl(d, \mathbb{R})$ such that the principal fundamental solution $X(\cdot)$ satisfies*
$$X(2p) = X(p)^2 = Q^{2p}.$$
The Floquet multipliers $\alpha_j \in \mathbb{C}$, the Floquet exponents $\lambda_j \in \mathbb{R}$, and the eigenvalues $\mu_j \in \mathbb{C}$ of Q are related by
$$|\alpha_j| = |\mu_j|^p \text{ and } \lambda_j = \frac{1}{p} \log |\alpha_j| = \log |\mu_j|.$$

7.1. Floquet Theory for Linear Difference Equations

Proof. By Lemma 7.1.3 a real matrix Q with $X(2p) = Q^{2p}$ exists. The $2p$-th powers of the eigenvalues of Q are the eigenvalues of $X(2p) = X(p)^2$, which are the squares of the eigenvalues α_j of $X(p)$, implying $|\alpha_j| = |\mu_j|^p$. Note that the algebraic multiplicities of the respective eigenvalues coincide as well. $\square$

Next we define a linear skew product system associated to (7.1.1). Denote by $\psi(n, n_0, x), n \in \mathbb{Z}$, the solution with initial condition $x_{n_0} = x$.

Definition 7.1.6. Consider $\mathbb{Z}_p := \{0, 1, \ldots, p-1\}$ and define the shift

$$\theta : \mathbb{Z} \times \mathbb{Z}_p \to \mathbb{Z}_p, \theta(n, \nu) = n + \nu \mod p \text{ for } n \in \mathbb{Z}, \nu \in \mathbb{Z}_p.$$

Furthermore, let $\varphi(n, \nu, x) := \psi(n + \nu, \nu, x), n \in \mathbb{Z}, \nu \in \mathbb{Z}_p, x \in \mathbb{R}^d$. Then a continuous linear skew product system is defined by

$$\Phi = (\theta, \varphi) : \mathbb{Z} \times \mathbb{Z}_p \times \mathbb{R}^d \longrightarrow \mathbb{Z}_p \times \mathbb{R}^d, \Phi(n, \nu, x) := (\theta(n, \nu), \varphi(n, \nu, x)).$$

In fact, the dynamical system property is clear for the base component θ and the p-periodicity of $A(\cdot)$ implies that

$$\psi(n + kp, \nu + kp, x) = \psi(n, \nu, x) \text{ for all } n, \nu, k \in \mathbb{Z}.$$

Hence one may always suppose that the second argument of ψ (the initial time) is in $\mathbb{Z}_p$. Then the cocycle property of the skew component φ follows from the 2-parameter cocycle property of ψ (cf. Theorem 6.4.1(iv)),

$$\varphi(n + m, \nu, x) = \psi(n + m + \nu, \nu, x) = \psi(n + m + \nu, m + \nu, \psi(m + \nu, \nu, x))$$
$$= \psi(n + \theta(m, \nu), \theta(m, \nu), \psi(m + \nu, \nu, x))$$
$$= \varphi(n, \theta(m, \nu), \varphi(m, \nu, x)).$$

The Lyapunov exponents or exponential growth rates of Φ are by definition

$$\lambda(x, \nu) := \limsup_{n \to \infty} \frac{1}{n} \log \|\varphi(n, \nu, x)\| \text{ for } (x, \nu) \in \mathbb{R}^d \times \mathbb{Z}_p.$$

The following theorem for periodic linear difference equations is a generalization of Theorem 1.5.6. It yields decompositions of the state space $\mathbb{R}^d$ into Lyapunov spaces corresponding to the Lyapunov exponents which coincide with the Floquet exponents. In contrast to the autonomous case, the Lyapunov spaces depend on the points in the base space $\mathbb{Z}_p$. Thus they change periodically over time.

Theorem 7.1.7. Let $\Phi = (\theta, \varphi) : \mathbb{Z} \times \mathbb{Z}_p \times \mathbb{R}^d \longrightarrow \mathbb{Z}_p \times \mathbb{R}^d$ be the linear skew product system associated by Definition 7.1.6 to the p-periodic linear difference equation (7.1.1). The Lyapunov exponents coincide with the Floquet exponents λ_j, $j = 1, \ldots, \ell \leq d$ and exist as limits. For each $\nu \in \mathbb{Z}_p$ there exists a decomposition

$$\mathbb{R}^d = L(\lambda_1, \nu) \oplus \ldots \oplus L(\lambda_\ell, \nu)$$

into linear subspaces $L(\lambda_j, \nu)$, called the Floquet or Lyapunov spaces, with the following properties:

(i) the Lyapunov spaces have dimensions independent of ν,

$$d_j := \dim L(\lambda_j, \nu) \text{ is constant for } \nu \in \mathbb{Z}_p;$$

(ii) they are invariant under multiplication by the principal fundamental matrix in the following sense:

$$X(n+\nu, \nu) L(\lambda_j, \nu) = L(\lambda_j, \theta(n, \nu)) \text{ for all } n \in \mathbb{Z} \text{ and } \nu \in \mathbb{Z}_p;$$

(iii) for every $\nu \in \mathbb{Z}_p$ the Lyapunov exponents satisfy

$$\lambda(x, \nu) = \lim_{n \to \pm\infty} \frac{1}{n} \log \|\varphi(n, \nu, x)\| = \lambda_j \text{ if and only if } 0 \neq x \in L(\lambda_j, \nu).$$

Proof. First we show that the Floquet exponents coincide with the Lyapunov exponents. Recall that $X(2p) = Q^{2p}$. For the autonomous linear difference equation $y_{n+1} = Q y_n$ Theorem 1.5.6 yields a decomposition of $\mathbb{R}^d$ into subspaces L_j which are characterized by the property that the Lyapunov exponents for $n \to \pm\infty$ are given by $\lambda_j = \log|\mu_j|$ where the μ_j are the eigenvalues of Q.

Now the proof is based on the fact that the matrix function $Z(n) := X(n) Q^{-n}, n \in \mathbb{Z}$, maps the solution $Q^n x_0$ of $y_{n+1} = Q y_n, y_0 = x_0 \in \mathbb{R}^d$, to the corresponding solution at time n of (7.1.1). This holds since

(7.1.4) $$X(n) x_0 = X(n) Q^{-n} Q^n x_0 = Z(n) (Q^n x_0).$$

Observe that $Z(0) = I_d$ and $Z(n)$ is $2p$-periodic, since

$$Z(n+2p) = X(n+2p) Q^{-(n+2p)} = X(n) X(2p) Q^{-2p} Q^{-n} = X(n) Q^{-n} = Z(n).$$

By periodicity, $Z(\cdot)$ and $Z(\cdot)^{-1}$ are bounded on $\mathbb{Z}$. Hence the exponential growth rates remain constant under multiplication by $Z(n)$; see Exercise 6.5.3.

Using also Proposition 7.1.5 we get a corresponding decomposition of $\mathbb{R}^d$ into subspaces, which is characterized by the property that a solution starting at time $n_0 = 0$ in the corresponding subspace $L(\lambda_j, 0) := L_j$ has exponential growth rates for $n \to \pm\infty$ equal to a given Floquet exponent λ_j. Then for every $\nu \in \mathbb{Z}$ the subspaces

$$L(\lambda_j, \nu) := X(\nu) L(\lambda_j, 0),$$

also yield a decomposition of $\mathbb{R}^d$. A solution of (7.1.1) starting at time $n_0 = \nu$ in the subspace $L(\lambda_j, \nu)$ has exponential growth rate for $n \to \pm\infty$ equal to λ_j.

7.1. Floquet Theory for Linear Difference Equations

In order to show the invariance property (ii) of the Lyapunov spaces we first note that by the 2-parameter cocycle property

$$X(n+\nu,\nu)L(\lambda_j,\nu) = X(n+\nu,\nu)X(\nu,0)L(\lambda_j,0) = X(n+\nu,0)L(\lambda_j,0)$$
$$= L(\lambda_j, n+\nu).$$

It remains to show that these subspaces change with period p; i.e., $L(\lambda_j, \nu + p) = L(\lambda_j, \nu)$ for all $\nu \in \mathbb{Z}$. For the proof consider solutions $x_n, n \in \mathbb{Z}$, and $z_n, n \in \mathbb{Z}$, of (7.1.1) corresponding to the initial conditions $x_\nu = x$ at time $n_0 = \nu$ and $z_{\nu+p} = x$ at time $n_0 = \nu+p$, respectively. Then by Lemma 7.1.4

$$x = x_\nu = X(\nu)x_0 \text{ and } x = z_{\nu+p} = X(\nu+p)z_0 = X(\nu)X(p)z_0.$$

Again by Lemma 7.1.4 this implies for all n,

$$z_{n+\nu+p} = X(n+\nu+p)z_0 = X(n+\nu)X(p)z_0 = X(n+\nu)X(\nu)^{-1}x$$
$$= X(n+\nu)x_0 = x_{n+\nu}.$$

Hence the solution $z_n, n \in \mathbb{Z}$, coincides with the time-p shift of the solution $x_n, n \in \mathbb{Z}$. This shows that the exponential growth rates for time tending to $\pm\infty$ coincide; i.e., $x \in L(\lambda_j, \nu + p)$ if and only if $x \in L(\lambda_j, \nu)$. □

Theorem 7.1.7 shows that for p-periodic matrix functions $A : \mathbb{Z} \longrightarrow Gl(d, \mathbb{R})$ the Floquet exponents and Floquet spaces replace the logarithms of the moduli of the eigenvalues and the Lyapunov spaces for constant matrices $A \in Gl(d, \mathbb{R})$; cf. Theorem 1.5.6. The number of Lyapunov exponents and the dimensions of the Lyapunov spaces are independent of $\nu \in \mathbb{Z}_p$, while the Lyapunov spaces themselves depend on the time parameter ν of the periodic matrix function $A(\cdot)$, and they form p-periodic orbits in the space of d_j-dimensional subspaces, the Grassmannians $\mathbb{G}_{d_j}$.

Remark 7.1.8. Transformations $Z(n)$ which together with $Z(n)^{-1}$ are bounded on $\mathbb{Z}$, are called Lyapunov transformations. A consequence of equation (7.1.4) is that the solutions at time n of the periodic linear difference equation are obtained by a Lyapunov transformation from the solutions of the autonomous difference equation $y_{n+1} = Qy_n$. This is a main assertion of Floquet theory. Theorem 7.1.7 gives more detailed information, since it describes the behavior of all Lyapunov spaces over the base space $\mathbb{Z}_p$ endowed with the shift θ. It is remarkable that the Lyapunov spaces have the same period p as the matrix function $A(\cdot)$, while the transformation $Z(\cdot)$ is $2p$-periodic.

Remark 7.1.9. Using sums of Lyapunov spaces, one can also construct flags of subspaces describing the Lyapunov exponents for every initial value. This follows from Theorem 1.5.8 and provides a generalization of this theorem.

The next corollary is a generalization of Theorem 4.2.1. It provides a topological characterization of the Lyapunov spaces constructed in Theorem 7.1.7 as chain components of an associated dynamical system. For this purpose we have to modify the construction of the skew product system: As base space we take $\mathbb{Z}_{2p} = \{0, 1, \ldots, 2p-1\}$. This is due to the fact that the Lyapunov transformation $Z(n), n \in \mathbb{Z}$, is $2p$-periodic.

Define a linear skew product flow (for notational simplicity we use the same notation as in Definition 7.1.6), by

$$\theta : \mathbb{Z} \times \mathbb{Z}_{2p} \to \mathbb{Z}_{2p}, \theta(n, \nu) = n + \nu \mod 2p \text{ for } n \in \mathbb{Z}, \nu \in \mathbb{Z}_{2p}$$

and

$$\Phi = (\theta, \varphi) : \mathbb{Z} \times \mathbb{Z}_{2p} \times \mathbb{R}^d \longrightarrow \mathbb{Z}_{2p} \times \mathbb{R}^d, \Phi(n, \nu, x) := (\theta(n, \nu), \varphi(n, \nu, x)).$$

Since the cocycle φ maps one-dimensional subspaces onto one-dimensional subspaces, the following projection of Φ is well defined:

(7.1.5)
$$\mathbb{P}\Phi(n, \nu, \mathbb{P}x) := (\theta(n, \nu), \mathbb{P}\varphi(n, \nu, x)) \text{ for } n \in \mathbb{Z}, \nu \in \mathbb{Z}_{2p}, 0 \neq x \in \mathbb{R}^d.$$

This is a continuous dynamical system on the compact metric space $\mathbb{Z}_{2p} \times \mathbb{P}^{d-1}$. We use the linear maps $\Phi(n, \nu) = \varphi(n + \nu, \nu, \cdot), n \in \mathbb{Z}, \nu \in \mathbb{Z}_{2p}$. Thus $\Phi(n, \nu)$ maps the fiber $\{\nu\} \times \mathbb{R}^d$ onto the fiber $\{\theta(n, \nu)\} \times \mathbb{R}^d$; it is represented by the principal fundamental matrix $X(n + \nu, \nu)$; cf. Theorem 6.4.1. The Lyapunov spaces $\{\nu\} \times L(\lambda_j, \nu)$ are contained in the fibers $\{\nu\} \times \mathbb{R}^d$, which in the following corollary are identified with $\mathbb{R}^d$.

Corollary 7.1.10. *Let $\mathbb{P}\Phi$ be the projection onto $\mathbb{Z}_{2p} \times \mathbb{P}^{d-1}$ given by (7.1.5) of the dynamical system $\Phi = (\theta, \varphi)$ corresponding to a p-periodic linear difference equation (7.1.1). Then the following assertions hold.*

(i) $\mathbb{P}\Phi$ has ℓ chain components $\mathcal{M}_1, \ldots, \mathcal{M}_\ell$, where ℓ is the number of Lyapunov exponents λ_j.

(ii) For each Lyapunov exponent λ_j one has (with an appropriate numbering) that

$$\mathcal{M}_j = \{(\nu, \mathbb{P}x) \mid 0 \neq x \in L(\lambda_j, \nu) \text{ and } \nu \in \mathbb{Z}_{2p}\}.$$

(iii) For the chain components $\mathcal{M}_j$, the sets

$$\mathcal{V}_j(\nu) := \{x \in \mathbb{R}^d \mid x = 0 \text{ or } (\nu, \mathbb{P}x) \in \mathcal{M}_j\}, \nu \in \mathbb{Z}_{2p},$$

coincide with the Lyapunov spaces $L(\lambda_j, \nu)$ and hence yield decompositions of $\mathbb{R}^d$ into linear subspaces

$$\mathbb{R}^d = \mathcal{V}_1(\nu) \oplus \ldots \oplus \mathcal{V}_\ell(\nu), \nu \in \mathbb{Z}_{2p}.$$

(iv) The Lyapunov spaces $L(\lambda_j, \nu)$ are invariant under the flow Φ, i.e., $\Phi(n, \nu)L(\lambda_j, \nu) = L(\lambda_j, \theta(n, \nu))$ for $j = 1, \ldots, \ell$, and their dimensions are constant.

Proof. For the autonomous linear equation $y_{n+1} = Qy_n$ we have a decomposition of $\mathbb{R}^d$ into the Lyapunov spaces $L(\lambda_j, 0)$ which by Theorem 4.2.1 correspond to the chain components. By (7.1.4) the $2p$-periodic matrix function $Z(n), n \in \mathbb{Z}$, maps the solution of $y_{n+1} = Qy_n, y(0) = x_0 \in \mathbb{R}^d$, to the solution of $x_{n+1} = A(n)x_n, x(0) = x_0$. These difference equations induce dynamical systems on $\mathbb{Z}_{2p} \times \mathbb{P}^{d-1}$ which are topologically conjugate: The map Z is well defined as a map from $\mathbb{Z}_{2p}$ to $Gl(d, \mathbb{R})$ since $Z(n), n \in \mathbb{Z}$, is $2p$-periodic. The conjugating map on $\mathbb{Z}_{2p} \times \mathbb{P}^{d-1}$ is given by $(\nu, y) \mapsto (\nu, \mathbb{P}Z(\nu)y)$ where $\mathbb{P}Z(\nu)$ is the map on $\mathbb{P}^{d-1}$ induced by the linear map $Z(\nu)$. Then it follows from Theorem 3.3.7 that the chain components are mapped onto each other.

For the autonomous equation, the chain components in $\mathbb{Z}_{2p} \times \mathbb{P}^{d-1}$ are given by the product of $\mathbb{Z}_{2p}$ with the chain components of $y_{n+1} = Qy_n$ in $\mathbb{P}^{d-1}$. In fact, take a point q in a chain component $\mathcal{M}$ in $\mathbb{P}^{d-1}$ and consider $(0, q) \in \mathbb{Z}_{2p} \times \mathbb{P}^{d-1}$. The ω-limit set $\omega(0, q)$ is contained in a chain component and its θ-component coincides with $\mathbb{Z}_{2p}$. Hence the chain component coincides with $\mathbb{Z}_{2p} \times \mathcal{M}$ and there are no other chain components. □

Remark 7.1.11. An alternative proof for determining the chain components of the autonomous equation considered in $\mathbb{Z}_{2p} \times \mathbb{P}^{d-1}$ can be based on Exercise 3.4.8. The component in $\mathbb{P}^{d-1}$ is independent of the component in $\mathbb{Z}_{2p}$ which is periodic. Hence the chain components in $\mathbb{Z}_{2p} \times \mathbb{P}^{d-1}$ are given by the product of $\mathbb{Z}_{2p}$ with the chain components in $\mathbb{P}^{d-1}$. For the periodic equation, the chain components do not have this product structure.

Stability

As an application of these results, consider the problem of stability of the zero solution of $x_{n+1} = A(n)x_n$ with period $p \in \mathbb{N}$. The following definition generalizes Definition 1.5.10.

Definition 7.1.12. The stable, center, and unstable subspaces associated with the periodic matrix function $A : \mathbb{Z} \longrightarrow Gl(d, \mathbb{R})$ are defined for $\nu \in \mathbb{Z}_p$ by

$$L^-(\nu) = \bigoplus_{\lambda_j < 0} L(\lambda_j, \nu), L^0(\nu) = L(0, \nu), \text{ and } L^+(\nu) = \bigoplus_{\lambda_j > 0} L(\lambda_j, \nu).$$

The collection $\{L^-(\nu) \mid \nu \in \mathbb{Z}_p\}$ is called the stable subbundle; analogously the center and unstable subbundles are defined. With these preparations we can state a result regarding stability of periodic linear difference equations.

Theorem 7.1.13. *The zero solution of the periodic linear difference equation $x_{n+1} = A(n)x_n$ is asymptotically stable if and only if all Floquet exponents are negative if and only if the stable subspace satisfies $L^-(\nu) = \mathbb{R}^d$ for some (and hence for all) $\nu \in \mathbb{Z}_p$.*

Proof. This follows from Theorem 7.1.7 and the construction of the Lyapunov spaces. □

7.2. Floquet Theory for Linear Differential Equations

This section presents general results on linear differential equations with periodic coefficients. Again, our development here is from the point of view of linear skew product flows, and we will determine the Lyapunov exponents and the corresponding Lyapunov spaces.

Definition 7.2.1. A periodic linear differential equation

$$\dot{x} = A(t)x \tag{7.2.1}$$

is given by a matrix function $A : \mathbb{R} \longrightarrow gl(d, \mathbb{R})$ that is locally integrable and periodic with period $T > 0$, i.e., $A(t+T) = A(t)$ for almost all $t \in \mathbb{R}$.

Recall from Theorem 6.1.1 that the principal fundamental solution $X(t) = X(t,0), t \in \mathbb{R}$, is the unique solution of the matrix differential equation

$$\dot{X}(t) = A(t)X(t) \text{ with initial value } X(0) = I.$$

Furthermore, the solutions of $\dot{x} = A(t)x$, $x(t_0) = x_0$, are given by $x(t) = X(t)X(t_0)^{-1}x_0$. The following lemma shows consequences of the periodicity assumption for $A(t)$ for the fundamental solution.

Lemma 7.2.2. *The principal fundamental solution $X(t)$ of $\dot{x} = A(t)x$ with T-periodic $A(\cdot)$ satisfies*

$$X(kT + t) = X(t)X(T)^k \text{ for all } t \in \mathbb{R} \text{ and all } k \in \mathbb{Z}.$$

Proof. The assertion is clear for $k = 0$. Suppose it holds for $k - 1 \geq 0$. Then

$$X(kT) = X((k-1)T + T) = X(T)X(T)^{k-1} = X(T)^k. \tag{7.2.2}$$

Define

$$Y(t) := X(t + kT)X(kT)^{-1}, t \in \mathbb{R}.$$

Then $Y(0) = I$ and differentiation yields using periodicity of $A(\cdot)$,

$$\frac{d}{dt}Y(t) = \dot{X}(kT + t)X(kT)^{-1} = A(kT + t)X(kT + t)X(kT)^{-1} = A(t)Y(t).$$

7.2. Floquet Theory for Linear Differential Equations

Since the solution of this initial value problem is unique, $Y(t) = X(t)$ and hence by (7.2.2),
$$X(t + kT) = X(t)X(kT) = X(t)X(T)^k \text{ for } t \in \mathbb{R}.$$
Similarly, one shows the assertion for $k < 0$. □

Consider a solution $x(t) = X(t)x_0$ of (7.2.1) and define $y_k := x(kT), k \in \mathbb{Z}$. This is a solution of an autonomous linear difference equation of the form
$$(7.2.3) \qquad y_{k+1} = X(T)y_k, k \in \mathbb{Z}.$$
The results from Section 1.5 on the relation between exponential growth rates and eigenvalues of $X(T)$ immediately apply to this equation. This motivates the following definition.

Definition 7.2.3. For the T-periodic differential equation (7.2.1), the eigenvalues $\alpha_j \in \mathbb{C}$ of $X(T)$ are called the Floquet multipliers, and the Floquet exponents are defined by
$$\lambda_j := \frac{1}{T} \log |\alpha_j|.$$

The matrix $X(T)$ is also called the monodromy matrix. We will need the following lemma which is derived using the Jordan canonical form and the scalar logarithm. The difference between the real and the complex situation becomes already evident by looking at $-1 = e^{i\pi}$.

Lemma 7.2.4. *For every invertible matrix $S \in Gl(d, \mathbb{C})$ there is a matrix $R \in gl(d, \mathbb{C})$ such that $S = e^R$. For every invertible matrix $S \in Gl(d, \mathbb{R})$ there is a real matrix $Q \in gl(d, \mathbb{R})$ such that $S^2 = e^Q$. The eigenvalues $\zeta \in \mathrm{spec}(S)$ are given by $\zeta = e^\mu, \mu \in \mathrm{spec}(R)$, in particular, $|\mu| = \log |\zeta|$, and the algebraic multiplicity of an eigenvalue ζ of S equals the sum of the algebraic multiplicities of the eigenvalues μ of R with $\zeta = e^\mu$. Analogous assertions hold for the eigenvalues of S^2 and Q.*

Proof. For the first statement observe that it suffices to consider a Jordan block, and write $S = \zeta I + N = \zeta(I + \frac{1}{\zeta}N)$ with nilpotent N, i.e., $N^m = 0$ for some $m \in \mathbb{N}$. Recall the series expansion $\log(1+z) = \sum_{j=1}^\infty \frac{(-1)^{j+1}}{j} z^j, |z| < 1$, and define
$$(7.2.4) \qquad R := (\log \zeta)I + \sum_{j=1}^m \frac{(-1)^{j+1}}{j\zeta^j} N^j.$$
Both summands commute and one finds
$$e^R = \exp(\log \zeta) \cdot \exp\left(\sum_{j=1}^m \frac{(-1)^{j+1}}{j\lambda^j} \left(\frac{1}{\zeta}N\right)^j \right) = \zeta(I + \frac{1}{\zeta}N) = S.$$

The proof is analogous to the arguments showing that the series expansions of the exponential function and the logarithm are inverse to each other (based on the Cauchy product of power series, here, however, only m summands occur in (7.2.4)).

For the second assertion observe that the matrices R and $\bar R$ commute, since their summands commute. Then, with $Q := R + \bar R \in gl(d, \mathbb{R})$ and $S = e^R = e^{\bar R}$, one finds $S^2 = e^R e^{\bar R} = e^{R+\bar R} = e^Q$. The proof above also shows that the eigenvalues of R and Q, respectively, are mapped onto the eigenvalues of e^R and e^Q, respectively, and that the assertions about the algebraic multiplicities hold (by considering all Jordan blocks). □

Remark 7.2.5. Another way to construct Q is to write a complex eigenvalue as $\mu = re^{i\varphi}, r > 0, \varphi \in (0, 2\pi)$. Then observe that the logarithm of

$$\begin{bmatrix} \operatorname{Re}\mu & \operatorname{Im}\mu \\ -\operatorname{Im}\mu & \operatorname{Re}\mu \end{bmatrix} = r \begin{bmatrix} \cos\varphi & -\sin\varphi \\ \sin\varphi & \cos\varphi \end{bmatrix} \text{ is } (\log r) \begin{bmatrix} 1 & 0 \\ 0 & 1 \end{bmatrix} + \begin{bmatrix} 0 & -\varphi \\ \varphi & 0 \end{bmatrix}$$

and discuss the Jordan blocks as above.

Remark 7.2.6. The real parts λ_j of the eigenvalues of R and Q, respectively, are uniquely determined by S, since $|\alpha_j| = e^{\lambda_j}$ for the eigenvalues α_j of S. The imaginary parts are unique up to addition of $2k\pi\iota, k \in \mathbb{Z}$. In particular, several eigenvalues of R and Q may be mapped to the same eigenvalue of e^R and e^Q, respectively.

Next we determine the relation between the Floquet exponents and the eigenvalues of Q.

Proposition 7.2.7. *There is a matrix $Q \in gl(d, \mathbb{R})$ such that the fundamental solution $X(\cdot)$ satisfies $X(2T) = e^{2TQ}$. The Floquet multipliers, i.e., the eigenvalues $\alpha_j \in \mathbb{C}$ of $X(T)$, the Floquet exponents $\lambda_j \in \mathbb{R}$, and the eigenvalues $\mu_j \in \mathbb{C}$ of Q are related by*

$$|\alpha_j| = e^{T \operatorname{Re}\mu_j} \text{ and } \lambda_j = \operatorname{Re}\mu_j = \frac{1}{T}\log|\alpha_j|.$$

Proof. By Lemma 7.2.4 a real matrix Q with $X(2T) = X(T)^2 = e^{2TQ}$ exists. The exponential function maps the eigenvalues of $2TQ$ to the eigenvalues of $X(2T) = X(T)^2$, which are the squares of the eigenvalues α_j of $X(T)$ and the imaginary parts of the eigenvalues of $2TQ$ do not contribute to the moduli of the α_j. □

We turn to define an associated linear skew product flow associated with (7.2.1). Denote by $\psi(t, t_0, x), t \in \mathbb{R}$, the solution with initial condition $x(t_0) = x$. Parametrize the unit sphere $\mathbb{S}^1$ by $t \in [0, T)$, i.e., identify it with $\mathbb{R}/(T\mathbb{Z})$.

7.2. Floquet Theory for Linear Differential Equations

Definition 7.2.8. Consider the unit sphere $\mathbb{S}^1$ parametrized by $t \in [0, T)$ and define the shift

$$\theta : \mathbb{R} \times \mathbb{S}^1 \to \mathbb{S}^1, \theta(t, \tau) = t + \tau \mod T \text{ for } t \in \mathbb{R}, \tau \in \mathbb{S}^1.$$

Furthermore, let $\varphi(t, \tau, x) := \psi(t + \tau, \tau, x), t \in \mathbb{R}, \tau \in \mathbb{S}^1, x \in \mathbb{R}^d$. Then a continuous linear skew product flow is defined by

$$\Phi = (\theta, \varphi) : \mathbb{R} \times \mathbb{S}^1 \times \mathbb{R}^d \longrightarrow \mathbb{S}^1 \times \mathbb{R}^d, \Phi(t, \tau, x) := (\theta(t, \tau), \varphi(t, \tau, x)).$$

In fact, the dynamical system property is clear for the base component θ on $B = \mathbb{S}^1$ and the T-periodicity of $A(\cdot)$ implies that

$$\psi(t + kT, \tau + kT, x) = \psi(t, \tau, x) \text{ for all } t, \tau \in \mathbb{R}, k \in \mathbb{Z}.$$

Hence one may always suppose that the second argument of ψ (the initial time) is in $\mathbb{S}^1$. Then the cocycle property of the skew component φ follows from the cocycle property of ψ,

$$\varphi(t + s, \tau, x) = \psi(t + s + \tau, \tau, x) = \psi(t + s + \tau, s + \tau, \psi(s + \tau, \tau, x))$$
$$= \psi(t + \theta(s, \tau), \theta(s, \tau), \psi(s + \tau, \tau, x))$$
$$= \varphi(t, \theta(s, \tau), \varphi(s, \tau, x)).$$

The Lyapunov exponents or exponential growth rates of Φ are by definition

$$\lambda(x, \tau) = \limsup_{t \to \infty} \frac{1}{t} \log \|\varphi(t, \tau, x)\| \text{ for } (x, \tau) \in \mathbb{R}^d \times \mathbb{S}^1.$$

The following theorem for periodic linear differential equations is a generalization of Theorem 1.4.3. It yields a decomposition into Lyapunov spaces corresponding to the Lyapunov exponents. In contrast to the autonomous case, the Lyapunov spaces depend on the points in the base space $\mathbb{R}$ modulo T which we have identified with the unit sphere $\mathbb{S}^1$. One may interpret this by saying that they change periodically over time.

Theorem 7.2.9. *Let $\Phi = (\theta, \varphi) : \mathbb{R} \times \mathbb{S}^1 \times \mathbb{R}^d \longrightarrow \mathbb{S}^1 \times \mathbb{R}^d$ be the linear skew product flow associated with the T-periodic linear differential equation (7.2.1). The Lyapunov exponents coincide with the Floquet exponents λ_j, $j = 1, \ldots, \ell \leq d$. For each $\tau \in \mathbb{S}^1$ there exists a decomposition*

$$\mathbb{R}^d = L(\lambda_1, \tau) \oplus \ldots \oplus L(\lambda_\ell, \tau)$$

into linear subspaces $L(\lambda_j, \tau)$, called the Floquet or Lyapunov spaces, with the following properties:

(i) The Lyapunov spaces have dimension independent of τ,

$$d_j := \dim L(\lambda_j, \tau) \text{ is constant for } \tau \in \mathbb{S}^1.$$

(ii) They are invariant under multiplication by the principal fundamental matrix $X(t,\tau)$ in the following sense:

$$X(t+\tau,\tau)L(\lambda_j,\tau) = L(\lambda_j,\theta(t,\tau)) \text{ for all } t \in \mathbb{R} \text{ and } \tau \in \mathbb{S}^1.$$

(iii) For every $\tau \in \mathbb{S}^1$ the Lyapunov exponents satisfy

$$\lambda(x,\tau) = \lim_{t \to \pm\infty} \frac{1}{t} \log \|\varphi(t,\tau,x)\| = \lambda_j \text{ if and only if } 0 \neq x \in L(\lambda_j,\tau).$$

Proof. By Proposition 7.2.7, the Floquet exponents λ_j coincide with the real parts of the eigenvalues of $Q \in gl(d,\mathbb{R})$ where $X(2T) = e^{2TQ}$. First we show that the Floquet exponents are the Lyapunov exponents. By Lemma 7.2.2 we can write

$$X(kT+s) = X(s)X(kT) \text{ for all } k \in \mathbb{Z} \text{ and } t,s \in \mathbb{R}.$$

For the autonomous linear differential equation $\dot{y} = Qy$ Theorem 1.4.3 yields a decomposition of $\mathbb{R}^d$ into subspaces $L(\lambda_j)$ which are characterized by the property that the Lyapunov exponents for $t \to \pm\infty$ are given by the real parts λ_j of the eigenvalues of Q. The matrix function $Z(t) := X(t)e^{-Qt}, t \in \mathbb{R}$, maps the solution $e^{Qt}x_0$ of $\dot{y} = Qy, y(0) = x_0 \in \mathbb{R}^d$, to the solution of $\dot{x} = A(t)x, x(0) = x_0$, since

$$(7.2.5) \qquad X(t)x_0 = X(t)e^{-Qt}e^{Qt}x_0 = Z(t)\left(e^{Qt}x_0\right).$$

Observe that $Z(0) = I$ and $Z(t)$ is $2T$-periodic, since by Proposition 7.2.7

$$Z(2T+t) = X(2T+t)e^{-(2T+t)Q} = X(t)X(2T)e^{-2TQ-tQ} = X(t)e^{-tQ} = Z(t).$$

Since $Z(\cdot)$ is continuous, it follows that $Z(t)$ and $Z(t)^{-1}$ are bounded on $\mathbb{R}$. The exponential growth rates remain constant under multiplication by the bounded matrix $Z(t)$ with bounded inverse $Z(t)^{-1}$; cf. Exercise 6.5.3. This shows that the Floquet exponents coincide with the Lyapunov exponents.

As a consequence we get a corresponding decomposition of $\mathbb{R}^d$ which is characterized by the property that a solution starting at time $t = 0$ in the corresponding subspace $L(\lambda_j, 0) := L(\lambda_j)$ has exponential growth rates for $t \to \pm\infty$ equal to the Floquet exponent λ_j. Then the subspaces

$$L(\lambda_j,\tau) := X(\tau)L(\lambda_j,0), \quad \tau \in \mathbb{R},$$

yield a decomposition of $\mathbb{R}^d$ into subspaces and their dimensions are constant for $\tau \in \mathbb{R}$ proving assertion (i).

Using $X(\tau) = X(\tau, 0)$ and the 2-parameter cocycle property for the principal fundamental solution (cf. Proposition 6.1.2(i)), one finds

$$(7.2.6) \quad X(t,\tau)L(\lambda_j,\tau) = X(t,\tau)X(\tau,0)L(\lambda_j,0) = X(t+\tau,0)L(\lambda_j,0)$$
$$= L(\lambda_j, t+\tau).$$

7.2. Floquet Theory for Linear Differential Equations

Hence the decomposition of $\mathbb{R}^d$ given by the subspaces $L(\lambda_j, \tau)$ is also characterized by the property that a solution starting at time $t = \tau$ in the corresponding subspace $L(\lambda_j, \tau)$ has exponential growth rate λ_j for $t \to \pm\infty$ and assertion (iii) follows.

Assertion (ii) follows from (7.2.6), if we can show that the Lyapunov spaces are T-periodic, hence they are well defined modulo T. The exponential growth rate of the solution $x(t, x_0)$ with $x(0) = x_0$ is equal to the exponential growth rate of the solution $z(t)$ of $\dot{x} = A(t)x$ with $z(T) = x_0$. In fact, for $t \in \mathbb{R}$,

$$x(t, x_0) = X(t)x_0 \text{ and } x_0 = z(T) = X(T)z(0), \text{ i.e., } z(0) = X(T)^{-1}x_0$$

implying

$$z(t) = X(t)z(0) = X(t)X(T)^{-1}x_0.$$

Hence by Lemma 7.2.2 we find for $t \in \mathbb{R}$,

$$z(t+T) = X(t+T)X(T)^{-1}x_0 = X(t)X(T)X(T)^{-1}x_0 = X(t)x_0 = x(t, x_0),$$

and the exponential growth rates for $t \to \pm\infty$ coincide. This shows that the decomposition into the subspaces $L(\lambda_j, \tau)$ is T-periodic. □

Remark 7.2.10. It is remarkable that the decomposition into the Lyapunov spaces has the same period T as the matrix function $A(\cdot)$, while the transformation $Z(\cdot)$ is only $2T$-periodic.

Remark 7.2.11. Recall the metric on the Grassmannians, introduced in Section 5.1. For each $j = 1, \ldots, \ell \leq d$ the map $L_j : \mathbb{S}^1 \longrightarrow \mathbb{G}_{d_j}$ defined by $\tau \longmapsto L(\lambda_j, \tau)$ is continuous, hence the Lyapunov spaces $L(\lambda_j, \tau)$ (sometimes called the Floquet spaces) of the periodic matrix function $A(t)$ change continuously with the base point τ. This follows from the construction of the spaces $L(\lambda_j, \tau)$. Observe that the Lyapunov spaces of the autonomous equation $\dot{x} = Qx$, naturally, are constant.

Remark 7.2.12. Using sums of Lyapunov spaces, one can also construct flags of subspaces describing the Lyapunov exponents for every initial value. This follows from Theorem 1.4.4 and provides a generalization of this theorem.

These results show that for periodic matrix functions $A : \mathbb{R} \longrightarrow gl(d, \mathbb{R})$ the Floquet exponents and Floquet spaces replace the real parts of eigenvalues and the Lyapunov spaces, concepts that are so useful in the linear algebra of (constant) matrices $A \in gl(d, \mathbb{R})$. The number of Lyapunov exponents and the dimensions of the Lyapunov spaces are independent of $\tau \in \mathbb{S}^1$, while the Lyapunov spaces themselves depend on the time parameter τ of the periodic matrix function $A(\cdot)$, and they form periodic orbits in the Grassmannians $\mathbb{G}_{d_j}$.

Remark 7.2.13. Transformations $Z(t)$ which together with $Z(t)^{-1}$ are bounded on $\mathbb{R}$, are known as Lyapunov transformations (or kinematic similarity transformation, Hahn [**61**, Definition 61.2]). Equation (7.2.5) shows that there is a periodic Lyapunov transformation of the solutions of a periodic linear differential equation to the solutions of an autonomous linear differential equation.

The next corollary is a generalization of Theorem 4.1.3. It shows that the Lyapunov spaces constructed in Theorem 7.2.9 can be characterized by the chain components of an associated skew product system. As in the discrete-time case (cf. Corollary 7.1.10), we have to consider twice the period. Here we parametrize the unit sphere $\mathbb{S}^1$ by $\tau \in [0, 2T)$, i.e., we identify $\mathbb{S}^1$ with $\mathbb{R}/(2T\mathbb{Z})$. Analogously to (7.1.5) we denote the linear skew product system corresponding to $\dot{x} = A(t)x$ again by $\Phi(t, \tau, x) = (\theta(t,\tau), \varphi(t,\tau,x)), t \in \mathbb{R}, \tau \in [0, 2T), x \in \mathbb{R}^d$. The dynamical system $\Phi = (\theta, \varphi)$ induces a projected flow $\mathbb{P}\Phi_t$ given by

(7.2.7) $\mathbb{P}\Phi(t, \tau, \mathbb{P}x) := (\theta(t,\tau), \mathbb{P}\varphi(t,\tau,x))$ for $t \in \mathbb{R}, \tau \in \mathbb{S}^1, 0 \neq x \in \mathbb{R}^d$.

This is a continuous dynamical system (in continuous time) on the compact metric space $\mathbb{S}^1 \times \mathbb{P}^{d-1}$ which we call a projective bundle. We use the linear maps $\Phi(t,\tau) = \varphi(t+\tau, \tau, \cdot), t \in \mathbb{R}, \tau \in \mathbb{S}^1$. Thus $\Phi(t+\tau, \tau)$ maps the fiber $\{\tau\} \times \mathbb{R}^d$ onto the fiber $\{\theta(t,\tau)\} \times \mathbb{R}^d, \tau \in [0, 2T)$; it is represented by the principal fundamental matrix $X(t+\tau,\tau)$; cf. Theorem 6.1.1. The Lyapunov spaces $\{\tau\} \times L(\lambda_j, \tau)$ are contained in the fibers $\{\tau\} \times \mathbb{R}^d$, which in the following corollary are identified with $\mathbb{R}^d$.

Corollary 7.2.14. *Let $\mathbb{S}^1$ be parametrized by $\tau \in [0, 2T)$ and consider the projection $\mathbb{P}\Phi$ to $\mathbb{S}^1 \times \mathbb{P}^{d-1}$ given by (7.2.7) of the linear skew product flow Φ associated to a T-periodic linear differential equation (7.2.1). Then the following assertions hold.*

(i) $\mathbb{P}\Phi$ has ℓ chain components $\mathcal{M}_1, \ldots, \mathcal{M}_\ell$, where ℓ is the number of Lyapunov exponents.

(ii) For each Lyapunov exponent λ_j one has (with an appropriate numbering) that

$$\mathcal{M}_j = \{(\tau, \mathbb{P}x) \mid 0 \neq x \in L(\lambda_j, \tau) \text{ and } \tau \in \mathbb{S}^1\}.$$

(iii) For the chain components $\mathcal{M}_j$ the sets

$$\mathcal{V}_j(\tau) := \{x \in \mathbb{R}^d \mid x = 0 \text{ or } (\tau, \mathbb{P}x) \in \mathcal{M}_j\}, \ \tau \in \mathbb{S}^1,$$

coincide with the Lyapunov spaces $L(\lambda_j, \tau)$ and hence yield decompositions of $\mathbb{R}^d$ into linear subspaces

(7.2.8) $\mathbb{R}^d = \mathcal{V}_1(\tau) \oplus \ldots \oplus \mathcal{V}_\ell(\tau), \ \tau \in \mathbb{S}^1.$

7.2. Floquet Theory for Linear Differential Equations

Proof. With Q as in Proposition 7.2.7, the autonomous linear equation $\dot{y} = Qy$ yields a decomposition of $\mathbb{R}^d$ into the Lyapunov spaces $L(\lambda_j, 0)$ which by Theorem 4.1.3 correspond to the chain components. By (7.2.5) the matrix function $Z(t)$ maps the solution of $\dot{y} = Qy, y(0) = x_0 \in \mathbb{R}^d$, to the solution of $\dot{x} = A(t)x, x(0) = x_0$. These differential equations induce dynamical systems on $\mathbb{S}^1 \times \mathbb{P}^{d-1}$ which are topologically conjugate: The map Z is well defined as a map from $\mathbb{S}^1$ parametrized by $\tau \in [0, 2T)$ to $Gl(d, \mathbb{R})$ since $Z(t), t \in \mathbb{R}$, is $2T$-periodic. The conjugating map on $\mathbb{S}^1 \times \mathbb{P}^{d-1}$ is given by $(\nu, y) \mapsto (\nu, \mathbb{P}Z(t)y)$ where $\mathbb{P}Z(t)$ is the map on $\mathbb{P}^{d-1}$ induced by the linear map $Z(t)$. Then it follows from Theorem 3.1.15(iii) that the chain components are mapped onto each other.

For the autonomous equation, the chain components in $\mathbb{S}^1 \times \mathbb{P}^{d-1}$ are given by the product of $\mathbb{S}^1$ with the chain components in $\mathbb{P}^{d-1}$. In fact, take a point p in a chain component $\mathcal{M}_j$ in $\mathbb{P}^{d-1}$ and consider $(0, p) \in \mathbb{S}^1 \times \mathbb{P}^{d-1}$. The ω-limit set $\omega(0, p)$ is contained in a maximal chain transitive set; its θ-component coincides with $\mathbb{S}^1$ while the component in $\mathbb{P}^{d-1}$ coincides with $\mathcal{M}_j$. Hence the maximal chain transitive sets coincide with $\mathbb{S}^1 \times \mathcal{M}_j$ and there are no other chain components. $\square$

Stability

As an application of these results, consider the problem of stability of the zero solution of $\dot{x}(t) = A(t)x(t)$ with period $T > 0$. The following definition generalizes Definition 1.4.7.

Definition 7.2.15. The stable, center, and unstable subspaces associated with the periodic matrix function $A : \mathbb{R} \longrightarrow gl(d, \mathbb{R})$ are defined for $\tau \in \mathbb{S}^1$ by

$$L^-(\tau) = \bigoplus_{\lambda_j < 0} L(\lambda_j, \tau), L^0(\tau) = L(0, \tau), \text{ and } L^+(\tau) = \bigoplus_{\lambda_j > 0} L(\lambda_j, \tau).$$

The collection $\{L^-(\nu) \,|\, \nu \in \mathbb{Z}_p\}$ is called the stable subbundle; analogously the center and unstable subbundles are defined. With these preparations we can state a result regarding stability of periodic linear differential equations.

Theorem 7.2.16. *The zero solution of the periodic linear differential equation $\dot{x} = A(t)x$ is asymptotically stable if and only if it is exponentially stable if and only if all Floquet exponents are negative if and only if $L^-(\tau) = \mathbb{R}^d$ for some (and hence for all) $\tau \in \mathbb{S}^1$.*

Proof. This follows from Theorem 1.4.8 and the construction of the Lyapunov spaces. $\square$

We briefly discuss periodic linear Hamiltonian systems.

Example 7.2.17. A matrix $H \in gl(2d, \mathbb{R})$ is called Hamiltonian if
$$H = \begin{bmatrix} H_{11} & H_{12} \\ H_{21} & H_{22} \end{bmatrix}$$
with $H_{11} = -H_{22}$ and H_{12} and H_{21} are symmetric. Equivalently,
$$JH = (JH)^\top = H^\top J^\top \text{ with } J = \begin{bmatrix} 0 & -I_d \\ I_d & 0 \end{bmatrix}.$$
With μ also $-\bar\mu$ is an eigenvalue. Let the images of $H : \mathbb{R} \to gl(2d, \mathbb{R})$ be Hamiltonian matrices $H(t), t \in \mathbb{R}$, and suppose that H is periodic and continuous. Then the differential equation
$$\begin{bmatrix} \dot{x} \\ \dot{y} \end{bmatrix} = \begin{bmatrix} H_{11}(t) & H_{12}(t) \\ H_{21}(t) & H_{22}(t) \end{bmatrix} \begin{bmatrix} x \\ y \end{bmatrix}$$
is called a linear periodic Hamiltonian system. One can show that for every Floquet exponent λ also its negative $-\lambda$ is a Floquet exponent. Hence the fixed point $0 \in \mathbb{R}^{2d}$ cannot be exponentially stable. Thus, for this class of differential equations weaker stability concepts are appropriate; cf. Section 7.5 for references.

7.3. The Mathieu Equation

To show the power of Floquet's approach we discuss the classical example of the Hill-Mathieu equation. This discussion also illustrates that considerable further work may be necessary in order to get explicit stability criteria for specific equations.

Let $q_1, q_2 : \mathbb{R} \to \mathbb{R}$ be T-periodic functions and suppose that q_1 is continuously differentiable and q_2 is continuous. Consider the periodic linear oscillator

(7.3.1) $$\ddot{y} + 2q_1(t)\dot{y} + q_2(t)y = 0.$$

Let Q_1 be an antiderivative of q_1 so that $\dot{Q}_1 = q_1$ and use the substitution $z(t) = y(t)\exp(Q_1(t)), t \in \mathbb{R}$. The chain rule yields
$$\dot{y} = \dot{z}\exp(-Q_1) - z\exp(-Q_1)\dot{Q}_1 = \exp(-Q_1)[\dot{z} - zq_1],$$
$$\ddot{y} = \ddot{z}\exp(-Q_1) - \dot{z}\exp(-Q_1)\dot{Q}_1 - \dot{z}\exp(-Q_1)\dot{Q}_1$$
$$+ z\exp(-Q_1)\dot{Q}_1\dot{Q}_1 - z\exp(-Q_1)\ddot{Q}_1$$
$$= \exp(-Q_1)\left[\ddot{z} - 2\dot{z}q_1 + zq_1q_1 - z\dot{q}_1\right].$$

Inserting this into (7.3.1) one obtains
$$0 = \ddot{y} + 2q_1\dot{y} + q_2 y$$
$$= \exp(-Q_1)\left[\ddot{z} - 2\dot{z}q_1 + zq_1q_1 - z\dot{q}_1 + 2q_1(\dot{z} - zq_1) + q_2 z\right]$$
$$= \exp(-Q_1)\left[\ddot{z} - zq_1q_1 - z\dot{q}_1 + q_2 z\right].$$

7.3. The Mathieu Equation

This is a special case of Hill's equation

(7.3.2) $$\ddot{z} + p(t)z = 0$$

with a T-periodic function p, here given by $p(t) := q_2(t) - q_1(t)^2 - \dot{q}_1(t)$, $t \in \mathbb{R}$. An example for (7.3.1) is

(7.3.3) $$\ddot{y} + 2k\dot{y} + (a + \varepsilon \cos 2t)y = 0 \text{ with } k > 0.$$

Here the substitution has the form $x(t) = y(t) \exp(kt), t \in \mathbb{R}$, and yields Mathieu's equation

(7.3.4) $$\ddot{x} + (\delta + \varepsilon \cos 2t)x = 0,$$

here with $\delta := a - k^2$. This is a linear oscillator with periodic restoring force. For simplicity, we only discuss this special case, although many arguments are also valid for more general equations of this type.

Remark 7.3.1. A physical interpretation of Mathieu's equation is obtained by a damped pendulum with oscillating pivot subject to gravity g and damping. Consider a mass m attached at the end of a massless pendulum of length l. Suppose the pivot point oscillates in the vertical direction according to $p(t) = A \cos \omega t$ with $A, \omega > 0$. Then the angle θ from the vertical to the pendulum obeys

(7.3.5) $$\frac{d^2}{dt^2}\theta(t) + k\dot{\theta}(t) + \frac{g + \ddot{p}(t)}{l} \sin \theta(t) = 0.$$

Here the term $k\dot{\theta}$ models the damping. We measure the angle θ such that when the pendulum is vertical, pointed downward, then $\theta = 0$. If we linearize the equation near the equilibrium $(\theta, \dot{\theta}) = (0, 0)$ (which is usually stable) we obtain the linearized equation

(7.3.6) $$\ddot{y}(t) + k\dot{y}(t) + \frac{g - A\omega^2 \cos \omega t}{l} y(t) = 0.$$

Introduce the new time t replacing $2t/\omega$ (i.e., we define $\tilde{y}(t) := y(\frac{2t}{\omega})$ and then omit the tilde). One obtains

(7.3.7) $$\ddot{y} + \frac{2k}{\omega}\dot{y} + \left(\frac{4g}{\omega^2 l} - \frac{4A}{l} \cos 2t\right) y = 0.$$

Define the new variable $x(t) = e^{\frac{k}{\omega}t}y(t)$ and abbreviate $\delta := \frac{4g}{\omega^2 l} - \frac{k^2}{\omega^2}$ and $\varepsilon := -\frac{4A}{l}$. Then one obtains an equation of the form (7.3.4). Note that for small damping $0 < k < \frac{4g}{l}$ one has $\delta > 0$.

If we linearize the equation near the equilibrium $(\theta, \dot{\theta}) = (\pi, 0)$ (which is usually unstable) one obtains the linearized equation

(7.3.8) $$\ddot{y}(t) + k\dot{y}(t) - \frac{g - A\omega^2 \cos \omega t}{l} y(t) = 0.$$

The same procedure as above leads an equation of the form (7.3.4) with $\delta := -\frac{4g}{\omega^2 l} - \frac{k^2}{\omega^2} < 0$ and $\varepsilon := \frac{4A}{l}$. (The reader is asked to verify the computations in this remark in Exercise 7.4.2.) We come back to the stability properties of the pendulum in Remark 7.3.3.

We proceed with the discussion of equations (7.3.3) and (7.3.4). With $y_1 := y, y_2 = \dot{y}$ the second order equation for y is equivalent to the system

$$\begin{bmatrix} \dot{y}_1 \\ \dot{y}_2 \end{bmatrix} = \begin{bmatrix} 0 & 1 \\ -a - \varepsilon \cos 2t & -2k \end{bmatrix} \begin{bmatrix} y_1 \\ y_2 \end{bmatrix}.$$

Similarly, with $x_1 = x, x_2 = \dot{x}$, the transformed equation (7.3.4) for x is equivalent to

$$\begin{bmatrix} \dot{x}_1 \\ \dot{x}_2 \end{bmatrix} = \begin{bmatrix} 0 & 1 \\ -(a - k^2) - \varepsilon \cos 2t & 0 \end{bmatrix} \begin{bmatrix} x_1 \\ x_2 \end{bmatrix}.$$

The solutions satisfy for $t \in \mathbb{R}$,

$$\begin{bmatrix} x_1(t) \\ x_2(t) \end{bmatrix} = e^{-kt} \begin{bmatrix} 1 & 0 \\ -k & 1 \end{bmatrix} \begin{bmatrix} y_1(t) \\ y_2(t) \end{bmatrix},$$

hence, for initial values $(x_{1,0}, x_{2,0}) = (y_{1,0}, y_{2,0})$, the Lyapunov exponents are related by

$$\lambda(x_{1,0}, x_{2,0}) = \lambda(y_{1,0}, y_{2,0}) - k.$$

The Floquet multipliers $\alpha_1, \alpha_2 \in \mathbb{C}$ of (7.3.4) are the eigenvalues of the principal fundamental solution $X(T)$ given by the solution of

$$\dot{X}(t) = A(t)X(t), X(0) = I, \text{ with } A(t) = \begin{bmatrix} 0 & 1 \\ -\delta - \varepsilon \cos 2t & 0 \end{bmatrix}.$$

Hence the sum of the Floquet multipliers satisfies

(7.3.9) $$\alpha_1 + \alpha_2 = \operatorname{tr} X(T)$$

and their product is given

(7.3.10) $$\alpha_1 \alpha_2 = \det X(T) = \det X(0) \exp\left(\int_0^\pi \operatorname{tr} A(s) ds\right) = 1.$$

Here we have used Liouville's formula; cf. Theorem 6.1.5.

By Proposition 7.2.7 and Theorem 7.2.9 the Floquet exponents (these are the Lyapunov exponents) are given by

$$\lambda_i = \frac{1}{\pi} \log |\alpha_i| \text{ for } i = 1, 2.$$

For the Floquet multipliers of (7.3.4), the following three cases may occur (we will use the results from Section 1.4.)

(i) Both α_1 and α_2, are real and $\alpha_1 \neq \alpha_2$. By (7.3.10) one of them, say α_1, is less than 1 and $\alpha_2 = 1/\alpha_1$ is greater than 1. In terms of the Floquet exponents, this means
$$\lambda_1 = \frac{1}{\pi} \log |\alpha_1| < 0 \text{ and } \lambda_2 = -\lambda_1 > 0.$$
Hence the origin is (exponentially) unstable for equation (7.3.4). For equation (7.3.3) the origin is exponentially stable if $\lambda_2 - k < 0$ (we also note that it is stable if $\lambda_2 - k = 0$).

(ii) The numbers $\alpha_1 \neq \alpha_2$ are complex conjugate. Then $|\alpha_1| = |\alpha_2| = 1$ and $\alpha_2 = 1/\alpha_1$. Thus we may assume $\alpha_1 = e^{i\theta}$ with $\theta \in (0, \pi)$ and $\alpha_2 = e^{-i\theta}$ and hence
$$\lambda_1 = \lambda_2 = 0.$$
Again, the system (7.3.4) is not exponentially stable. In contrast, the origin is exponentially stable for (7.3.3), since $\lambda_2 - k = -k < 0$. Since, by Proposition 7.2.7, the Floquet multipliers $\alpha_{1,2}$ are the eigenvalues of the matrix Q, Theorem 1.4.10 implies that the origin is stable for the autonomous equation $\dot{x} = Qx$. Then (7.2.5) shows that the origin is also stable for the periodic equation (7.3.4).

(iii) If (i) and (ii) do not hold, it follows that $\alpha_1 = \alpha_2$ is real. This implies $\alpha_1 = \alpha_2 = 1$ or $\alpha_1 = \alpha_2 = -1$ and in both cases one has
$$\lambda_1 = \lambda_2 = 0.$$
Again the system (7.3.4) is not exponentially stable, while (7.3.3) is exponentially stable, since $\lambda_{1,2} - k = -k < 0$. Concerning stability of (7.3.4), two cases are possible: either $\lambda = 0$ has geometric multiplicity 2 (thus there are two one-dimensional Jordan blocks) as an eigenvalue of Q which implies that $\dot{x} = Qx$ is stable; or, $\lambda = 0$ has geometric multiplicity 1 (which means that we have a single Jordan block) and $\dot{x} = Qx$ is unstable. Again, equation (7.2.5) shows that this entails the same properties for the periodic differential equation (7.3.4).

While this discussion sheds some light on the possible cases, it does not determine the stability properties for given parameters ε and δ. For this purpose, we discuss the parameter dependence of the eigenvalues α_1, α_2 of the matrices
$$X(T) = X(T; \delta, \varepsilon) \text{ for } \delta, \varepsilon \geq 0.$$
The eigenvalues of a matrix depend continuously on the entries, and the solution $X(T; \delta, \varepsilon)$ depends continuously on the parameters δ, ε; see Theorem 6.1.4. Hence the eigenvalues of $X(T; \delta, \varepsilon)$ depend continuously on the parameters δ and ε. By the discussion above, we know that exponential instability can only occur if one of the eigenvalues has modulus greater than 1. Now, if there is a complex conjugate pair of eigenvalues, they must lie

on the unit circle in $\mathbb{C}$, and hence a small perturbation cannot lead to an eigenvalue with modulus greater than 1. Otherwise, there are two real eigenvalues α_1 and α_2, and the system is exponentially unstable if and only if one of them is outside the interval $[-1, 1]$ (and then the other one is in the interval $(-1, 1)$). Hence a transition from stability to exponential instability or vice versa can only occur via transition through one of the eigenvalue pairs $\alpha_1 = \alpha_2 = -1$ or $\alpha_1 = \alpha_2 = 1$.

The first case is equivalent to the existence of a 2π-periodic solution, the second case is equivalent to the existence of a π-periodic solution. For the equation $\ddot{z} + \delta z = 0$ with $\varepsilon = 0$ it is easy to characterize these cases: The corresponding system

$$\begin{bmatrix} \dot{z}_1 \\ \dot{z}_2 \end{bmatrix} = \begin{bmatrix} 0 & 1 \\ -\delta & 0 \end{bmatrix} \begin{bmatrix} z_1 \\ z_2 \end{bmatrix}$$

is time independent and has the principal fundamental solution

$$X(t) = \begin{bmatrix} \cos\sqrt{\delta}t & \frac{1}{\sqrt{\delta}}\sin\sqrt{\delta}t \\ -\sqrt{\delta}\sin\sqrt{\delta}t & \cos\sqrt{\delta}t \end{bmatrix}, t \in \mathbb{R}.$$

Thus the eigenvalues of

$$X(\pi) = \begin{bmatrix} \cos\sqrt{\delta}\pi & \frac{1}{\sqrt{\delta}}\sin\sqrt{\delta}\pi \\ -\sqrt{\delta}\sin\sqrt{\delta}\pi & \cos\sqrt{\delta}\pi \end{bmatrix}$$

are

$$\alpha_{1,2} = \cos\sqrt{\delta}\pi \pm \sqrt{-\sin^2\sqrt{\delta}\pi}.$$

Hence the eigenvalues are real if and only if $\sqrt{\delta}$ is an integer, and -1 is an eigenvalue if and only if $\sqrt{\delta}$ is odd, i.e., $\delta = (2n+1)^2$, $n \in \mathbb{N}$. Analogously, 1 is an eigenvalue if and only if δ is even, i.e., $\delta = (2n)^2$. As can be read off the fundamental solution, the origin is stable (but, naturally, not exponentially stable) in these cases.

For $\varepsilon > 0$ we also observe that the parameter pairs (δ, ε) for which there are eigenvalue pairs $\alpha_1 = \alpha_2 = 1$ and $\alpha_1 = \alpha_2 = -1$, respectively, are the only parameter values where the stability properties may change. It will turn out that in the (ε, δ)-space these critical parameter values are given by curves separating stable and unstable regions emanating from the points with $\varepsilon = 0$ and $\delta = (2n)^2$ and $\delta = (2n+1)^2$, $n \in \mathbb{N}$, respectively. We do this by constructing Fourier expansions of the corresponding π-periodic and 2π-periodic solutions, respectively.

First, we consider the critical parameters with eigenvalue 1, i.e., the case where a π-periodic solution $x(t), t \in \mathbb{R}$, exists. Since solutions are continuously differentiable, a standard result of analysis (see, e.g., Amann

7.3. The Mathieu Equation

and Escher [5, Chapter VI, Theorem 7.21]) shows that it has an absolutely and uniformly convergent Fourier series,

$$x(t) = \sum_{n=0}^{\infty} a_n \cos(2nt) + \sum_{n=0}^{\infty} b_n \sin(2nt), t \in [0, \pi],$$

with coefficients $a_n, b_n \in \mathbb{R}$. Since the minimal period is equal to π, it follows that $a_1 \neq 0$ or $b_1 \neq 0$. Inserting this series into the differential equation, one obtains for all t, that

$$0 = \ddot{x}(t) + (\delta + \varepsilon \cos 2t)x(t)$$
$$= \sum_{n=0}^{\infty}(\delta - 4n^2)a_n \cos(2nt) + \sum_{n=0}^{\infty}(\delta - 4n^2)b_n \sin(2nt)$$
$$+ \varepsilon \sum_{n=0}^{\infty} a_n \cos(2nt)\cos(2t) + \varepsilon \sum_{n=0}^{\infty} b_n \sin(2nt)\cos(2t).$$

Using the trigonometric identities

$$\cos(2nt)\cos(2t) = \frac{1}{2}\left[\cos(2(n+1)t) + \cos(2(n-1)t)\right],$$
$$\sin(2nt)\cos(2t) = \frac{1}{2}\left[\sin(2(n+1)t) + \sin(2(n-1)t)\right],$$

and sorting by the sine and cosine functions one computes

$$0 = \left(\delta a_0 + \frac{\varepsilon}{2}a_1\right)\cos 0 + \left((\delta - 4)a_1 + \frac{\varepsilon}{2}(2a_0 + a_2)\right)\cos(2t)$$
$$+ \sum_{n=1}^{\infty}(\delta - 4n^2)a_n \cos(2nt) + \left((\delta - 4)b_1 + \frac{\varepsilon}{2}b_2\right)\sin(2t)$$
$$+ \sum_{n=2}^{\infty}\left((\delta - 4n^2)b_n + \frac{\varepsilon}{2}(b_{n-1} + b_{n+1})\right)\sin(2nt).$$

For $n, m = 0, 1, 2, \ldots$, using the orthogonality relations

$$\int_0^\pi \cos(2nt)\cos(2mt)dt = \delta_{n,m}, \quad \int_0^\pi \sin(2nt)\sin(2mt)dt = \delta_{n,m},$$
$$\int_0^\pi \cos(2nt)\sin(2mt)dt = 0,$$

one finds that all Fourier coefficients of the sine and cosine functions must vanish. Hence at least one of the following two (infinite) linear systems must

have a nontrivial solution:
$$\begin{bmatrix} 0 \\ 0 \\ 0 \\ 0 \\ \cdot \\ \cdot \\ \cdot \end{bmatrix} = \begin{bmatrix} \delta & \frac{\varepsilon}{2} & & & & 0 \\ \varepsilon & \delta - 4 \cdot 1^2 & \frac{\varepsilon}{2} & & & \\ & \frac{\varepsilon}{2} & \delta - 4 \cdot 2^2 & \frac{\varepsilon}{2} & & \\ & & \frac{\varepsilon}{2} & \delta - 4 \cdot 3^2 & \frac{\varepsilon}{2} & \\ & & & \cdot & \cdot & \cdot \\ 0 & & & & \cdot & \cdot & \cdot \end{bmatrix} \begin{bmatrix} a_0 \\ a_1 \\ a_2 \\ a_3 \\ \cdot \\ \cdot \\ \cdot \end{bmatrix}$$

and

$$\begin{bmatrix} 0 \\ 0 \\ 0 \\ \cdot \\ \cdot \\ \cdot \end{bmatrix} = \begin{bmatrix} \delta - 4 \cdot 1^2 & \frac{\varepsilon}{2} & & & \\ \frac{\varepsilon}{2} & \delta - 4 \cdot 2^2 & \frac{\varepsilon}{2} & & \\ & \frac{\varepsilon}{2} & \delta - 4 \cdot 3^2 & \frac{\varepsilon}{2} & \\ & & \cdot & \cdot & \cdot \\ & & & \cdot & \cdot & \cdot \end{bmatrix} \begin{bmatrix} b_1 \\ b_2 \\ b_3 \\ \cdot \\ \cdot \\ \cdot \end{bmatrix}.$$

Solvability of these equations can be characterized by the condition that (appropriately defined) determinants for the corresponding infinite matrices vanish. For a numerical approach, we only consider finite submatrices depending on δ and ε. Then at least one of the two resulting $n \times n$-matrices must have vanishing determinant. Thus, for fixed n, we obtain a nonlinear equation for the relation between the critical parameters (ε, δ). Numerical computation shows that, starting from the critical parameter values $(\varepsilon, \delta) = (0, (2m+1)^2), m \in \mathbb{N}$, this determines curves separating stable and unstable regions. An analogous computation for 2π-periodic solutions determines further curves of critical parameters emanating from the points $(\varepsilon, \delta) = (0, (2m)^2), m \in \mathbb{N}$.

Figure 7.1 shows the resulting stability diagram. It has been computed using the system above with $n = 20$ equations. A Newton method is used to determine δ for a grid of ε. This determines numerically the curves starting at the points with $\varepsilon = 0$ and $\delta = 4, 9, 16,$ and 25. The stability regions in the (ε, δ)-plane are shaded.

Remark 7.3.2. An alternative computation of the curves in Figure 7.1 can be based on the observation that δ in the diagonal of matrices above can be interpreted as an eigenvalue of the matrix obtained by subtracting δI_n. Then standard numerical eigenvalue algorithms (e.g. shifted QR-methods) may be employed.

Remark 7.3.3. The discussion above has consequences for the damped pendulum with oscillating pivot (see Remark 7.3.1). It shows that equation (7.3.7) obtained by linearization at the lower position (which has $\delta > 0$) may be exponentially unstable for small periodic oscillations (i.e., small ε).

7.4. Exercises

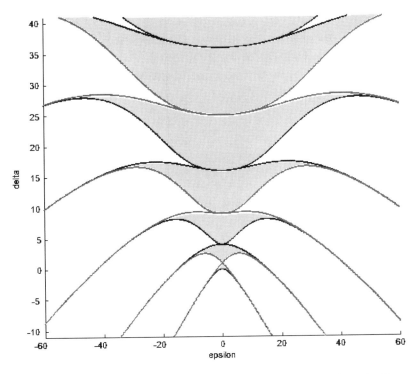

Figure 7.1. Stability diagram for the Mathieu equation (7.3.4). Regions of stability are shaded.

Naturally, for the problem without oscillating pivot (i.e., $A = \varepsilon = 0$) the origin is always exponentially stable. It is worth noting that the local stability properties near the equilibrium of the original nonlinear differential equation (7.3.5) can be derived from the stability properties of the linearized equation using the theory of stable/unstable manifolds. In a similar vein, one also sees that for the inverted pendulum with $\delta < 0$ and oscillating pivot and damping $k \geq 0$ the linearized equation (7.3.7) (which has $\delta < 0$) may be stable (look at the small region below the ε-axis in Figure 7.1). Here, naturally, for the problem without oscillating pivot (i.e., $A = \epsilon = 0$) the origin is not stable.

7.4. Exercises

Exercise 7.4.1. Let $x(t), t \in \mathbb{R}$, be a solution of the Mathieu equation $\ddot{x} + (\delta + \varepsilon \cos 2t)x = 0$. (i) Now show that $y(t) := x(-t), t \in \mathbb{R}$, is also a solution. (ii) Prove that the Mathieu equation has an even and an odd solution (recall that a function $f : \mathbb{R} \to \mathbb{R}$ is even if $f(t) = f(-t), t \in \mathbb{R}$, and it is odd if $f(t) = -f(-t), t \in \mathbb{R}$). (iii) Show that the curves in the stability diagram of the Mathieu equation are symmetric with respect to the δ-axis.

Exercise 7.4.2. Check the computations in Remark 7.3.1. In particular, derive a formula relating the Lyapunov (Floquet) exponents of the normalized equation (7.3.4) to those of the pendulum (7.3.7) and (7.3.8) linearized about the lower and the upper equilibrium, respectively. Use this to show that the claims in Remark 7.3.1 actually hold true.

Exercise 7.4.3. Let $A : \mathbb{R} \to gl(d, \mathbb{R})$ be continuous and T-periodic, $T > 0$ with principal fundamental solution $X(t, t_0)$ of the associated differential equation. Assume that the trace $\operatorname{tr} A(t) = 0$ for all $t \in [0, T]$. Show:

(i) The matrices $X(t, t_0)$ are volume preserving, i.e., $\det X(t, t_0) = 1$ for all $t_0, t \in \mathbb{R}$.

(ii) If $|\operatorname{tr} X(T, 0)| > 2$ there are unbounded solutions. If $|\operatorname{tr} X(T, 0)| < 2$, all solutions are bounded.

(iii) Consider the Mathieu equation

$$\ddot{y}(t) = -\omega^2(1 + \varepsilon \cos t)y(t),$$

where $\omega > 0$, $\varepsilon \in \mathbb{R}$ are given parameters. Why are all solutions of this differential equation bounded for sufficiently small $|\varepsilon|$, if $\omega \neq k/2$ for $k = 1, 2, 3, \ldots$?
Hint: Let $X_\varepsilon(t, t_0)$ be a corresponding principal fundamental matrix. Show that $\lim_{\varepsilon \to 0} X_\varepsilon(\omega, 0) = X_0(\omega, 0)$. Determine $X_0(\omega, 0)$ and apply (ii).

Exercise 7.4.4. Show that the Mathieu equation (7.3.4) is a periodic linear Hamiltonian system by constructing a corresponding continuous quadratic form $H(\cdot, \cdot, t)$ in two variables; cf. Example 7.2.17.
Hint: Use a physical interpretation taking into account kinetic and potential energy.

Exercise 7.4.5. Consider the 2-periodic difference equation $x_{k+1} = A(k)x_k$ in $\mathbb{R}^2$ with

$$A(k) := \begin{bmatrix} 0 & 1 + \frac{(-1)^k}{2} \\ 1 + \frac{(-1)^{k+1}}{2} & 0 \end{bmatrix}.$$

Show that the eigenvalues of $A(0)$ and $A(1)$ have modulus less than 1, while one Floquet exponent is positive and hence the system is unstable.

Exercise 7.4.6. Consider a T-periodic linear differential equation. Show that a T-periodic solution exists if and only if there is a Floquet exponent $\lambda = 0$. Formulate and prove also the analogous result for difference equations.

Exercise 7.4.7. Consider $x_{n+1} = A(n)x_n$ in $\mathbb{R}$ where $A(n) = 2$ for n even and $A(n) = -1$ for n odd. Determine the Floquet exponent and show that there is no $Q \in \mathbb{R}$ with $Q^2 = X(2)$, but there is Q such that $Q^4 = $

$X(2)^2$. Show that the transformation $Z(n) = X(n)Q^{-n}$ to the associated autonomous equation $y_{n+1} = By_n$ is 4-periodic and not 2-periodic.

Exercise 7.4.8. Show that for every dimension d, there is a matrix $A \in gl(d, \mathbb{R})$ that is not the square of any other matrix in $gl(d, \mathbb{R})$.

Exercise 7.4.9. Explore Remark 7.3.3, i.e., use the stability discussion of the Mathieu equation to determine stability of the two linear oscillators (with $\delta > 0$ and $\delta < 0$, respectively). Use Figure 7.1 for numerical results (or write your own program).

7.5. Orientation, Notes and References

Orientation. This chapter presents periodic matrix functions and the dynamical systems generated by them: periodic linear difference equations and periodic linear differential equations. They can be treated using Floquet theory, which is based on an analysis of principal fundamental solutions which allow us to reduce the analysis to autonomous equations. The main result is that the Lyapunov exponents (coinciding with the Floquet exponents) exist as limits and that they lead to decompositions of the state space $\mathbb{R}^d$ into linear subspaces. This decomposition changes in a periodic way. In other words, for every point in the base space (given by $\mathbb{Z}_p$, that is, $\mathbb{Z}$ mod p in the discrete-time case, and by $\mathbb{S}^1$ identified with $\mathbb{R}$ mod T in the continuous-time case) there is such a decomposition invariant under the flow and the dimensions of the corresponding subspaces are independent of the base points. The discussion of the classical Mathieu differential equation has illustrated that definite stability results for specific equations and parameters may require considerable effort in addition to the general theory.

In Chapter 9 we will give up the periodicity requirement and analyze Lyapunov decompositions over the base space. Here the topological considerations in Chapter 4 are taken up: chain transitivity in projective space will play a decisive role. The results in Chapter 3 on chain components will have to be extended by introducing Morse decompositions and attractors. Since these notions have abstract topological flows on compact metric spaces as their appropriate framework, we will develop their theory in Chapter 8 in this setting. The results will be used in Chapter 9 for linear flows.

Notes and references. Our exposition of Floquet theory for linear difference equations in Section 7.1 is partly based on Elaydi [43]. Exercise 7.4.5 is taken from [43, Example 4.12]. It seems that Floquet theory for linear difference equations is much less popular than for differential equations, although the constructions are quite analogous and, in fact, much simpler since the principal fundamental solution is given explicitly. The development in Section 7.1 made essential use of the assumption that the matrices $A(n)$

and hence the principal fundamental solution $X(p) = \prod_{j=1}^{p} A(p-j)$ are invertible. If this is not the case, one must treat the generalized eigenspace for the eigenvalue 0 of the matrix $X(p)$ separately; cf. Hilger [**66**].

Further details supporting our discussion in Section 7.2 and additional results can be found in Amann [**4**], Guckenheimer and Holmes [**60**], Hahn [**61**], and Wiggins [**140**]. Partly, we follow the careful exposition in Chicone [**26**, Section 2.4]. In particular, the proof of Lemma 7.2.4 is taken from [**26**, Theorem 2.47]; see also Amann [**4**, Lemma 20.7]. Meyer and Hall [**104**, Chapter III] present a detailed discussion of linear Hamiltonian differential equations including Floquet theory, as discussed in Example 7.2.17.

If the entries of a periodic matrix function change slowly enough, one may expect that the eigenvalues of the individual (constant) matrices still determine the stability behavior. This is in fact true; cf. Hahn [**61**, Section 62]. If the entries of $A(t)$ defining the right-hand side of a linear differential equation has two or more different periods, the matrix function is called quasi-periodic. In this case, one may identify the base space with a k-torus $\mathbb{T}^k$, instead of the unit circle $\mathbb{S}^1 = \mathbb{T}^1$ as in the periodic case. More generally, the entries of the matrix function may be almost periodic functions, which can be interpreted as the combination of countably many periods. A condensed survey of the linear skew product systems for linear differential equations with almost periodic coefficients is included in Fabbri, Johnson and Zampogni [**45**].

The discussion of Hill's and Mathieu's equation in Section 7.3 is classical and has a long history. Hill's paper [**67**] (on the perigee of the moon, i.e., the point on its orbit which is closest to earth) appeared in 1886. In order to deal with the infinite-dimensional linear equations for the Fourier coefficients, one can define determinants of infinite matrices and study convergence for finite-dimensional submatrices. This has generated a huge body of literature, probably starting with Poincaré [**113**]. We refer the reader to Magnus and Winkler [**98**] and Gohberg, Goldberg and Krupnik [**52**]; see also Mennicken [**102**]. A classical reference for numerically computed stability diagrams is Stoker [**131**, Chapters VI.3 and 4]. In our discussion of the Mathieu equation, we also found the lecture notes [**138**] by Michael Ward helpful.

Chapter 8

Morse Decompositions of Dynamical Systems

In this short chapter we come back to the global theory of general continuous flows on compact metric spaces, as already considered in Chapter 3. The purpose is to prepare the analysis of linear flows in Chapter 9 within a topological framework. For this endeavor, we introduce the notions of Morse decompositions and attractors of continuous flows on compact metric spaces and relate them to chain transitivity discussed in Chapter 3. We will show that Morse decompositions can be constructed by sequences of attractors and that the finest Morse decomposition, if it exists, yields the chain components. Recall that Chapter 4 characterized the Lyapunov spaces of autonomous linear differential equations as those subspaces which, when projected to projective space, coincide with the chain components of the induced flow. We will aim at a similar characterization for general linear flows, which needs the additional material in the present chapter. The theory will be developed for the continuous-time case only and also our applications in Chapter 9 will be confined to this case.

Section 8.1 introduces isolated invariant sets and Morse decompositions. Section 8.2 shows that Morse decompositions correspond to sequences of attractor-repeller pairs and Section 8.3 establishes the relation between the finest Morse decomposition, attractor-repeller sequences and the chain components.

8.1. Morse Decompositions

This section introduces Morse decompositions and explains some of their properties. The basic idea is to describe the global structure of a flow by specifying invariant subsets of the state space and an order between them capturing the limit behavior of the flow, forward and backward in time. Here one may start with a very rough picture, and then more and more details of the dynamics are revealed by refining the picture.

First we need the definition of isolated invariant sets; this will allow us to separate invariant sets.

Definition 8.1.1. For a flow Φ on a compact metric space X, a compact subset $K \subset X$ is called invariant if $\Phi(t, x) \in K$ for all $x \in K$ and all $t \in \mathbb{R}$. It is called isolated invariant if it is invariant and there exists a neighborhood N of K, i.e., a set N with $K \subset \text{int } N$, such that $\Phi(t, x) \in N$ for all $t \in \mathbb{R}$ implies $x \in K$.

The next example illustrates the difference between invariant sets and isolated invariant sets.

Example 8.1.2. Consider on the interval $[0, 1] \subset \mathbb{R}$ the ordinary differential equation
$$\dot{x} = \begin{cases} x^2 \sin(\frac{\pi}{x}) & \text{for } x \in (0, 1], \\ 0 & \text{for } x = 0. \end{cases}$$
Then the points $x_0 = 0$ and $x_n = \frac{1}{n}, n \geq 1$, are equilibria, since $\sin(\frac{\pi}{x_n}) = \sin(n\pi) = 0$ for $n \geq 1$. Hence every set $\{x_n\}, n \in \mathbb{N}_0$, is invariant for the associated flow. These invariant sets are isolated for $n \geq 1$, while the set $\{0\}$ is not isolated: Every neighborhood N of $\{0\}$ contains an entire trajectory.

Definition 8.1.3. A Morse decomposition of a flow Φ on a compact metric space X is a finite collection $\{\mathcal{M}_i \mid i = 1, \ldots, \ell\}$ of nonvoid, pairwise disjoint, and compact isolated invariant sets such that

(i) for all $x \in X$ the limit sets satisfy $\omega(x), \alpha(x) \subset \bigcup_{i=1}^{\ell} \mathcal{M}_i$;

(ii) suppose there are $\mathcal{M}_{j_0}, \mathcal{M}_{j_1}, \ldots, \mathcal{M}_{j_n}$ and $x_1, \ldots, x_n \in X \setminus \bigcup_{i=1}^{\ell} \mathcal{M}_i$ with $\alpha(x_i) \subset \mathcal{M}_{j_{i-1}}$ and $\omega(x_i) \subset \mathcal{M}_{j_i}$ for $i = 1, \ldots, n$; then $\mathcal{M}_{j_0} \neq \mathcal{M}_{j_n}$.

The elements of a Morse decomposition are called Morse sets.

Thus the Morse sets contain all limit sets and "cycles" (sometimes called "homoclinic structures") are not allowed. We notice the preservation of the concept of Morse decompositions under conjugacies between dynamical systems.

Proposition 8.1.4. *(i) Topological conjugacies on a compact metric space X map Morse decompositions onto Morse decompositions.*

(ii) For a compact invariant subset $Y \subset X$, a Morse decomposition $\{\mathcal{M}_i \mid i = 1, \ldots, \ell\}$ in X yields a Morse decomposition in Y given by
$$\{\mathcal{M}_i \cap Y \mid i = 1, \ldots, \ell\},$$
where only those indices i with $\mathcal{M}_i \cap Y \neq \varnothing$ are considered.

8.1. Morse Decompositions

Proof. Assertion (i) follows from Proposition 3.1.15(i) which shows the preservation of α- and ω-limit sets. Assertion (ii) is immediate from the definitions. $\square$

Since we consider conjugate flows as essentially 'the same', Proposition 8.1.4(i) means that Morse decompositions are intrinsic properties of flows.

The next proposition discusses the flow between Morse sets and gives a first hint that Morse decompositions actually may characterize the global behavior of a dynamical system. Recall that an order on a set A is a reflexive, antisymmetric and transitive relation $\preceq$, i.e., (i) $a \preceq a$, (ii) $a \preceq b$ and $b \preceq a$ is equivalent to $a = b$, and (ii) $a \preceq b$ and $b \preceq c$ implies $a \preceq c$.

Proposition 8.1.5. *For a Morse decomposition* $\{\mathcal{M}_i \mid i = 1, \ldots, \ell\}$ *an order is defined by the relation* $\mathcal{M}_i \preceq \mathcal{M}_j$ *if there are indices* $j_0, \ldots, j_n$ *with* $\mathcal{M}_i = \mathcal{M}_{j_0}, \mathcal{M}_j = \mathcal{M}_{j_n}$ *and points* $x_{j_i} \in X$ *with*

$$\alpha(x_{j_i}) \subset \mathcal{M}_{j_i-1} \text{ and } \omega(x_{j_i}) \subset \mathcal{M}_{j_i} \text{ for } i = 1, \ldots, n.$$

Proof. Reflexivity and transitivity are immediate. If $\mathcal{M}_i \preceq \mathcal{M}_j$ and $\mathcal{M}_j \preceq \mathcal{M}_i$, then $\mathcal{M}_i = \mathcal{M}_j$ follows from property (ii) of Morse decompositions. $\square$

This result says that the flow of a dynamical system goes from a lesser (with respect to the order $\preceq$) Morse set to a greater Morse set for trajectories that do not start in one of the Morse sets.

We enumerate the Morse sets in such a way that $\mathcal{M}_i \preceq \mathcal{M}_j$ implies $i \leq j$. Then Morse decompositions describe the flow via its movement from Morse sets with lower indices toward those with higher ones. But note that $i < j$ does not imply $\mathcal{M}_i \preceq \mathcal{M}_j$ and, in particular, it does not imply the existence of $x \in X$ with $\alpha(x) \subset \mathcal{M}_i$ and $\omega(x) \subset \mathcal{M}_j$.

A Morse decomposition $\{\mathcal{M}_1, \ldots, \mathcal{M}_\ell\}$ is called finer than a Morse decomposition $\{\mathcal{M}'_1, \ldots, \mathcal{M}'_{\ell'}\}$, if for all $j \in \{1, \ldots, \ell'\}$ there is $i \in \{1, \ldots, \ell\}$ such that $\mathcal{M}_i \subset \mathcal{M}'_j$. A finest Morse decomposition has the property that any finer Morse decomposition is the same. The intersection of two (or finitely many) Morse decompositions $\{\mathcal{M}_1, \ldots, \mathcal{M}_\ell\}$ and $\{\mathcal{M}'_1, \ldots, \mathcal{M}'_{\ell'}\}$ defines a Morse decomposition

$$\{\mathcal{M}_i \cap \mathcal{M}'_j \mid i, j\},$$

where only the indices $i \in \{1, \ldots, \ell\}, j \in \{1, \ldots, \ell'\}$ with $\mathcal{M}_i \cap \mathcal{M}'_j \neq \varnothing$ are allowed. Note that, in general, intersections of infinitely many Morse decompositions do not define a Morse decomposition. In particular, there need not exist a finest Morse decomposition; cf. Example 8.1.8. If there exists a finest Morse decomposition, it is unique.

For one-dimensional systems the Morse decompositions are easy to describe.

Proposition 8.1.6. *Let $\varphi : \mathbb{R} \times X \to X$ be a continuous flow on a compact interval $X = [a, b] \subset \mathbb{R}$, generated by a differential equation $\dot{x} = f(x)$ with a Lipschitz continuous map $f : X \to \mathbb{R}$ satisfying $f(a) = f(b) = 0$. Then any Morse set consists of fixed points and intervals between them. The finest Morse decomposition (if it exists) consists of single fixed points or intervals of fixed points.*

Proof. Let $x \in [a, b]$. Then one of the following three cases hold: (i) $f(x) > 0$, (ii) $f(x) = 0$, or (iii) $f(x) < 0$. If $f(x) = 0$, then the ω-limit set $\omega(x) = \{x\}$ is contained in a Morse set. If $f(x) > 0$, then $f(y) > 0$ for all y in a neighborhood of x. Hence $x < z \leq b$ for every $z \in \omega(x)$. But $\omega(x)$ cannot consists of more than one point, and hence is an equilibrium. Similarly, one argues for $\alpha(x)$. Concluding, one sees that either x is an equilibrium and hence in a Morse set, or it is in an interval between equilibria, either contained in a Morse set or not. □

The following one-dimensional examples illustrate the concept of Morse decompositions.

Example 8.1.7. Consider the dynamical system discussed in Example 3.1.4 which is generated by

$$\dot{x} = x(x-1)(x-2)^2(x-3)$$

on the compact interval $X := [0, 3]$. This flow has, e.g., the following Morse decompositions

$$\mathcal{M}_1 := \{0\} \preceq \mathcal{M}_2 := [1, 3],$$
$$\mathcal{M}_1 := \{0\} \preceq \mathcal{M}_3 := \{1\} \succeq \mathcal{M}_2 := [2, 3],$$
$$\mathcal{M}_1 := \{0\} \preceq \mathcal{M}_3 := [1, 2] \succeq \mathcal{M}_2 := \{3\},$$
$$\mathcal{M}_1 := \{0\} \cup [2, 3] \preceq \mathcal{M}_1 := \{1\}.$$

It has the finest Morse decomposition $\{0\} \preceq \{1\} \succeq \{2\} \succeq \{3\}$. This also illustrates that the order is not linear (not all pairs of Morse sets are related) and hence the numbering of the Morse sets according to their order is not unique. In particular, minimal (and maximal) Morse sets need not be unique.

Example 8.1.8. Consider again the dynamical system defined in Example 8.1.2. For every $n \in \mathbb{N}$ the two sets

$$\mathcal{M}_1^n := \left\{\frac{1}{2n}\right\}, \mathcal{M}_2^n := \left[0, \frac{1}{2n+1}\right] \cup \left[\frac{1}{2n-1}, 1\right]$$

form a Morse decomposition of the associated flow. Note that the intersection of all these Morse decompositions which is the family of sets

$\{\{0\}, \{\frac{1}{n}\}$ for $n \in \mathbb{N}\}$ is not a Morse decomposition, since Morse decompositions are finite. This system does not have a finest Morse decomposition, since the countably many sets $\{\frac{1}{n}\}$ for $n \in \mathbb{N}$ would have to be included as Morse sets.

Example 8.1.9. Consider the dynamical system defined in Example 3.2.2 given by $\dot{x} = \sin^2 x$ on $X = \mathbb{S}^1$, which we parametrize by $[0, 2\pi)$; thus 2π is identified with 0. The only Morse decomposition is the trivial one $\mathcal{M} = \{\mathbb{S}^1\}$. In fact, property (i) of Morse decompositions is trivially satisfied and

$$\omega(x) = \begin{cases} \{\pi\} & \text{for } x \in (0, \pi], \\ \{0\} & \text{for } x \in (\pi, 0], \end{cases} \text{ and } \alpha(x) = \begin{cases} \{0\} & \text{for } x \in [0, \pi), \\ \{\pi\} & \text{for } x \in [\pi, 0). \end{cases}$$

Hence there are Morse sets $\mathcal{M}$ and $\mathcal{M}'$, with $0 \in \mathcal{M}$ and $\pi \in \mathcal{M}'$. By the no-cycle condition (ii) of Definition 8.1.3 the points 0 and π are in the same Morse set, hence $\mathcal{M} = \mathcal{M}'$. Then the no-cycle condition again (now with $n = 0$) implies that $\mathcal{M} = \mathbb{S}^1$.

8.2. Attractors

In this section, attractors and complementary repellers are defined and it is shown that Morse decompositions can be constructed from sequences of attractors and their complementary repellers. While the term 'attractor' has an intuitive appeal, there are many ways in the mathematical literature to make this idea precise. The notion employed here which is based on ω-limit sets of neighborhoods (recall the definition of such limit sets in Definition 3.1.1) will be given first.

Definition 8.2.1. For a flow Φ on a compact metric space X a compact invariant set A is an attractor if it admits a neighborhood N such that $\omega(N) = A$. A repeller is a compact invariant set R that has a neighborhood N^* with $\alpha(N^*) = R$.

We also allow the empty set $\varnothing$ as an attractor and a repeller. A neighborhood N as in Definition 8.2.1 is called an attractor neighborhood and N^* is called a repeller neighborhood. Every attractor is compact and invariant, and a repeller is an attractor for the time-reversed flow. Furthermore, if A is an attractor in X and $Y \subset X$ is a compact invariant set, then $A \cap Y$ is an attractor for the flow restricted to Y.

Example 8.2.2. Consider again the dynamical system discussed in Examples 3.1.4 and 8.1.7. This system has, besides the empty set and the entire space $[0, 3]$, three attractors, namely $\{1\}$, $[1, 2]$, and $[1, 3]$. The fact that these sets are indeed attractors follows directly from the determination of the limit sets in Example 3.1.4. To see that there are no other attractors one argues as in Example 3.1.4. Similarly, the nontrivial repellers of this system are seen to be $\{0\}$, $[2,3]$, $\{3\}$, $\{0\} \cup [2,3]$, and $\{0\} \cup \{3\}$.

Example 8.2.3. Consider again the system discussed in Examples 3.2.2 and 8.1.9 on $\mathbb{S}^1$, given by $\dot{x} = \sin^2 x$. For this flow, the only attractors are $\varnothing$ and $\mathbb{S}^1$: Let $A \subset \mathbb{S}^1$ be an attractor with $\omega(N) = A$ for a neighborhood $N(A)$. For each point $x \in \mathbb{S}^1$ the limit set $\omega(x)$ contains at least one of the two fixed points 0 or π, which implies that each attractor has to contain at least one of the fixed points. Consider the point π and let $N(\pi)$ be any neighborhood. We have $[\pi, 2\pi] \subset \omega(N) \subset A$. Repeating this argument for the fixed point 0, we see that $[0, \pi] \subset A$, and hence $A = \mathbb{S}^1$.

We write
$$\Phi(I, Y) := \{\Phi(t, x) \mid t \in I \text{ and } x \in Y\} \text{ for } I \subset \mathbb{R} \text{ and } Y \subset X.$$

The following lemma shines new light on attractor neighborhoods.

Lemma 8.2.4. *For every attractor neighborhood N of an attractor A there is a time $t^* > 0$ with $\mathrm{cl}(\Phi([t^*, \infty), N) \subset \mathrm{int}\, N$.*

Proof. We may assume that N is closed. First we show that there is $t^* > 0$ with $\Phi([t^*, \infty), N) \subset \mathrm{int}\, N$. Otherwise, there are (t_n, x_n) with $t_n \to \infty$ and $x_n \in N$ such that $\Phi(t_n, x_n) \notin \mathrm{int}\, N$. This contradicts the assumption that N is an attractor neighborhood implying $\lim_{n \to \infty} \Phi(t_n, x_n) \in A$.

In order to show that $\mathrm{cl}(\Phi([t^*, \infty), N) \subset \mathrm{int}\, N$ note that the same argument shows that for every converging sequence $\Phi(t_n, x_n) \in \Phi([t^*, \infty), N)$ with $t_n \to \infty$ it follows that $\lim_{n \to \infty} \Phi(t_n, x_n) \in \mathrm{int}\, N$. If $\Phi(t_n, x_n) \to z$ with a bounded sequence (t_n), then a subsequence converges to $z = \Phi(t, x) \in \mathrm{int}\, N$. $\square$

The following lemma shows that every attractor comes with a repeller.

Lemma 8.2.5. *For an attractor A, the set $A^* = \{x \in X \mid \omega(x) \cap A = \varnothing\}$ is a repeller, called the complementary repeller of A, and (A, A^*) is called an attractor-repeller pair.*

Proof. Let N be a compact attractor neighborhood of A. Choose $t^* > 0$ such that $\mathrm{cl}(\Phi([t^*, \infty), N) \subset N$ and define an open set V by
$$V = X \setminus \mathrm{cl}(\Phi([t^*, \infty), N).$$
Then $X = N \cup V$. Furthermore $\Phi((-\infty, -t^*], V) \subset X \setminus N$ and therefore V is a neighborhood of $\alpha(V) \subset X \setminus N \subset V$. Hence $\alpha(V)$ is a repeller with repeller neighborhood V and by invariance $\alpha(V) \subset A^*$. For the converse inclusion, note that $x \in A^*$ implies $x \notin N$, thus $x \in V$. Furthermore, $\omega(x) \cap A = \varnothing$ implies for all t that $\omega(\Phi(t, x)) \cap A = \varnothing$, hence $\Phi(t, x) \in A^* \subset V$. Thus $x = \Phi(-t, \Phi(t, x))$ for all $t \geq 0$ and it follows that $x \in \alpha(V)$. $\square$

8.2. Attractors

Note that an attractor A and its complementary repeller A^* are disjoint. There are always the trivial attractor-repeller pairs $A = X, A^* = \varnothing$ and $A = \varnothing, A^* = X$.

Example 8.2.6. Consider again the dynamical system discussed in Examples 3.1.4, 8.1.7 and 8.2.2. The nontrivial attractor-repeller pairs of this system are $A_1 = \{1\}$ with $A_1^* = \{0\} \cup [2,3]$, $A_2 = [1,2]$ with $A_2^* = \{0\} \cup \{3\}$, and $A_3 = [1,3]$ with $A_3^* = \{0\}$.

As noted above, a repeller A^* is an attractor of the time-reversed system. A consequence of the following proposition is, in particular, that for an attractor-repeller pair (A, A^*) the complementary repeller of the attractor A^* of the time-reversed system is A.

Proposition 8.2.7. *If (A, A^*) is an attractor-repeller pair and $x \notin A \cup A^*$, then $\alpha(x) \subset A^*$ and $\omega(x) \subset A$.*

Proof. By definition of A^* it follows that $\omega(x) \cap A \neq \varnothing$. Thus there is $t_0 > 0$ with $\Phi(t_0, x) \in N$, where N is a neighborhood of the attractor A with $\omega(N) = A$, and hence $\omega(x) \subset A$. Now suppose that there is $y \in \alpha(x) \setminus A^*$. Thus by definition of A^* one has $\omega(y) \cap A \neq \varnothing$. Using $\omega(y) \subset \alpha(x)$ one finds that there are $t_n \to \infty$ with $\Phi(-t_n, x) \to A$, and thus for n large enough, $\Phi(-t_n, x) \in N$. Clearly, $\Phi(t_n, \Phi(-t_n, x)) \to x$ and hence $\omega(N) = A$ implies that $x \in A$, contradicting the choice of x. Thus $\alpha(x) \subset A^*$. $\square$

Trajectories starting in a neighborhood of an attractor leave the neighborhood in backwards time.

Lemma 8.2.8. *A compact invariant set A is an attractor if and only if there exists a compact neighborhood N of A such that $\Phi((-\infty, 0], x) \not\subset N$ for all $x \in N \setminus A$.*

Proof. The necessity of the condition is clear because $\Phi((-\infty, 0], x) \subset N$ implies $x \in \omega(N)$. Conversely, let N be a compact neighborhood of A such that $\Phi((-\infty, 0], x) \not\subset N$ for all $x \in N \setminus A$. Thus there exists a $t^* > 0$ such that $\Phi([-t^*, 0], x) \not\subset N$ for all x in the compact set $N \cap \mathrm{cl}(X \setminus N)$. Since A is a compact invariant set and Φ is continuous, we can choose a neighborhood V of A such that $\Phi([0, t^*], V) \subset N$. Then $\Phi([0, \infty), V) \subset N$ implying that $\omega(V) = A$ and A is an attractor. In fact, if $\Phi([0, \infty), V) \not\subset N$, there are $t > 0$ and $y \in V$ such that $\Phi(t, y) \notin N$. It follows that $t > t^*$ and hence
$$t^{**} := \inf\{t > 0 \mid \Phi(t, y) \notin N\} \geq t^*.$$
Then $x := \Phi(t^{**}, y) \in N \cap \mathrm{cl}(X \setminus N)$ and $\Phi([-t, 0], x) \subset N$ for all $t \in [0, t^*]$. This contradicts $\Phi([-t^*, 0], x) \not\subset N$. $\square$

This implies the following characterization of attractor-repeller pairs.

Lemma 8.2.9. *A pair (A, A^*) of disjoint compact invariant sets is an attractor-repeller pair if and only if (i) $x \in X \setminus A^*$ implies $\Phi([0,\infty), x) \cap N \neq \varnothing$ for every neighborhood N of A, and (ii) $x \in X \setminus A$ implies $\Phi((-\infty, 0], x) \cap N^* \neq \varnothing$ for every neighborhood N^* of A^*.*

Proof. Certainly, these conditions are necessary. Conversely, suppose that (i) holds and let W be a compact neighborhood of A with $W \cap A^* = \varnothing$. Then (ii) implies that $\Phi((-\infty, 0], x) \not\subset W$ for all $x \in W \setminus A$. By Lemma 8.2.8 this implies that A is an attractor. Moreover, it follows from (i) that $\omega(x) \cap A \neq \varnothing$ for all $x \in X \setminus A^*$. Hence $A^* = \{x \in X \mid \omega(x) \cap A = \varnothing\}$ is the complementary repeller of A. □

The following result characterizes Morse decompositions via attractor-repeller sequences. It is the main result on the relation between these notions.

Theorem 8.2.10. *For a flow Φ on a compact metric space X a finite collection of subsets $\{\mathcal{M}_1, \ldots, \mathcal{M}_n\}$ defines a Morse decomposition if and only if there is a strictly increasing sequence of attractors*

$$\varnothing = A_0 \subset A_1 \subset A_2 \subset \ldots \subset A_n = X$$

such that

(8.2.1) $\qquad \mathcal{M}_{n-i} = A_{i+1} \cap A_i^* \text{ for } 0 \leq i \leq n-1.$

Proof. (i) Suppose that $\{\mathcal{M}_1, \ldots, \mathcal{M}_n\}$ is a Morse decomposition. Define a strictly increasing sequence of invariant sets by $A_0 := \varnothing$ and

$$A_k := \{x \in X \mid \alpha(x) \subset \mathcal{M}_n \cup \ldots \cup \mathcal{M}_{n-k+1}\} \text{ for } k = 1, \ldots, n.$$

First we show that the sets A_k are closed. Since for every $x \in X$ there is j with $\alpha(x) \subset \mathcal{M}_j$, it follows that $A_n = X$, and hence A_n is closed. Proceeding by induction, assume that A_{k+1} is closed and consider $x_i \in A_k$ with $x_i \to x$. We have to show that $\alpha(x) \subset \mathcal{M}_n \cup \ldots \cup \mathcal{M}_{n-k+1}$. The induction hypothesis implies that $x \in A_{k+1}$ and hence we have $\alpha(x) \subset \mathcal{M}_n \cup \ldots \cup \mathcal{M}_{n-k+1} \cup \mathcal{M}_{n-k}$. Thus either the assertion holds or $\alpha(x) \subset \mathcal{M}_{n-k}$.

In order to see that the latter case cannot occur, we will proceed by contradiction and assume that $\alpha(x) \subset \mathcal{M}_{n-k}$. Let V be an open neighborhood of $\mathcal{M}_{n-k}$ such that $V \cap \mathcal{M}_j = \varnothing$ for $j \neq n-k$. There are a sequence $t_\nu \to \infty$ and $z \in \mathcal{M}_{n-k}$ such that $\Phi(-t_\nu, x) \in V$ and $d(\Phi(-t_\nu, x), z) \leq \nu^{-1}$ for all $\nu \geq 1$. Hence for every ν there is a $m_\nu \geq \nu$ such that $\Phi(-t_\nu, x_{m_\nu}) \in V$ and $d(\Phi(-t_\nu, x_{m_\nu}), z) \leq 2\nu^{-1}$. Because $\alpha(x_i) \subset \mathcal{M}_n \cup \ldots \cup \mathcal{M}_{n-k+1}$ for all i, there are $\tau_\nu < t_\nu < \sigma_\nu$ such that $\Phi(-\sigma_\nu, x_{m_\nu})$ and $\Phi(-\tau_\nu, x_{m_\nu}) \in \partial V$ and $\Phi(-t, x_{m_\nu}) \in \text{cl } V$ for all $t \in [\tau_\nu, \sigma_\nu]$. Invariance of $\mathcal{M}_{n-k}$ implies that $t_\nu - \tau_\nu \to \infty$ as $\nu \to \infty$. We may assume that there is $y \in \partial V$ with $\Phi(-\sigma_\nu, x_{m_\nu}) \to y$ for $\nu \to \infty$. Then it follows that $\Phi([0, \infty), y) \subset \text{cl } V$ and hence by the choice of V one has $\omega(y) \subset \mathcal{M}_{n-k}$. Because A_{k+1} is closed and invariant, we have $y \in A_{k+1}$ and so $\alpha(y) \subset \mathcal{M}_n \cup \ldots \cup \mathcal{M}_{n-k}$. The ordering of the Morse sets implies that $y \in \mathcal{M}_{n-k}$, contradicting $y \in \partial V$.

Now assume that A_k is not an attractor. Then Lemma 8.2.8 implies that for every neighborhood N of A_k there is $x \in N \setminus A_k$ with $\Phi((-\infty, 0], x) \subset N$. Then there is $j \geq n - k + 1$ with $\alpha(x) \subset \mathcal{M}_j$. On the other hand, $x \notin A_k$ implies $\alpha(x) \not\subset \mathcal{M}_n \cup \ldots \cup \mathcal{M}_{n-k+1}$, hence $\alpha(x) \in \mathcal{M}_i$ for some $i < n-k+1$. This contradiction implies that A_k is an attractor.

It remains to show that $\mathcal{M}_{n-i} = A_{i+1} \cap A_i^*$. Clearly, $\mathcal{M}_{n-i} \subset A_{i+1}$. Suppose that $x \in \mathcal{M}_{n-i} \setminus A_i^*$. Then $\omega(x) \subset A_i$ and therefore $\omega(x) \subset \mathcal{M}_j$ for some $j \geq n - i + 1$. This contradiction proves $\mathcal{M}_{n-i} \subset A_{i+1} \cap A_i^*$. If conversely, $x \in A_{i+1} \cap A_i^*$, then $\alpha(x) \subset \mathcal{M}_n \cup \ldots \cup \mathcal{M}_{n-i}$. From $x \in A_i^*$ we conclude
$$\omega(x) \cap (\mathcal{M}_n \cup \ldots \cup \mathcal{M}_{n-i+1}) \subset \omega(x) \cap A_i = \varnothing$$
and hence $\omega(x) \subset \mathcal{M}_1 \cup \ldots \cup \mathcal{M}_{n-i}$. Now the definition of a Morse decomposition implies $x \in \mathcal{M}_{n-i}$.

(ii) Conversely, let the sets $\mathcal{M}_j$, $i = 1, \ldots, n$, be defined by an increasing sequence of attractors as indicated in (8.2.1). Clearly these sets are compact and invariant. If $i < j$, then $\mathcal{M}_{n-i} \cap \mathcal{M}_{n-j} = A_{i+1} \cap A_i^* \cap A_{j+1} \cap A_j^* = A_{i+1} \cap A_j^* \subset A_j \cap A_j^* = \varnothing$; hence the sets $\mathcal{M}_i$ are pairwise disjoint.

We claim that for $x \in X$ either $\Phi(\mathbb{R}, x) \subset \mathcal{M}_j$ for some j or else there are indices $i \leq j$ such that $\alpha(x) \subset \mathcal{M}_{n-j}$ and $\omega(x) \subset \mathcal{M}_{n-i+1}$. In fact, there is a smallest integer i such that $\omega(x) \subset A_i$, and there is a largest integer j such that $\alpha(x) \subset A_j^*$. Clearly, $i > 0$ and $j < n$. Now $\omega(x) \not\subset A_{i-1}$, i.e., $x \in A_{i-1}^*$. Thus by invariance $\Phi(\mathbb{R}, x) \subset A_{i-1}^*$ and $\omega(x) \subset A_{i-1}^*$. On the other hand, $\alpha(x) \not\subset A_{j+1}^*$. If $\Phi(t, x) \notin A_{j+1}$ for some $t \in \mathbb{R}$, then by Proposition 8.2.7 $\alpha(x) \subset A_{j+1}^*$, a contradiction. Hence it follows that $\Phi(\mathbb{R}, x) \subset A_{j+1}$. Now $j \geq i - 1$, because otherwise $j + 1 \leq i - 1$ and thus $A_{j+1} \subset A_{i-1}$, which implies $\Phi(\mathbb{R}, x) \subset A_{i-1}^* \cap A_{i-1} = \varnothing$. If $j = i - 1$, then $\Phi(\mathbb{R}, x) \subset A_{i-1}^* \cap A_i = \mathcal{M}_{n-i-1}$. If $j > i - 1$, then $\mathcal{M}_{n-i+1} \neq \mathcal{M}_{n-j}$ and we know $\omega(x) \subset A_{i-1}^* \cap A_i = \mathcal{M}_{n-i+1}$ and $\alpha(x) \subset A_j^* \cap A_{j+1} = \mathcal{M}_{n-j}$. This proves the claim. The claim also shows that the sets $\mathcal{M}_j$ are isolated invariant and cycles cannot occur, since, as seen above, one always has $\omega(x) \subset \mathcal{M}_{n-i+1}$ and $\alpha(x) \subset \mathcal{M}_{n-j}$ with $n - i + 1 \geq n - j$. □

One obtains more information on the minimal and the maximal element in the order of Morse sets.

Corollary 8.2.11. *For a given Morse decomposition, every maximal Morse set (with respect to the order of Morse sets) is an attractor, and every minimal Morse set is a repeller.*

Proof. Let $\mathcal{M}$ be a maximal Morse set of a Morse decomposition. Then we can number the sets in the Morse decomposition such that it is given by $\{\mathcal{M}_1, \ldots, \mathcal{M}_n\}$ with $\mathcal{M}_n = \mathcal{M}$. By Theorem 8.2.10, there is an attractor sequence with $\mathcal{M} = \mathcal{M}_n = A_1 \cap A_0^* = A_1$ since $A_0 = \varnothing$. Analogously, one sees the minimal Morse set is $\mathcal{M}_1 = A_n \cap A_{n-1}^* = A_{n-1}^*$. □

Example 8.2.12. We illustrate Theorem 8.2.10 and Corollary 8.2.11 by looking again at Example 8.1.7. For this system a strictly increasing sequence of attractors with their corresponding repellers is

$$A_0 = \varnothing \subset A_1 = \{1\} \subset A_2 = [1,2] \subset A_3 = [1,3] \subset A_4 = [0,3],$$
$$A_0^* = [0,3] \supset A_1^* = \{0\} \cup [2,3] \supset A_2^* = \{0\} \cup \{3\} \supset A_3^* = \{0\} \supset A_4^* = \varnothing.$$

The associated Morse decomposition is

$$\mathcal{M}_4 = A_1 \cap A_0^* = \{1\}, \; \mathcal{M}_3 = A_2 \cap A_1^* = \{2\},$$
$$\mathcal{M}_2 = A_3 \cap A_2^* = \{3\}, \; \mathcal{M}_1 = A_4 \cap A_3^* = \{0\}.$$

Note that $\mathcal{M}_4$ is an attractor and $\mathcal{M}_1$ is a repeller.

8.3. Morse Decompositions, Attractors, and Chain Transitivity

The previous section has characterized Morse decompositions via sequences of attractor-repeller pairs. In the present section, we will make contact with the concepts and results from Chapter 3 based on chains. Thus we proceed to analyze the relation between finest Morse decompositions, chain recurrence and attractors leading to the main result of this chapter in Theorem 8.3.3.

Throughout this section, we consider a flow Φ on a compact metric space X. The following version of limit sets via chains is helpful. It generalizes the notion of ω-limit sets.

Definition 8.3.1. For $Y \subset X$ and $\varepsilon, T > 0$ let

$$\Omega(Y, \varepsilon, T) = \{z \in X \mid \text{there is an } (\varepsilon, T)\text{-chain from some } y \in Y \text{ to } z\},$$

and define the chain limit set of Y as

$$\Omega(Y) = \bigcap_{\varepsilon, T > 0} \Omega(Y, \varepsilon, T).$$

Note that $\Omega(Y) = \{z \in X \mid \text{for all } \varepsilon, T > 0 \text{ there are } y \in Y \text{ and an } (\varepsilon, T)\text{-chain from } y \text{ to } z\}$ and $\omega(Y) \subset \Omega(Y)$. For a one-point set $Y = \{x\}$, we also write $\Omega(x)$. Arguing as for $\omega(Y)$ in Proposition 3.1.5 one sees that $\Omega(Y)$ is invariant under the flow.

A first relation between chains and attractors is given by the following proposition.

8.3. Morse Decompositions, Attractors, and Chain Transitivity

Proposition 8.3.2. *For $Y \subset X$ the set $\Omega(Y)$ is the intersection of all attractors containing $\omega(Y)$.*

Proof. Suppose that A is any attractor containing $\omega(Y)$. Let V be an open neighborhood of A disjoint from A^* and let $t > 0$ be such that $\operatorname{cl} \Phi(t, V) \subset V$. Let
$$0 < \varepsilon < \inf \{d(y, z) \mid y \notin V \text{ and } z \in \operatorname{cl} \Phi(t, V)\}.$$
Choose $T > t$ such that $\Phi(T, Y) \subset \Phi(t, \operatorname{cl} V)$. Then every (ε, T)-chain from Y must end in V. Therefore, if $\omega(Y) \subset A$, then also $\Omega(Y) \subset A$ and hence $\Omega(Y)$ is contained in the intersection of all attractors containing $\omega(Y)$.

For $\varepsilon, T > 0$ let $N := \operatorname{cl}(\Omega(Y, \varepsilon, T))$. We will show that $A := \omega(N)$ is an attractor with attractor neighborhood N. Note that $\omega(N) \subset \Omega(Y, \varepsilon, T) \subset \operatorname{int} N$, where the second inclusion follows, because $\Omega(Y, \varepsilon, T)$ is open and contained in N. Now let $z \in \omega(N)$. Then there are $t_n \to \infty$ and $x_n \in N$ with $\Phi(t_n, x_n) \to z$. Choose $n_0 \in \mathbb{N}$, $\delta > 0$ and $p \in \Omega(Y, \varepsilon, T)$ with
$$d(p, x_{n_0}) < \delta, \ t_{n_0} > T, \text{ and } d(\Phi(t_{n_0}, x_{n_0}), z) < \frac{\varepsilon}{2}$$
and
$$d(\Phi(t_{n_0}, y), \Phi(t_{n_0}, x_{n_0})) < \frac{\varepsilon}{2} \text{ for all } y \text{ with } d(y, x_{n_0}) < \delta.$$
By definition of p there is an (ε, T)-chain from some $y \in Y$ to p and we obtain
$$d(\Phi(t_{n_0}, p), z) \leq d(\Phi(t_{n_0}, p), \Phi(t_{n_0}, x_{n_0})) + d(\Phi(t_{n_0}, x_{n_0}), z) < \frac{\varepsilon}{2} + \frac{\varepsilon}{2} = \varepsilon.$$
Thus concatenation yields an (ε, T)-chain from y to z, hence $z \in \Omega(Y, \varepsilon, T)$. We have shown that $A := \omega(N)$ is a closed invariant set with attractor neighborhood N, hence an attractor. By invariance of $\Omega(Y)$ we have $A = \omega(\operatorname{cl}(\Omega(Y, \varepsilon, T))) \supset \Omega(Y) \supset \omega(Y)$. Direct inspection shows that $\Omega(Y) = \omega(\Omega(Y))$ in fact equals the intersection of these attractors containing $\omega(Y)$. $\square$

This proposition implies, in particular, that a chain transitive flow has only the trivial attractors $A = \varnothing, X$ because for every $Y \subset X$ one has that $\Omega(Y) = X$. See also Examples 8.1.9 and 8.2.3 for a simple illustration.

We obtain the following fundamental relation between the chain recurrent set and attractors.

Theorem 8.3.3. *The chain recurrent set $\mathcal{R}$ satisfies*

(8.3.1) $$\mathcal{R} = \bigcap \{A \cup A^* \mid A \text{ is an attractor}\}.$$

In particular, there exists a finest Morse decomposition $\{\mathcal{M}_1, \ldots, \mathcal{M}_n\}$ if and only if the chain recurrent set $\mathcal{R}$ has only finitely many connected components. In this case, the Morse sets coincide with the chain components of

$\mathcal{R}$ and the flow restricted to every Morse set is chain transitive and chain recurrent.

Proof. If A is an attractor and $x \in X$ either $\omega(x) \subset A$ or $x \in A^*$. If $x \in \mathcal{R}$, then $x \in \Omega(x)$ and Proposition 8.3.2 shows that x is contained in every attractor containing $\omega(x)$. Hence $x \in A \cup A^*$. Conversely, if x is in the intersection in (8.3.1), then x is in every attractor A containing $\omega(x)$. Again Proposition 8.3.2 shows that $x \in \Omega(x)$, that is $x \in \mathcal{R}$.

If there exists a finest Morse decomposition, then the flow restricted to a corresponding Morse set must be chain transitive. In fact, suppose that a Morse set $\mathcal{M}_k$ is not chain transitive. Then there exists $x \in \mathcal{M}_k$ with $\Omega(x) \neq \mathcal{M}_k$. By Theorem 8.2.10 the Morse sets can be written via a sequence of attractors as

$$\mathcal{M}_{n-i} = A_{i+1} \cap A_i^* \text{ for } 0 \leq i \leq n-1.$$

By invariance of $\mathcal{M}_k$ one has $\omega(x) \subset \mathcal{M}_k$. Now Proposition 8.3.2 shows that there is an attractor $\hat{A}$ containing $\omega(x)$ and $\mathcal{M}_k \not\subset \hat{A}$. The attractor sequence $\varnothing \subset \hat{A} \subset X$ gives rise to the Morse decomposition

$$\hat{\mathcal{M}}_1 = X \cap \hat{A}^* = \hat{A}^*, \hat{\mathcal{M}}_2 = \hat{A} \cap X = \hat{A}.$$

This is a contradiction to the assumption that we started with the finest Morse decomposition, since $\hat{\mathcal{M}}_2$ is not contained in any Morse set $\mathcal{M}_j$. We have shown, that the Morse sets in the finest Morse decomposition are chain transitive and hence connected by Proposition 3.2.5. Thus every Morse set in the finest Morse decomposition is contained in a connected component of the chain recurrent set $\mathcal{R}$.

Conversely, the finitely many connected components $\mathcal{M}_i$ of $\mathcal{R}$ define a Morse decomposition, because they are ordered isolated invariant sets. In fact, this is the finest Morse decomposition: Using again the characterization of Morse decompositions via increasing attractor sequences in Theorem 8.2.10, one sees that a finer Morse decomposition would imply the existence of an attractor A such that $A \cap \mathcal{M}_i$ is a proper subset of $\mathcal{M}_i$ for some i, and hence this would be an attractor of the flow restricted to $\mathcal{M}_i$. This is a contradiction, since the flow restricted to a chain component $\mathcal{M}_i$ is by Proposition 3.2.4 chain transitive. □

8.4. Exercises

Exercise 8.4.1. Let $n \in \mathbb{N}$. Construct an example with a Morse decomposition such that there are n Morse sets which are maximal with respect to the order of Morse sets.

Exercise 8.4.2. In Example 8.1.7 several Morse decompositions for a system on the interval $X = [0, 3]$ have been given. Construct for each of them an increasing sequence of attractors as in Theorem 8.2.10. Compare with Example 8.2.12.

Exercise 8.4.3. Let $\mathcal{M}_1, \ldots, \mathcal{M}_\ell$ be a Morse decomposition. Show that for every $x \in X$ there is $i \in \{1, \ldots, \ell\}$ with $\omega(x) \subset \mathcal{M}_i$.

Exercise 8.4.4. Give an example where the chain components are not determined by a finest Morse decomposition.

Exercise 8.4.5. Let Φ be a flow on a compact metric space X. Assume that there exists a finest Morse decomposition. Prove that for every set $Y \subset X$ the chain limit set

$$\Omega(Y) = \left\{ z \in X \;\middle|\; \begin{array}{l} \text{for all } \varepsilon, T > 0 \text{ there are } y \in Y \\ \text{and an } (\varepsilon, T)\text{-chain from } y \text{ to } z \end{array} \right\}$$

is an attractor. Give an example where there is no finest Morse decomposition and $\Omega(Y)$ is not an attractor.

8.5. Orientation, Notes and References

Orientation. This chapter has shown that one may construct the chain components of a continuous flow on a compact metric space by finding the finest Morse decomposition, if it exists. This is the content of Theorem 8.3.3. Theorem 8.2.10 shows that Morse decompositions, in turn, are associated with attractor-repeller sequences. Hence, if one wants to construct the finest Morse decomposition, one may refine attractor-repeller sequences. We will follow this path in the next chapter, where we construct the chain components of the flow in the projective bundle associated to a continuous linear flow. This yields a generalization of the Lyapunov spaces for autonomous linear equations.

Notes and references. The theory of continuous flows on compact metric spaces based on chains, attractors and Morse decompositions is essentially due to C. Conley; cf. [**34, 35**].

Morse decompositions are closely related to the existence of a complete Lyapunov function; see Robinson [**117**] or Alongi and Nelson [**2**]. Thus (this is Conley's Fundamental Theorem) the flow outside of the chain recurrent set $\mathcal{R}$ is gradient-like in the sense that there is a Lyapunov function which is a strictly decreasing along trajectories outside $\mathcal{R}$ and constant on the components of $\mathcal{R}$. The important notion of Conley index associates to every Morse set an index which reflects the topological properties of the flow. Easton [**42**] develops the topological theory for continuous maps on compact metric spaces.

The characterization of Morse decompositions in Theorem 8.2.10 is sometimes taken as a definition; see Salamon [**120**] or Salamon and Zehnder [**121**]. There may exist countably many attractors; see Akin [**3**, Proposition 4.8] or Robinson [**117**, Lemma 10.1.7]. The intersection of all Morse decompositions for a flow need not consist of only countably many sets as in Example 8.1.8. It may form a Cantor set; see Akin [**3**, p. 25] (and use Theorems 8.2.10 and 8.3.3).

Rasmussen [**114**] developed a theory of Morse decompositions for nonautonomous systems (in continuous and discrete time). There are extensions of the theory of Morse decompositions and chain transitivity to infinite-dimensional systems (see Rybakowski [**118**, Definition III.1.5 and Theorem III.1.8]), and to semi-dynamical systems; Patrão and San Martin [**110, 111**].

Chapter 9

Topological Linear Flows

In this chapter we consider as a second class of time-varying matrices robust linear systems or, equivalently, bilinear control systems; cf. Example 6.3.4. They find an appropriate framework in the general class of topological linear flows or dynamical systems on vector bundles. The techniques from the topological theory of flows on compact metric spaces developed in Chapter 3 and complemented in Chapter 8 will play a dominant role, since they will allow us to construct first appropriate subspace decompositions and then a spectral theory for (generalized) Lyapunov exponents. In particular, we will construct the finest Morse decomposition for the induced system in projective space and characterize the Lyapunov spectrum. The theory will be developed for the continuous-time case only.

Section 9.1 presents the main results: Theorem 9.1.5 is a spectral decomposition theorem involving the notion of the Morse spectrum and Theorem 9.1.12 gives an application to stability. The ensuing sections develop the theory which is needed for the proof of Theorem 9.1.5: In Section 9.2 the finest Morse decomposition is constructed which yields a decomposition into subbundles; this result, Theorem 9.2.5, is due to Selgrade. The properties of the Morse spectrum are derived in Section 9.3, and the relation to the Lyapunov exponents is clarified in Section 9.4. Here the proof of Theorem 9.1.5 is also completed. Finally, Section 9.5 proves that robust linear systems (or, equivalently, bilinear control systems) define topological linear flows on an appropriate vector bundle and presents several examples.

9.1. The Spectral Decomposition Theorem

In this section we define topological linear flows which are continuous linear skew product flows over a compact metric space. Then we introduce the Morse spectrum and formulate the main result of this chapter, which furnishes a decomposition according to exponential growth rates. The result is compared to the spectral decomposition theorems derived for autonomous and periodic differential equations. An application to stability is given.

We start by defining topological linear dynamical systems. Their state space has the form $\mathcal{V} = B \times \mathbb{R}^d$ where B is a compact metric space and $\mathbb{R}^d$ is endowed with an inner product. One should imagine this space, called a vector bundle, as a bundle of copies of $\mathbb{R}^d$ indexed by the elements of B. Each fiber $\mathcal{V}_b := \{b\} \times \mathbb{R}^d, b \in B$, has the linear structure of a vector space. Frequently, it will be convenient to identify a fiber $\mathcal{V}_b$ with $\mathbb{R}^d$.

Definition 9.1.1. Let B be a compact metric space. A continuous flow $\Phi = (\theta, \varphi) : \mathbb{R} \times B \times \mathbb{R}^d \to B \times \mathbb{R}^d$ is a topological linear dynamical system (or continuous linear skew product flow) if for all $t \in \mathbb{R}$ and $(b, x), (b, y) \in B \times \mathbb{R}^d$ one has $\Phi(t, b, x) = (\theta(t)b, \varphi(t, b, x))$ and

$$\varphi(t, b, \alpha x + \beta y) = \alpha \varphi(t, b, x) + \beta \varphi(t, b, y) \text{ for } \alpha, \beta \in \mathbb{R}.$$

Thus the map Φ is linear in every fiber $\{b\} \times \mathbb{R}^d$, and it is convenient to write the skew-component or cocycle, i.e., the maps on the fibers, as

$$\Phi(t, b) = \varphi(t, b, \cdot) : \{b\} \times \mathbb{R}^d \to \{\theta(t, b)\} \times \mathbb{R}^d.$$

Where convenient, we also write the base flow as $\theta_t b$ or $\theta(t)b$. Topological linear dynamical systems will, in particular, provide an appropriate framework for families of linear time-varying differential equations of the form

$$(9.1.1) \qquad \dot{x}(t) = A(u(t))x(t) := \left[A_0 + \sum_{i=1}^{m} u_i(t) A_i\right] x(t), \ u \in \mathcal{U}.$$

Here $A_0, \ldots, A_m \in gl(d, \mathbb{R})$, $\mathcal{U} = \{u : \mathbb{R} \longrightarrow U \,|\, \text{integrable on every bounded interval}\}$ and $U \subset \mathbb{R}^m$ is compact and convex. Equation (9.1.1) defines a linear skew product flow, if we consider as base component the shift $\theta : \mathbb{R} \times \mathcal{U} \longrightarrow \mathcal{U}$, $\theta(t, u(\cdot)) := u(\cdot + t)$, and the skew-component consists of the solutions $\varphi(t, u, x)$, $t \in \mathbb{R}$, of the differential equation. The (nontrivial) proof that Φ is continuous when $\mathcal{U}$ is endowed with an appropriate metric making it into a compact metric space, and hence Φ is a topological linear flow will be deferred to Section 9.5. In an application context, there are two alternative interpretations of system (9.1.1): It may be considered as a robust linear system, when we interpret the functions $u \in \mathcal{U}$ as perturbations acting on the system; here only the range $U \subset \mathbb{R}^m$ of the perturbations is known and arbitrary perturbation functions $u \in \mathcal{U}$ are considered. In this case, one is

9.1. The Spectral Decomposition Theorem

usually interested in the worst case behavior, for instance, in the question, whether or not one can guarantee stability for all $u \in \mathcal{U}$. Alternatively, we can interpret the functions $u \in \mathcal{U}$ as controls which can be chosen so as to achieve a desired behavior, for instance, such that for some $u \in \mathcal{U}$ all corresponding solutions tend to zero as time tends to infinity. Note that here the controls act on the coefficients and not additively. Thus this is not a linear control system, which has the form

$$\dot{x}(t) = Ax(t) + Bu(t)$$

with matrices A and B of appropriate dimensions.

We need some preparations in order to formulate the main theorem of this chapter concerning the Lyapunov exponents and corresponding decompositions of the state space of topological linear dynamical systems.

Definition 9.1.2. Let $\mathcal{W}$ be a closed subset of a vector bundle $\mathcal{V} = B \times \mathbb{R}^d$ that intersects each fiber in a linear subspace $\mathcal{W}_b := \mathcal{W} \cap \left[\{b\} \times \mathbb{R}^d \right], b \in B$. Then $\mathcal{W}$ is a called a subbundle if $\dim \mathcal{W}_b = \dim \mathcal{W}_{b'}$ for all $b, b' \in B$.

Thus a subbundle of a vector bundle is a subset such that over every point in the base space one has a subspace of $\mathbb{R}^d$ of constant dimension. Subbundles are an appropriate generalization of the concept of linear subspaces to vector bundles.

As discussed in Chapter 6, the Lyapunov exponents of nonautonomous differential equations may not always come with corresponding subspace decompositions into Lyapunov spaces. In order to deal with this problem, we will generalize the concept of Lyapunov exponents. Here we rely heavily on the concepts introduced in Chapter 4 where, for autonomous linear differential equations, we could characterize the Lyapunov spaces by their projections onto projective space: They coincide with the chain components of the induced flow. In the following we will take up this characterization which also applies to periodic systems as seen in Chapter 7.

The projective bundle $\mathbb{P}\mathcal{V} := B \times \mathbb{P}^{d-1}$ is endowed with the metric which is the maximum of the metric on B and the metric on $\mathbb{P}^{d-1}$ introduced in (4.1.3),

$$d(p, p') := \min\left(\left\| \frac{x}{\|x\|} - \frac{x'}{\|x'\|} \right\|, \left\| \frac{x}{\|x\|} + \frac{x'}{\|x'\|} \right\| \right),$$

where $x \in \mathbb{R}^d$ is any element projecting to p, i.e., $x \in \mathbb{P}^{-1}(p)$ and $x' \in \mathbb{P}^{-1}(p')$ and $\mathbb{P} : \mathbb{R}^d \setminus \{0\} \to \mathbb{P}^{d-1}$ denotes the projection onto $\mathbb{P}^{d-1}$. As in Chapter 4, the continuous linear flow $\Phi = (\theta, \varphi)$ induces a continuous projective flow on the projective bundle, given by

$$\mathbb{P}\Phi(t, b, p) = (\theta(t, b), \mathbb{P}(\varphi(t, b, x))) \text{ for } (t, b, p) \in \mathbb{R} \times B \times \mathbb{P}^{d-1} \text{ and } x \in \mathbb{P}^{-1}(p).$$

Next, we generalize the notion of Lyapunov exponents or exponential growth rates. For $(b, x) \in B \times \mathbb{R}^d$ the Lyapunov exponents satisfy

$$\lambda(b,x) := \limsup_{t \to \infty} \frac{1}{t} \log \|\varphi(t,b,x)\| = \limsup_{t \to \infty} \frac{1}{t} \left(\log \|\varphi(t,b,x)\| - \log \|x\|\right).$$

Now, instead of looking at a single trajectory, we look at pieces of finitely many trajectories which, when projected to $B \times \mathbb{P}^{d-1}$, are chains. Since it is convenient to start with a chain in the projective bundle we are led to the following concept for an (ε, T)-chain ζ in $B \times \mathbb{P}^{d-1}$ given by

$$T_0, \ldots, T_{n-1} > T, \ (b_0, p_0), \ldots, (b_{n-1}, p_{n-1}) \in B \times \mathbb{P}^{d-1}$$

and satisfying $d(\mathbb{P}\Phi(T_i, b_i, p_i), (b_{i+1}, p_{i+1})) < \varepsilon$ for $i = 0, 1, \ldots, n-1$. The total time of the chain ζ is $\tau(\zeta) := \sum_{i=0}^{n-1} T_i$.

Definition 9.1.3. The finite time exponential growth rate of a chain ζ as above (or chain exponent) is, for any $x_i \in \mathbb{P}^{-1}(p_i)$,

$$\lambda(\zeta) = \frac{1}{\tau(\zeta)} \sum_{i=0}^{n-1} \left(\log \|\varphi(T_i, b_i, x_i)\| - \log \|x_i\|\right).$$

The choice of $x_i \in \mathbb{P}^{-1}(p_i)$ does not matter, since by linearity for $\alpha \neq 0$,

$$\log \|\varphi(T_i, b_i, \alpha x_i)\| - \log \|\alpha x_i\| = \log \|\varphi(T_i, b_i, x_i)\| - \log \|x_i\|.$$

In particular, one may take all x_i with $\|x_i\| = 1$ and then the terms $\log \|x_i\|$ vanish. We obtain the following concept of exponential growth rates.

Definition 9.1.4. The Morse spectrum of a chain transitive set $\mathcal{M} \subset B \times \mathbb{P}^{d-1}$ of the projective flow $\mathbb{P}\Phi$ is

$$\Sigma_{Mo}(\mathcal{M}) = \left\{ \begin{array}{c} \lambda \in \mathbb{R} \mid \text{there exist } \varepsilon^k \to 0, \ T^k \to \infty \text{ and} \\ (\varepsilon^k, T^k)\text{-chains } \zeta^k \text{ in } \mathcal{M} \text{ with } \lim_{k \to \infty} \lambda(\zeta^k) = \lambda \end{array} \right\}.$$

Thus the Morse spectrum is given by limit points of exponential growth rates of pieces of finitely many trajectories in $\mathbb{R}^d$ which, when projected to $B \times \mathbb{P}^{d-1}$, are chains in the chain component $\mathcal{M}$. The following theorem is the spectral theorem for linear flows. It generalizes the autonomous and the periodic cases. It also provides decompositions of $\mathbb{R}^d$ into subspaces, which change with $b \in B$ according to the dynamics in the base space. All points in the base space are considered, thus also irregular situations (cf. the counterexamples in Chapter 6), are included. Associated to the subspaces are exponential growth rates which, in contrast to the periodic case, do not yield numbers, but intervals. In fact, instead of the Lyapunov spectrum, i.e., the set of Lyapunov exponents, the generalized version called Morse spectrum is taken, which, however, contains the Lyapunov spectrum and is 'not much larger'.

9.1. The Spectral Decomposition Theorem

Theorem 9.1.5. *Let $\Phi = (\theta, \varphi) : \mathbb{R} \times B \times \mathbb{R}^d \to B \times \mathbb{R}^d$ be a topological linear flow on the vector bundle $\mathcal{V} = B \times \mathbb{R}^d$ with chain transitive base flow θ on the compact metric space B and denote by $\Phi(t,b) := \varphi(t,b,\cdot), (t,b) \in \mathbb{R} \times B$, the corresponding cocycle. Then for every $b \in B$ the following assertions hold:*

(i) There are decompositions

(9.1.2) $$\mathbb{R}^d = V_1(b) \oplus \ldots \oplus V_\ell(b), b \in B,$$

of $\mathbb{R}^d$ into linear subspaces $V_j(b)$ which are invariant under the flow Φ, i.e., $\Phi(t,b) V_j(b) = V_j(\theta_t b)$ for all $j = 1, \ldots, \ell$, and their dimensions d_j are constant; they form subbundles $\mathcal{V}_j$ of $B \times \mathbb{R}^d$.

(ii) There are compact intervals $[\kappa_i^*, \kappa_i], i = 1, \ldots, \ell$ with $\kappa_i^* < \kappa_{i+1}^*$ and $\kappa_i < \kappa_{i+1}$ for all i such that for each $b \in B, x \in \mathbb{R}^d \setminus \{0\}$ the Lyapunov exponents $\lambda^\pm(b,x)$ forward and backward in time satisfy for some $i \leq j$,

$$\lambda^+(b,x) := \limsup_{t \to \infty} \frac{1}{t} \log \|\varphi(t,b,x)\| \in [\kappa_j^*, \kappa_j],$$

$$\lambda^-(b,x) := \liminf_{t \to -\infty} \frac{1}{t} \log \|\varphi(t,b,x)\| \in [\kappa_i^*, \kappa_i].$$

Each boundary point κ_i^*, κ_i equals the Lyapunov exponent $\lambda^+(b,x) = \lambda^-(b,x)$ of some $b \in B, x \in \mathbb{R}^d \setminus \{0\}$ and these exponents are limits.

(iii) The projections $\mathbb{P}\mathcal{V}_j$ of the subbundles $\mathcal{V}_j, j = 1, \ldots, \ell$, to the projective bundle $B \times \mathbb{P}^{d-1}$ are the chain components $\mathcal{M}_j$ of the projective flow $\mathbb{P}\Phi$, hence they form the finest Morse decomposition and they are linearly ordered by $\mathcal{M}_1 \preceq \ldots \preceq \mathcal{M}_\ell$.

(iv) The intervals in (ii) are the Morse spectral intervals of the subbundles $\mathcal{V}_j$, i.e., $[\kappa_j^*, \kappa_j] = \Sigma_{Mo}(\mathbb{P}\mathcal{V}_j), j = 1, \ldots, \ell$.

(v) For each $j = 1, \ldots, \ell$ the maps $V_j : B \to \mathbb{G}_{d_j}$ to the Grassmannian are continuous.

The subspaces $V_j(b)$ may be considered as generalizations of Lyapunov spaces. The subbundles $\mathcal{V}_j$ are called the Selgrade bundles of the flow Φ and we may write the decompositions (9.1.2) as

$$B \times \mathbb{R}^d = \mathcal{V}_1 \oplus \ldots \oplus \mathcal{V}_\ell.$$

Such a decomposition into subbundles is called a Whitney sum. The proof of this theorem will be based on the results in Sections 9.2–9.4.

To understand the relation of this theorem to our previous results, we apply it to the cases of constant and periodic matrix functions.

Example 9.1.6. Let $A \in gl(d, \mathbb{R})$ and consider the dynamical system $\varphi : \mathbb{R} \times \mathbb{R}^d \longrightarrow \mathbb{R}^d$ generated by the solutions of the autonomous linear differential equation $\dot{x} = Ax$. The flow φ can be considered as the skew-component of a topological linear dynamical system over the base flow given by $B = \{b\}$ and $\theta : \mathbb{R} \times B \longrightarrow B$ defined as the constant map $\theta_t b = b$ for all $t \in \mathbb{R}$. Since the base flow is trivially chain transitive, Theorem 9.1.5 yields the decomposition of $\mathbb{R}^d$ into Lyapunov spaces projecting to the chain components $\mathcal{M}_i$ in $\mathbb{P}^{d-1}$. In order to recover the results from Theorem 4.1.3 it only remains to show that the Morse spectrum over $\mathcal{M}_i$ coincides with the Lyapunov exponent λ_i. Consider for $\delta > 0$ a time $\hat{T} > 0$ large enough such that for all $t \geq \hat{T}$ and all $x \in L_i = L(\lambda_i)$ with $\|x\| = 1$,

$$\left| \frac{1}{t} \log \|\varphi(t,x)\| - \lambda_i \right| < \delta.$$

Then for every (ε, T)-chain ζ in $\mathcal{M}_i = \mathbb{P}L_i$ with $T \geq T'$ one has

$$|\lambda(\zeta) - \lambda_i| = \left| \frac{1}{\tau(\zeta)} \sum_{i=0}^{n-1} \log \|\varphi(T_i, x_i)\| - \lambda_i \right|$$

$$= \left| \sum_{i=0}^{n-1} \frac{T_i}{\tau(\zeta)} \left[\frac{1}{T_i} \log \|\varphi(T_i, x_i)\| - \lambda_i \right] \right| < \sum_{i=0}^{n-1} \frac{T_i}{\tau(\zeta)} \delta = \delta.$$

Since this holds for every $\delta > 0$, it follows that in the limit for $\varepsilon^k \to 0, T^k \to \infty$ the Morse spectrum over $\mathcal{M}_i$ coincides with $\{\lambda_i\}$. Note that here no use is made of explicit solution formulas.

For Floquet theory, one can argue analogously.

Example 9.1.7. Let $A : \mathbb{R} \longrightarrow gl(d, \mathbb{R})$ be a continuous, T-periodic matrix function. Define the base flow as follows: Let $B = \mathbb{S}^1$ be parametrized by $\tau \in [0, 2T)$ and take θ is the shift $\theta(t, \tau) = t + \tau$. Then θ is a chain transitive dynamical system (cf. Exercise 3.4.2), and the solutions $\varphi(\cdot, \tau, x)$ of $\dot{x} = A(t)x$ define a topological linear dynamical system $\Phi : \mathbb{R} \times \mathbb{S}^1 \times \mathbb{R}^d \longrightarrow \mathbb{S}^1 \times \mathbb{R}^d$ via $\Phi(t, \tau, x) = (\theta(t, \tau), \varphi(t, \tau, x))$. With this setup, Theorem 9.1.5 recovers the results of Floquet Theory in Section 7.2. In fact, by Corollary 7.2.14 the chain components $\mathcal{M}_i$ in $\mathbb{S}^1 \times \mathbb{R}^d$ are the projections of the Lyapunov spaces $L(\lambda_i, \tau), \tau \in \mathbb{S}^1$. It only remains to show that the Morse spectrum over $\mathcal{M}_i$ coincides with the Lyapunov exponent λ_i. Consider for $\delta > 0$ a time $\hat{T} > 0$ large enough such that for all $t \geq \hat{T}$ and all $x \in L(\lambda_i, \tau), \tau \in \mathbb{S}^1$, with $\|x\| = 1$,

$$\left| \frac{1}{t} \log \|\varphi(t, \tau, x)\| - \lambda_i \right| < \delta.$$

Then for every (ε, T')-chain ζ with $T' \geq \hat{T}$ one has

$$|\lambda(\zeta) - \lambda_i| = \left| \frac{1}{\tau(\zeta)} \sum_{i=0}^{n-1} \log \|\varphi(T_i, \tau_i, x_i)\| - \lambda_i \right| < \delta.$$

9.1. The Spectral Decomposition Theorem

Since this holds for every $\delta > 0$, it follows that, taking the limit for $\varepsilon^k \to 0, T^k \to \infty$, the Morse spectrum over $\mathcal{M}_i$ coincides with $\{\lambda_i\}$.

Part (iii) of Theorem 9.1.5 shows how the subbundles $\mathcal{V}_j$ are constructed: One has to analyze the chain components of the projective flow. In Section 9.2 it is shown that the set of points x in $\mathbb{R}^d$ which project to a chain component, form a subbundle and these subbundles yield a decomposition of the bundle $\mathcal{V}$ (Selgrade's Theorem).

Stability

The problem of stability of the zero solution of topological linear dynamical systems can now be analyzed in analogy to the case of a constant matrix or a periodic matrix function. The following definition generalizes the last part of Definition 1.4.7 taking into account that the Morse spectrum of a Selgrade bundle is, in general, an interval.

Definition 9.1.8. The stable, center, and unstable subbundles associated with a topological linear flow Φ on a vector bundle $\mathcal{V} = B \times \mathbb{R}^d$ are defined as

$$\mathcal{V}^- := \bigoplus_{j:\max \Sigma_{Mo}(\mathcal{V}_j)<0} \mathcal{V}_j, \quad \mathcal{V}^0 := \bigoplus_{j:0\in \Sigma_{Mo}(\mathcal{V}_j)} \mathcal{V}_j, \quad \mathcal{V}^+ := \bigoplus_{j:\min \Sigma_{Mo}(\mathcal{V}_j)>0} \mathcal{V}_j.$$

Remark 9.1.9. Note that the center bundle may be the sum of several Selgrade bundles, since the corresponding Morse spectral intervals may overlap. The presence of a nontrivial center subbundle $\mathcal{V}^0$ will make the analysis of the long time behavior complicated. If $\mathcal{V}^0 = \varnothing$, one says that the linear flow admits an exponential dichotomy which is a uniform hyperbolicity condition.

We consider the following notions of stability.

Definition 9.1.10. Let $\Phi = (\theta, \varphi)$ be a topological linear flow on $B \times \mathbb{R}^d$. The zero solution $\varphi(t, b, 0) = 0$ for all $b \in B$ and $t \in \mathbb{R}$ is

stable if for all $\varepsilon > 0$ there exists a $\delta > 0$ such that $\|\varphi(t, b, x_0)\| < \varepsilon$ for all $t \geq 0$ and all $b \in B$ whenever $\|x_0\| < \delta$;

asymptotically stable if it is stable and there exists a $\gamma > 0$ such that $\lim_{t \to \infty} \varphi(t, b, x_0) = 0$ whenever $\|x_0\| < \gamma$;

exponentially stable if there are constants $\alpha \geq 1$ and $\beta > 0$ such that for all $b \in B$ and $x \in \mathbb{R}^d$,

$$\|\varphi(t, b, x)\| \leq \alpha e^{-\beta t} \|x\| \text{ for all } t \geq 0.$$

In order to characterize the stability properties of topological linear flows the following result, known as Fenichel's uniformity lemma, is needed.

Lemma 9.1.11. *Let $\Phi = (\theta, \varphi)$ be a linear flow on the vector bundle $\mathcal{V} = B \times \mathbb{R}^d$ and assume that*

$$\lim_{t \to \infty} \|\varphi(t, b, x)\| = 0 \text{ for all } (b, x) \in B \times \mathbb{R}^d.$$

Then there are $\alpha \geq 1$ and $\beta > 0$ such that

$$\|\Phi(t, b)\| \leq \alpha e^{-\beta t} \text{ for all } t \geq 0 \text{ and all } b \in B,$$

where $\|\Phi(t, b)\| := \sup \{\|\Phi(t, b)x\| \mid \|x\| = 1\}$ is the operator norm.

Proof. By assumption for every point in the unit sphere bundle $B \times \mathbb{S}^{d-1} = \{(b, x) \in B \times \mathbb{R}^d \mid \|x\| = 1\}$ there is a time $T(b, x) > 0$ with $\|\Phi(T(b, x), b)x\| < \frac{1}{2}$. Thus for every $(b, x) \in B \times \mathbb{S}^{d-1}$ there is a neighborhood $N(b, x)$ in $B \times \mathbb{S}^{d-1}$ such that

$$\|\Phi(T(b, x), b')x'\| < \frac{1}{2} \text{ for all } (b', x') \in N(b, x).$$

Using compactness of $B \times \mathbb{S}^{d-1}$ and linearity one finds a finite covering $N_1, \ldots, N_n$ of $B \times \mathbb{S}^{d-1}$ and times $T_1, \ldots, T_n > 0$ such that for all $(b', x') \in B \times \mathbb{R}^d$ with $\left(b', \frac{x'}{\|x'\|}\right) \in N_i, i = 1, \ldots, n$,

$$\|\Phi(T_i, b')x'\| < \frac{1}{2} \|x'\|.$$

Now fix $(b, x) \in B \times \mathbb{S}^{d-1}$. Choose a (nonunique) sequence of integers $i_j \in \{1, \ldots, n\}, j \in \mathbb{N}$, in the following way:

$$i_1 : (b, x) \in N_{i_1},$$
$$i_2 : \frac{\Phi(T_{i_1}, b)x}{\|\Phi(T_{i_1}, b)x\|} \in N_{i_2},$$
$$\vdots$$
$$i_j : \frac{\Phi(T_{i_1} + T_{i_2} + \ldots + T_{i_{j-1}}, b)x}{\|\Phi(T_{i_1} + T_{i_2} + \ldots + T_{i_{j-1}}, b)x\|} \in N_{i_j}.$$

Let $\tau_j = T_{i_1} + T_{i_2} + \ldots + T_{i_j}$. Then we write $t > 0$ as $t = \tau_j + r$ for some $j \geq 0$ and $0 \leq r \leq T := \max_{i=1,\ldots,n} T_i$. We obtain

$\|\Phi(t, b)x\|$
$\leq \|\Phi(r, \theta(\tau_j, b))\| \, \|\Phi(\tau_j, b)x\|$
$= \|\Phi(r, \theta(\tau_j, b))\| \, \|\Phi(T_{i_j}, \theta(T_{i_1} + \ldots + T_{i_{j-1}}, b))\Phi(T_{i_1} + \ldots + T_{i_{j-1}}, b)x\|$
$\leq \|\Phi(r, \theta(\tau_j, b))\| \, \frac{1}{2} \, \|\Phi(T_{i_1} + T_{i_2} + \ldots + T_{i_{j-1}}, b)x\|$
$\leq \|\Phi(r, \theta(\tau_j, b))\| \left(\frac{1}{2}\right)^j \|x\|.$

9.1. The Spectral Decomposition Theorem

Note that $\log 2 > \frac{1}{2}$ and $t = \tau_j + r \leq jT + r$ implies $j \geq \frac{t-r}{T}$. Hence it follows that

$$\left(\frac{1}{2}\right)^j = \mathrm{e}^{-j\log 2} \leq \mathrm{e}^{-\frac{t-r}{2T}} = \mathrm{e}^{-\frac{t}{2T}} \cdot \mathrm{e}^{\frac{r}{T}} \leq \mathrm{e} \cdot \mathrm{e}^{-t/T}.$$

Let $\beta := \frac{1}{2T}$ and $\alpha := \mathrm{e} \cdot \max_{0 \leq r \leq T} \max_{b \in B} \|\Phi(r,b)\|$. Then we conclude, as claimed, that $\|\Phi(t,b)x\| \leq \alpha \mathrm{e}^{-\beta t} \|x\|$ for every $(b,x) \in B \times \mathbb{R}^d$ and $t > 0$. □

With these preparations Theorem 9.1.5 implies the following main result regarding stability of topological linear flows. It generalizes Theorem 1.4.8 which considers autonomous linear differential equations and Theorem 7.2.16 which considers periodic linear differential equations.

Theorem 9.1.12. *Let $\Phi = (\theta, \varphi)$ be a topological linear flow on $B \times \mathbb{R}^d$. Then for the zero solution $\varphi(t,b,0) = 0, b \in B, t \in \mathbb{R}$, the following statements are equivalent:*

(i) the zero solution is asymptotically stable,

(ii) the zero solution is exponentially stable,

(iii) all Morse spectral intervals $\Sigma_{Mo}(\mathcal{V}_j)$ are contained in $(-\infty, 0)$,

(iv) the stable subbundle $\mathcal{V}^-$ satisfies $\mathcal{V}^- = B \times \mathbb{R}^d$.

Proof. First observe that by linearity asymptotic (and exponential) stability of the zero solution $\varphi(t,b,0) = 0, b \in B, t \in \mathbb{R}$, for all $(b, x_0) \in B \times \mathbb{R}^d$ with $\|x_0\| < \gamma$ implies asymptotic and exponential stability, respectively, for all points $(b,x) \in B \times \mathbb{R}^d$: In fact, for exponential stability it follows that for $x \in \mathbb{R}^d$ the point $x_0 := \frac{\gamma}{2} \frac{x}{\|x\|}$ satisfies $\|x_0\| < \gamma$ and hence

$$\|\varphi(t,b,x)\| = \|\Phi(t,b)x\| = \left\|\frac{2\|x\|}{\gamma} \Phi(t,b) \left(\frac{\gamma}{2} \frac{x}{\|x\|}\right)\right\| = \frac{2\|x\|}{\gamma} \|\varphi(t,b,x_0)\|$$

$$\leq \frac{2\|x\|}{\gamma} \alpha \|x_0\| \mathrm{e}^{-\beta t} = \alpha \|x\| \mathrm{e}^{-\beta t},$$

and analogously for asymptotic stability. By Theorem 9.1.5, properties (ii), (iii) and (iv) are equivalent and imply (i). Conversely, assume asymptotic stability of the zero solution, i.e.,

$$\lim_{t \to \infty} \|\varphi(t,b,x)\| = 0 \text{ for all } (b,x) \in B \times \mathbb{R}^d.$$

Then Lemma 9.1.11 implies that there are $\alpha \geq 1$ and $\beta > 0$ such that for all $t \geq 0$ and all $b \in B$,

$$\|\Phi(t,b)\| \leq \alpha \mathrm{e}^{-\beta t}, \text{ and hence } \|\varphi(t,b,x)\| \leq \alpha \mathrm{e}^{-\beta t} \|x\| \text{ for all } x \in \mathbb{R}^d. \quad \square$$

In general, Lyapunov exponents for topological linear flows are difficult to compute explicitly—numerical methods are usually the way to go. In Section 9.5 we will discuss several examples.

In the next section we begin to prove Theorem 9.1.5.

9.2. Selgrade's Theorem

This section characterizes the chain components of projective linear flows and shows that they yield a decomposition of the vector bundle into invariant subbundles. This is the content of Selgrade's Theorem.

Throughout this section, $\Phi = (\theta, \varphi) : \mathbb{R} \times B \times \mathbb{R}^d \longrightarrow B \times \mathbb{R}^d$ is a linear skew product flow with continuous base flow $\theta : \mathbb{R} \times B \longrightarrow B$ on a compact metric base space B which is chain transitive. Furthermore, $\mathbb{R}^d$ is endowed with the Euclidean scalar product.

We start with the following lemma which provides the projective space with a metric which is equivalent to the metric $d(\cdot, \cdot)$ in (4.1.3).

Lemma 9.2.1. *(i) A metric on $\mathbb{P}^{d-1}$ is defined by*

$$\bar{d}(\mathbb{P}x, \mathbb{P}y) := \sqrt{1 - \frac{\langle x, y \rangle^2}{\|x\|^2 \|y\|^2}}.$$

(ii) There exists a constant $c_0 > 0$ such that

$$c_0 d(\mathbb{P}x, \mathbb{P}y) \leq \bar{d}(\mathbb{P}x, \mathbb{P}y) \leq d(\mathbb{P}x, \mathbb{P}y) \text{ for all } 0 \neq x, y \in \mathbb{R}^d.$$

Proof. (i) It suffices to consider unit vectors $x, y \in \mathbb{R}^d$ with $d(\mathbb{P}x, \mathbb{P}y) = \|x - y\|$. In the Euclidean space $\mathbb{R}^d$ one has $|\langle x, y \rangle| \leq \|x\| \|y\|$ and equality holds if and only if x and y are linearly dependent, hence $1 - \langle x, y \rangle^2 = 0$ if and only if $\mathbb{P}x = \mathbb{P}y$. Clearly, $1 - \langle x, y \rangle^2 = 1 - \langle y, x \rangle^2$. Verification of the triangle inequality is a bit more complicated: Observe that multiplication by an orthogonal matrix O does not change $\bar{d}$, since $\langle Ox, Oy \rangle = \langle x, y \rangle$ for $x, y \in \mathbb{R}^d$. Hence it suffices to show that

$$\bar{d}(\mathbb{P}x, \mathbb{P}y) \leq \bar{d}(\mathbb{P}x, \mathbb{P}z) + \bar{d}(\mathbb{P}z, \mathbb{P}y)$$

for unit vectors $z = \mathbf{e}_1, x = x_1 \mathbf{e}_1 + x_2 \mathbf{e}_2, y = y_1 \mathbf{e}_1 + y_2 \mathbf{e}_2 + y_3 \mathbf{e}_3$ and the canonical basis vectors $\mathbf{e}_1, \mathbf{e}_2, \mathbf{e}_3$. This computation is left as Exercise 9.6.3.

(ii) One computes for x, y with $\|x\| = \|y\| = 1$:

$$(9.2.1) \quad \|x - y\|^2 \|x + y\|^2 = \langle x - y, x - y \rangle \langle x + y, x + y \rangle = 4(1 - \langle x, y \rangle^2).$$

Since $\|x + y\| \leq 2$, this implies

$$d(\mathbb{P}x, \mathbb{P}y)^2 = \|x - y\|^2 \geq \frac{1}{4} \|x - y\|^2 \|x + y\|^2 = 1 - \langle x, y \rangle^2 = \bar{d}(\mathbb{P}x, \mathbb{P}y)^2.$$

For the converse inequality, it suffices to show that for all sequences (x_n, y_n) in $\mathbb{R}^d \times \mathbb{R}^d$ with $\|x_n\| = \|y_n\| = 1$ and $d(\mathbb{P}x_n, \mathbb{P}y_n) \to 0$ it follows that $\bar{d}(\mathbb{P}x_n, \mathbb{P}y_n) \to 0$.

9.2. Selgrade's Theorem

We may assume that $\|x_n - y_n\| \to 0$. Then $\langle x_n, y_n \rangle \to 1$: In fact, if this does not hold, one finds the contradiction that there are converging subsequences $x_{n_k}, y_{n_k} \to x$ with

$$\lim_{k \to \infty} \langle x_{n_k}, y_{n_k} \rangle = \langle x, x \rangle < 1 = \|x\|.$$

Hence it follows that

$$\bar{d}(\mathbb{P} x_n, \mathbb{P} y_n) = \sqrt{1 - \langle x_n, y_n \rangle} \to 0. \qquad \square$$

The metric introduced above can be extended to a metric on the projective bundle $B \times \mathbb{P}^{d-1}$. For the proof of Selgrade's Theorem we will need the characterization of Morse decompositions via attractors from Theorem 8.2.10 and Proposition 8.3.2. We begin with the following lemma which shows that attractors in the projective bundle define linear subspaces. (Note that Lemmas 9.2.2 and 9.2.3 do not require that the base space is chain transitive.)

Lemma 9.2.2. *Consider the set up of Theorem 9.1.5 with $\mathcal{V} = B \times \mathbb{R}^d$ and the cocycle Φ on $\mathbb{R}^d$. Let $\mathcal{W}$ be an invariant subbundle of $\mathcal{V}$ and consider an attractor A in $\mathbb{P}\mathcal{W}$. Then we have*

(i) Let $(b, x), (b, x') \in \mathcal{W}$ with $x, x' \neq 0$ be given with $(b, \mathbb{P} x) \in A$ and $(b, \mathbb{P} x') \notin A$. Then

(9.2.2) $$\lim_{t \to -\infty} \|\Phi(t, b) x\| / \|\Phi(t, b) x'\| = 0.$$

(ii) The set $\mathbb{P}^{-1} A$ intersects each fiber $\mathcal{W}_b$ in a linear space.

Proof. (i) Define the two-dimensional subspace L in the fiber over b by $L = \{(b, cx + c'x') \mid c, c' \in \mathbb{R}\}$. First assume that $(b, \mathbb{P} x) \in A$ is a boundary point of $A \cap \mathbb{P} L$ in $\mathbb{P} L$. Choose ε with $0 < \varepsilon < d_H(A, A^*)$, where $A^* := \{(b, p) \in \mathbb{P}\mathcal{W} \mid \omega(b, p) \cap A = \varnothing\}$ is the complementary repeller of A and d_H denotes the Hausdorff metric; cf. (3.2.1). By Lemma 9.2.1 it follows for $\delta := (c_0 \varepsilon)^2 > 0$ that for all $(b, x_0), (b, x_1) \in \mathcal{V}$ with $x_0, x_1 \neq 0$,

(9.2.3) $$\langle x_0, x_1 \rangle^2 / \|x_0\|^2 \|x_1\|^2 \geq 1 - \delta \text{ implies } d(\mathbb{P} x_0, \mathbb{P} x_1) \leq \varepsilon.$$

Now suppose that (9.2.2) does not hold. Then there exist a sequence $t_k \to -\infty$ and a constant $K > 0$ such that for all $k \in \mathbb{N}$,

(9.2.4) $$\|\Phi(t_k, b) x'\| / \|\Phi(t_k, b) x\| \leq K.$$

For $c \in \mathbb{R}$,

$$\Phi(t_k, b)(cx' + x) = c\Phi(t_k, b)x' + \Phi(t_k, b)x = cx'_k + x_k$$

with $x'_k := \Phi(t_k, b) x'$ and $x_k := \Phi(t_k, b) x$. The boundedness assumption (9.2.4) implies that also $\left\langle \frac{x'_k}{\|x_k\|}, \frac{x_k}{\|x_k\|} \right\rangle$ remains bounded for $k \to \infty$. We compute

$$
\begin{aligned}
\frac{\langle \Phi(t_k,b)(cx'+x), \Phi(t_k,b)x\rangle^2}{\|\Phi(t_k,b)(cx'+x)\|^2 \|\Phi(t_k,b)x\|^2} &= \frac{\langle cx'_k + x_k, x_k\rangle^2}{\|cx'_k + x_k\|^2 \|x_k\|^2} \\
&= \frac{c^2\langle x'_k, x_k\rangle^2 + 2c\langle x'_k, x_k\rangle \|x_k\|^2 + \|x_k\|^4}{c^2\|x'_k\|^2 \|x_k\|^2 + 2c\langle x'_k, x_k\rangle \|x_k\|^2 + \|x_k\|^4} \\
&= \frac{c^2 \left\langle \frac{x'_k}{\|x_k\|}, \frac{x_k}{\|x_k\|} \right\rangle^2 + 2c \left\langle \frac{x'_k}{\|x_k\|}, \frac{x_k}{\|x_k\|} \right\rangle + 1}{c^2 \frac{\|x'_k\|^2}{\|x_k\|^2} + 2c \left\langle \frac{x'_k}{\|x_k\|}, \frac{x_k}{\|x_k\|} \right\rangle + 1}.
\end{aligned}
$$

(9.2.5)

It follows that for $|c|$ small enough this expression is $\geq 1 - \delta$ for all $k \in \mathbb{N}$. Hence, by (9.2.3), for all $k \in \mathbb{N}$,

$$d(\mathbb{P}\Phi(t_k,b)(cx'+x), A) \leq \varepsilon$$

and thus $\alpha(b,\mathbb{P}(cx'+x)) \not\subset A^*$. By the choice of ε, this implies $(b,\mathbb{P}(cx'+x)) \in A$ for $|c|$ sufficiently small. This contradicts the assumption that $(b, \mathbb{P}x)$ is a boundary point of $A \cap \mathbb{P}L$ in $\mathbb{P}L$. Thus (9.2.2) holds in this case.

(ii) Suppose that $A \cap \mathbb{P}L \neq \varnothing$ and does not coincide with $\mathbb{P}L$. Then, without loss of generality, we may assume that $(b, \mathbb{P}x)$ is a boundary point of $A \cap \mathbb{P}L$ in $\mathbb{P}L$. Then there is $x' \in L$ such that $(b, \mathbb{P}x') \notin A$ and any point in $\mathbb{P}L \setminus \{(b, \mathbb{P}x)\}$ is given by $(b, \mathbb{P}(x' + cx))$ for some $c \in \mathbb{R}$. Then (9.2.2) and the same computation as in (9.2.5) imply that the limit for $t \to -\infty$ of

$$\frac{\langle \Phi(t,b)(x'+cx), \Phi(t,b)x'\rangle^2}{\|\Phi(t,b)(x'+cx)\|^2 \|\Phi(t,b)x'\|^2}$$

is equal to 1. By Lemma 9.2.1 this implies

$$\begin{aligned}
0 &= \lim_{t \to -\infty} \overline{d}\left(\mathbb{P}\Phi(t,b)(x'+cx), \mathbb{P}\Phi(t,b)x'\right) \\
&= \lim_{t \to -\infty} d\left(\mathbb{P}\Phi(t,b)(x'+cx), \mathbb{P}\Phi(t,b)x'\right).
\end{aligned}$$

Since $(b, \mathbb{P}x') \notin A$ the limit points for $t \to -\infty$ of

$$\mathbb{P}\Phi(t,b,x') = (\theta(t,b), \mathbb{P}\Phi(t,b)x)$$

are contained in the complementary repeller A^*; see Proposition 8.2.7. Hence this also holds for $(b, x'+cx)$ implying $(b, \mathbb{P}(x'+cx)) \notin A$. Therefore $A \cap \mathbb{P}L$ consists of a single point. We have shown that for any two-dimensional subspace L in $\mathcal{W}_b$ the set $A \cap \mathbb{P}L$ is empty, equals $\mathbb{P}L$, or consists of a single point. This implies that $\mathbb{P}^{-1}A$ intersects each fiber in a linear subspace. Furthermore, it also shows that assertion (i) is only nontrivial if $(b, \mathbb{P}x)$ is a boundary point of $A \cap \mathbb{P}L$ as assumed in the proof of (i). □

9.2. Selgrade's Theorem

Next we use this result to analyze the behavior of subspaces under the flow. Recall from Definition 8.3.1 that $\Omega(b) = \{b' \in B \,|\, \text{for all } \varepsilon, T > 0 \text{ there is an } (\varepsilon, T)\text{-chain from } b \text{ to } b'\}$.

Lemma 9.2.3. *With the notation of Lemma 9.2.2 suppose that $L_b \subset \mathcal{W}_b$ is a linear subspace and $b' \in \Omega(b)$. Then*

$$L_{b'} = \{v' \in \mathcal{W}_{b'} \,|\, v' = (b', x') \text{ with } x' \neq 0 \text{ implies } \mathbb{P}v' \in \Omega(\mathbb{P}L_b)\}$$

is a linear subspace of $\mathcal{W}_{b'}$ and $\dim L_{b'} \geq \dim L_b$.

Proof. By Proposition 8.3.2, the set $\Omega(\mathbb{P}L_b)$ is the intersection of attractors. Hence Lemma 9.2.2 implies that it intersects each fiber in a projective linear space. Therefore $L_{b'}$ is a linear subspace of $\mathcal{V}_{b'}$. Now let us define the set $L_{b'}(\varepsilon, T)$ to be the closure of all points $v \in \mathcal{V}_{b'}$ such that there exists an (ε, T)-chain from some point in $\mathbb{P}L_b$ to $\mathbb{P}\Phi(-T, v)$. Thus for every $v' = (b', x') \in L_{b'}(\varepsilon, T)$ with $x' \neq 0$ there exists an (ε, T)-chain from some point in $\mathbb{P}L_b$ to $\mathbb{P}v'$. This implies

$$(9.2.6) \qquad \bigcap_{n \in \mathbb{N}} L_{b'}(1/n, n) \subset L_{b'}.$$

The following construction shows that $L_{b'}(\varepsilon, T)$ contains a linear subspace of dimension at least that of L_b. First note that $\Omega(b)$ is invariant and hence $\theta_{-T-1}b' \in \Omega(b)$. Then there exists an (ε, T)-chain $b_0, \ldots, b_k$ with times $T_0, \ldots, T_{k-1} > T$ from b to $\theta_{-T-1}b'$. Given any $v = (b, x) \in L_b$ define the sequence $v_0 = v, \ldots, v_k$ in $\mathcal{V}$ such that $v_0 = (b, x)$, $v_j = (b_j, \varphi(T_0 + \ldots + T_{j-1}, x_0, b_0))$ for $j = 1, \ldots, k$. Since there are only trivial jumps in the second component, this sequence defines an (ε, T)-chain from $\mathbb{P}v \in \mathbb{P}L_b$ to $\mathbb{P}v_k = (\theta_{-T-1}b', \mathbb{P}\varphi(T_0 + \ldots + T_{k-1}, x_0, b_0))$, and therefore $v' = \Phi(T+1, v_k) \in L_{b'}(\varepsilon, T)$. Furthermore, the points $v' \in \pi^{-1}(b')$ obtained in this way form a linear subspace of the same dimension as L_b, since in the second component we employ the linear isomorphism

$$\Phi(T_0 + \ldots + T_{k-1} + T + 1) = \varphi(T_0 + \ldots + T_{k-1} + T + 1, b, \cdot) : \mathcal{V}_b \to \mathcal{V}_{b'}.$$

We conclude that the set $\mathcal{L}(\varepsilon, T)$ of m-dimensional subspaces of $\mathcal{V}_b$ contained in $L_{b'}(\varepsilon, T)$ is nonempty for $m = \dim L_b$. This is a compact subset of the Grassmannian $\mathbb{G}_m$. Since the intersection of a decreasing sequence of nonempty compact sets is nonempty, it follows that the intersection of the sets $\mathcal{L}(1/n, n)$ is nonempty. Together with (9.2.6) this proves the statement of the lemma. □

These lemmas will imply that attractors in $\mathbb{P}\mathcal{W}$ generate subbundles in $\mathcal{W}$ if the flow on the base space is chain transitive.

Proposition 9.2.4. *Let A be an attractor in $\mathbb{P}\mathcal{W}$ for an invariant subbundle $\mathcal{W}$ of a vector bundle $\mathcal{V} = B \times \mathbb{R}^d$ with chain transitive base space B. Then*

$$\mathbb{P}^{-1}A := \{v = (b, x) \in \mathcal{W} \mid x = 0 \text{ or } (b, \mathbb{P}x) \in A\}$$

is an invariant subbundle contained in $\mathcal{W}$.

Proof. We have to show that $\mathbb{P}^{-1}A$ is a closed subset of $\mathcal{W}$ that intersects each fiber in a linear subspace of constant dimension. Closedness is clear by definition of A and closedness of $\mathcal{W}$ in $\mathcal{V}$. By Lemma 9.2.2(ii) the intersections of $\mathbb{P}^{-1}A$ with fibers are linear subspaces $L_b, b \in B$. Because A is an attractor containing $\omega(\mathbb{P}L_b)$, it follows by Proposition 8.3.2 that $\Omega(\mathbb{P}L_b) \subset A$ and hence for every $b' \in \Omega(b)$,

$$\{v = (b, x) \in \mathcal{W}_{b'} \mid x = 0 \text{ or } (b, \mathbb{P}x) \in \Omega(\mathbb{P}L_b)\} \subset L_{b'}.$$

Therefore we obtain from Lemma 9.2.3 that $\dim L_{b'} \geq \dim L_b$. Chain transitivity of B implies that $\Omega(b) = B$ and hence $\dim L_b$ is constant for $b \in B$. □

This result allows us to characterize the chain components of a projective linear flow over a chain transitive base space. The following theorem due to Selgrade shows that a finest Morse decomposition of the projected flow exists and that it provides a decomposition of the vector bundle $\mathcal{V} = B \times \mathbb{R}^d$ into invariant subbundles.

Theorem 9.2.5 (Selgrade). *Let $\Phi = (\theta, \varphi) : \mathbb{R} \times B \times \mathbb{R}^d \to B \times \mathbb{R}^d$ be a topological linear flow on the vector bundle $\mathcal{V} = B \times \mathbb{R}^d$ with chain transitive base flow θ on the compact metric space B. Then the projected flow $\mathbb{P}\Phi$ has a finite number of chain components $\mathcal{M}_1, \ldots, \mathcal{M}_\ell$, $\ell \leq d$. These components form the finest Morse decomposition for $\mathbb{P}\Phi$, and they are linearly ordered $\mathcal{M}_1 \preceq \ldots \preceq \mathcal{M}_\ell$. Their lifts $\mathcal{V}_i := \mathbb{P}^{-1}\mathcal{M}_i \subset B \times \mathbb{R}^d$ are subbundles, called the Selgrade bundles. They form a continuous bundle decomposition*

(9.2.7) $$B \times \mathbb{R}^d = \mathcal{V}_1 \oplus \ldots \oplus \mathcal{V}_\ell.$$

Since the $\mathcal{V}_i$ are subbundles, the dimensions of the subspaces $\mathcal{V}_{i,b}$ are independent of $b \in B$. The subbundle decomposition, also called a Whitney sum, means that for every $b \in B$ one has a decomposition of $\mathbb{R}^d$ into the ℓ linear subspaces $\mathcal{V}_{i,b} := \{x \in \mathbb{R}^d \mid (b, x) \in \mathcal{V}_i\}$,

$$\mathbb{R}^d = \mathcal{V}_{1,b} \oplus \ldots \oplus \mathcal{V}_{\ell,b}.$$

The subbundles are invariant under the flow Φ, since they are the preimages of the Morse sets $\mathcal{M}_i$. For the fibers, this means that $\mathcal{V}_{i,\theta_t b} = \{\varphi(t, x, b) \in \mathbb{R}^d \mid (b, x) \in \mathcal{V}_i\}$. Hence this result is a generalization of the decomposition (7.2.8) in the context of Floquet theory derived from a finest Morse decomposition in $\mathbb{S}^1 \times \mathbb{P}^{d-1}$. In the next section we will discuss the corresponding exponential growth rates.

9.2. Selgrade's Theorem

Proof of Theorem 9.2.5. Note first that there is always a Morse decomposition of $\mathbb{P}\Phi$: Define $A_0 = \varnothing$, $A_1 = \mathbb{P}\mathcal{V}$ and $\mathcal{M}_1 = A_1 \cap A_0^*$; then a Morse decomposition is given by $\{\mathcal{M}_1\}$. Next we claim that for every invariant subbundle $\mathcal{W}$ of $\mathcal{V}$ the following holds: for every Morse decomposition $\{\mathcal{M}_1, \ldots, \mathcal{M}_n\}$ of $\mathbb{P}\mathcal{W}$ corresponding to an attractor sequence $\varnothing = A_0 \subset A_1 \subset \ldots \subset A_n = \mathbb{P}\mathcal{W}$ the sets $\mathbb{P}^{-1}\mathcal{M}_{n-i} = \mathbb{P}^{-1}A_{i+1} \cap \mathbb{P}^{-1}A_i^*$, $i = 0, \ldots, n-1$, define a Whitney decomposition of $\mathcal{W}$ into subbundles. For $n = 1$, this is obviously true. So we assume that the assertion is true for all invariant subbundles $\mathcal{W}$ and their Morse decompositions corresponding to an attractor sequences of length $n - 1$, and we prove it for n.

Thus let $\mathcal{W}$ be an invariant subbundle, and consider an attractor sequence of length n. Because $\mathcal{M}_n = A_1$, it is an attractor, and by Proposition 9.2.4, $\mathbb{P}^{-1}A_1$ and $\mathbb{P}^{-1}A_1^*$ are invariant subbundles. It is easily seen that $\{\mathcal{M}_1, \ldots, \mathcal{M}_{n-1}\}$ is a Morse decomposition of A_1^*. Hence by the induction assumption $\mathbb{P}^{-1}\mathcal{M}_j$, $j = 2, \ldots, n$ form a Whitney decomposition of $\mathbb{P}^{-1}A_1^*$,

$$\mathbb{P}^{-1}A_1^* = \mathbb{P}^{-1}\mathcal{M}_2 \oplus \ldots \oplus \mathbb{P}^{-1}\mathcal{M}_n.$$

It remains to show that $\mathbb{P}^{-1}A_1$ and $\mathbb{P}^{-1}A_1^*$ form a Whitney decomposition of $\mathcal{W}$. Write $A = A_1$, choose $b \in B$, and assume that the corresponding fibers of $\mathbb{P}^{-1}A$ and $\mathbb{P}^{-1}A^*$ have dimensions r and s, respectively. Because A and A^* are disjoint, it suffices to prove that $r + s \geq \dim \mathcal{W} =: d_1$. Fix $b \in B$ and consider a subspace F_b of $\mathcal{W}_b$ complementary to A_b^*, hence $\dim F_b = d_1 - s$. By the definition of A^*, one has $\omega(\mathbb{P}F_b) \subset A$. Because A is an attractor, Proposition 8.3.2 implies that $\Omega(\mathbb{P}F_b) \subset A$. Now chain transitivity in B and Lemma 9.2.3 show that for each $b' \in B$ the set $\Omega(\mathbb{P}F_b)$ meets $\mathbb{P}\mathcal{V}_{b'}$ in a subspace of dimension at least $d_1 - s$. Therefore $r \geq d_1 - s$ as claimed, and we obtain that

$$\mathcal{W} = \mathbb{P}^{-1}A_1 \oplus \mathbb{P}^{-1}A_1^* = \mathbb{P}^{-1}\mathcal{M}_1 \oplus \mathbb{P}^{-1}\mathcal{M}_2 \oplus \ldots \oplus \mathbb{P}^{-1}\mathcal{M}_n.$$

In particular, this holds for the vector bundle $\mathcal{V}$.

In order to see that there exists a finest Morse decomposition for the flow on $\mathbb{P}\mathcal{V}$, let $\{\mathcal{M}_1, \ldots, \mathcal{M}_n\}$ and $\{\mathcal{M}'_1, \ldots, \mathcal{M}'_m\}$ be two Morse decompositions corresponding to the attractor sequences $\varnothing = A_0 \subset A_1 \subset \ldots \subset A_n = \mathbb{P}\mathcal{V}$ and $\varnothing = A'_0 \subset A'_1 \subset \ldots \subset A'_m = \mathbb{P}\mathcal{V}$. By Proposition 9.2.4, all $\mathbb{P}^{-1}A_i$, $\mathbb{P}^{-1}A'_j$ are subbundles of $\mathcal{V}$; hence by a dimension argument $n, m \leq d$. By the first part of this proof it follows that $\mathbb{P}^{-1}\mathcal{M}_{n-i} = \mathbb{P}^{-1}A_{i+1} \cap \mathbb{P}^{-1}A_i^*$ and $\mathbb{P}^{-1}\mathcal{M}'_{m-j} = \mathbb{P}^{-1}A'_{j+1} \cap \mathbb{P}^{-1}A'^*_j$ are subbundles. Their intersections $\mathcal{M}_{ij} := \mathcal{M}_i \cap \mathcal{M}'_j$ define a Morse decomposition. This shows that refinements of Morse decompositions of $(\mathbb{P}\mathcal{V}, \mathbb{P}\Phi)$ lead to finer Whitney decompositions of $\mathcal{V}$. Therefore, again by a dimension argument, there exists a finest Morse decomposition $\{\mathcal{M}_1, \ldots, \mathcal{M}_\ell\}$ with $1 \leq \ell \leq d$. $\square$

Remark 9.2.6. The proof also shows that for every $j \in \{1, \ldots, \ell-1\}$ the subbundles $\mathcal{V}_1 \oplus \ldots \oplus \mathcal{V}_j$ and $\mathcal{V}_{j+1} \oplus \ldots \oplus \mathcal{V}_\ell$ yield an attractor $A := \mathbb{P}(\mathcal{V}_1 \oplus \ldots \oplus \mathcal{V}_j)$ with complementary repeller $A^* := \mathbb{P}(\mathcal{V}_1 \oplus \ldots \oplus \mathcal{V}_j)$.

9.3. The Morse Spectrum

This section comes back to the problem of characterizing exponential growth rates and the corresponding subspaces. It characterizes the Morse spectrum for linear flows on vector bundles and shows that it consists of compact intervals corresponding to the Selgrade bundles.

As before, the Lyapunov exponents of a topological linear flow are defined as the exponential growth rates of trajectories. Unfortunately, the set of Lyapunov exponents of a topological linear flow Φ is rather difficult to handle as we have seen in the discussion in Chapter 6. We will consider a generalized version of Lyapunov exponents motivated by Selgrade's theorem, Theorem 9.2.5, where the Selgrade bundles are constructed from the chain components in the projective bundle. Recall that the Selgrade bundles are generalizations of the Lyapunov spaces for linear flows in $\mathbb{R}^d$ which determine the Lyapunov exponents for linear autonomous differential equations; in the autonomous case all Lyapunov exponents can be obtained by the exponential growth rates of trajectories which project to the chain components. Instead of looking at exponential growth rates defined via trajectories, we will consider exponential growth rates defined via pieces of trajectories in the Selgrade bundles which project to chains in the chain components.

Recall from Section 9.1 the definition of the Morse spectrum. Let ζ be an (ε, T)-chain in $B \times \mathbb{P}^{d-1}$ given by $T_0, \ldots, T_{n-1} > T, (b_0, p_0), \ldots, (b_{n-1}, p_{n-1}) \in B \times \mathbb{P}^{d-1}$ and satisfying $d(\mathbb{P}\Phi(T_i, b_i, p_i), (b_{i+1}, p_{i+1})) < \varepsilon$ for $i = 0, 1, \ldots, n-1$. The total time of the chain ζ is $\tau(\zeta) = \sum_{i=0}^{n-1} T_i$ and the chain exponent of ζ is

$$\lambda(\zeta) = \frac{1}{\tau(\zeta)} \sum_{i=0}^{n-1} (\log \|\varphi(T_i, b_i, x_i)\| - \log \|x_i\|),$$

where $x_i \in \mathbb{P}^{-1}(p_i)$. By Selgrade's Theorem, Theorem 9.2.5, the chain components $\mathcal{M}_i$ of the projective flow determine the Selgrade bundles $\mathcal{V}_i = \mathbb{P}^{-1}\mathcal{V}_i$, and hence we let

$$(9.3.1) \quad \Sigma_{Mo}(\mathcal{V}_i) := \left\{ \begin{array}{l} \lambda \in \mathbb{R} \,|\, \text{there are } \varepsilon^k \to 0, T^k \to \infty \text{ and } (\varepsilon^k, T^k)\text{-} \\ \text{chains } \zeta^k \text{ in } \mathbb{P}\mathcal{V}_i = \mathcal{M}_i \text{ with } \lim \lambda(\zeta^k) = \lambda \end{array} \right\}.$$

We start the analysis of the Morse spectrum with the following lemma providing, in particular, a uniform bound on chain exponents.

9.3. The Morse Spectrum

Lemma 9.3.1. *(i) For every topological linear flow Φ on $B \times \mathbb{R}^d$ there are $c, \mu > 0$ such that for all $b \in B$ and all $t \geq 0$,*

$$c^{-1}e^{-\mu t} < \inf_{\|x\|=1} \|\varphi(t,b,x)\| \leq \sup_{\|x\|=1} \|\varphi(t,b,x)\| < ce^{\mu t}.$$

(ii) For all $t_1, t_2 \geq 1$ and $(b,x) \in B \times \mathbb{R}^d$ with $x \neq 0$ one has with $M := \log c + \mu$,

$$\left| \frac{1}{t_1+t_2} \log \|\varphi(t_1+t_2,b,x)\| - \frac{1}{t_1} \log \|\varphi(t_1,b,x)\| \right| \leq 2M \frac{t_2}{t_1+t_2}.$$

(iii) For all $\varepsilon > 0, T \geq 1$ and every (ε, T)-chain ζ the chain exponent satisfies $\lambda(\zeta) \leq M$.

Proof. (i) Let $\|\Phi(t,b)\| := \sup_{\|x\|=1} \|\varphi(t,b,x)\|$ and define

$$c := \max\{\|\Phi(t,b)\| \mid b \in B \text{ and } 0 \leq t \leq 1\} < \infty \text{ and } \mu := \log c.$$

Note that every $t \geq 0$ can be written as $t = n + \tau$ for some $n \in \mathbb{N}_0$ and $0 \leq \tau < 1$. The cocycle property then implies that

$$\|\Phi(t,b)\| = \|\Phi(\tau, \theta(n)b) \cdot \Phi(h, \theta(n-1)b) \ldots \Phi(h,b)\| \leq c^{n+1} \leq ce^{\mu t}.$$

The estimate from below follows analogously.

(ii) Define the finite time Lyapunov exponent

$$\lambda(t,b,x) := \frac{1}{t} \left(\log \|\varphi(t,b,x)\| - \log \|x\| \right).$$

Then the cocycle property shows

(9.3.2)
$$\begin{aligned}
&t_2 \lambda(t_2, \Phi(t_1,b,x)) \\
&= \log \|\varphi(t_2, \Phi(t_1,b,x))\| - \log \|\varphi(t_1,b,x)\| \\
&= \log \|\varphi(t_1+t_2,b,x)\| - \log \|\varphi(t_1,b,x)\| \\
&= \log \|\varphi(t_1+t_2,b,x)\| - \log \|x\| - [\log \|\varphi(t_1,b,x)\| - \log \|x\|] \\
&= (t_1+t_2) \lambda(t_1+t_2,b,x) - t_1 \lambda(t_1,b,x).
\end{aligned}$$

Assertion (i) and linearity imply that for $b \in B, x \in \mathbb{R}^d$ and $t \geq 1$,

$$\lambda(t,b,x) = \frac{1}{t} [\log \|\varphi(t,b,x)\| - \log \|x\|] \leq \log c + \mu = M.$$

Hence it follows that

$$|\lambda(t_1+t_2,b,x) - \lambda(t_1,b,x)|$$
$$= \left| \frac{1}{t_1+t_2} [t_2 \lambda(t_2, \Phi(t_1,b,x)) + t_1 \lambda(t_1,b,x)] - \lambda(t_1,b,x) \right|$$
$$\le \frac{t_2}{t_1+t_2} |\lambda(t_2, \Phi(t_1,b,x))| + \left|\frac{t_1}{t_1+t_2} - 1\right| |\lambda(t_1,b,x)|$$
$$\le \frac{t_2}{t_1+t_2} M + \frac{t_2}{t_1+t_2} M = 2M \frac{t_2}{t_1+t_2}.$$

(iii) One computes

$$\lambda(\zeta) = \frac{1}{\tau(\zeta)} \sum_{i=0}^{n-1} (\log \|\varphi(T_i, b_i, x_i)\| - \log \|x_i\|)$$
$$\le \frac{1}{\tau(\zeta)} \sum_{i=0}^{n-1} (\log c + \mu T_i) \le \log c + \mu = M. \qquad \square$$

The following lemma shows that the chain exponent of concatenated chains is a convex combination of the individual chain exponents.

Lemma 9.3.2. *Let ξ, ζ be (ε, T)-chains in $B \times \mathbb{P}^{d-1}$ with total times $\tau(\xi)$ and $\tau(\zeta)$, respectively, such that the initial point of ξ coincides with the final point of ζ. Then the chain exponent of the concatenated chain $\zeta \circ \xi$ is given by*

$$\lambda(\xi \circ \zeta) = \frac{\tau(\xi)}{\tau(\xi)+\tau(\zeta)} \lambda(\xi) + \frac{\tau(\zeta)}{\tau(\xi)+\tau(\zeta)} \lambda(\zeta).$$

Proof. Let the chains ξ and ζ be given by $(b_0, \mathbb{P}x_0), \ldots, (b_k, \mathbb{P}x_k) \in B \times \mathbb{P}^{d-1}$ and $(a_0, \mathbb{P}y_0) = (b_k, \mathbb{P}x_k), \ldots, (a_n, \mathbb{P}y_n)$ with times $S_0, \ldots, S_{k-1}$ and $T_0, \ldots, T_{n-1}$, respectively. Thus one computes

$$(\tau(\xi) + \tau(\zeta)) \lambda(\xi \circ \zeta)$$
$$= \sum_{i=0}^{k-1} [\log \|\Phi(S_i, x_i)\| - \log \|x_i\|] + \sum_{i=0}^{n-1} [\log \|\Phi(T_i, y_i)\| - \log \|y_i\|]$$
$$= \tau(\xi)\lambda(\xi) + \tau(\zeta)\lambda(\zeta). \qquad \square$$

It is actually sufficient to consider periodic chains in the definition of the Morse spectrum, i.e., the exponents of chains from a point to itself. The proof is based on Lemma 3.2.8, which gives a uniform upper bound for the time needed to connect any two points in a chain component.

9.3. The Morse Spectrum

Proposition 9.3.3. *Let $\mathcal{V}_i \subset B \times \mathbb{R}^d$ be a Selgrade bundle with $\mathcal{M}_i = \mathbb{P}\mathcal{V}_i$. Then the Morse spectrum over $\mathcal{V}_i$ defined in (9.3.1) satisfies*

$$\Sigma_{Mo}(\mathcal{V}_i) = \left\{ \begin{array}{c} \lambda \in \mathbb{R} \mid \text{ there are } \varepsilon^k \to 0, \, T^k \to \infty \text{ and periodic} \\ (\varepsilon^k, T^k)\text{-chains } \zeta^k \text{ in } \mathcal{M}_i \text{ with } \lambda(\zeta^k) \to \lambda \text{ as } k \to \infty \end{array} \right\}.$$

Proof. Let $\lambda \in \Sigma_{Mo}(\mathcal{V}_i)$ and fix $\varepsilon, T > 0$. It suffices to prove that for every $\delta > 0$ there exists a periodic (ε, T)-chain ζ' with $|\lambda - f(\zeta')| < \delta$. By Lemma 3.2.8 there exists $\bar{T}(\varepsilon, T) > 0$ such that for all $(b, p), (b', p') \in \mathcal{M}_i$ there is an (ε, T)-chain ξ in $\mathcal{M}_i$ from (b, p) to (b', p') with total time $\tau(\xi) \leq \bar{T}(\varepsilon, T)$. For $S > T$ choose an (ε, S)-chain ζ with $|\lambda - \lambda(\zeta)| < \frac{\delta}{2}$ given by, say, $(b_0, p_0), \ldots, (b_m, p_m)$ with times $S_0, \ldots, S_{m-1} > S$ and with total time $\tau(\zeta)$. Concatenate this with an (ε, T)-chain ξ from (b_m, p_m) to (b_0, p_0) with times $T_0, \ldots, T_{m-1} > T$ and total time $\tau(\xi) \leq \bar{T}(\varepsilon, T)$. The periodic (ε, T)-chain $\zeta' = \xi \circ \zeta$ has the desired approximation property: Since the chain ξ depends on ζ, also τ depends on ζ. However, the total length of ξ is bounded as $\tau(\xi) \leq \bar{T}(\varepsilon, T)$. Lemma 9.3.2(ii) implies

$$|\lambda(\zeta) - \lambda(\xi \circ \zeta)| = \left| \lambda(\zeta) - \frac{\tau(\xi)}{\tau(\xi) + \tau(\zeta)} \lambda(\xi) - \frac{\tau(\zeta)}{\tau(\xi) + \tau(\zeta)} \lambda(\zeta) \right|$$

$$\leq \frac{\tau(\xi)}{\tau(\xi) + \tau(\zeta)} |\lambda(\zeta)| + \frac{\tau(\xi)}{\tau(\xi) + \tau(\zeta)} |\lambda(\xi)|.$$

By Lemma 9.3.2(i) there is a uniform bound for $|\lambda(\xi)|$ and $|\lambda(\zeta)|$ for all considered chains ξ and ζ. Since $\tau(\xi)$ remains bounded for chains ζ with total time $\tau(\zeta)$ tending to ∞, the right-hand side tends to 0 as $S \to \infty$ and hence $\tau(\zeta) \to \infty$. $\square$

The following result describes the behavior of the Morse spectrum under time reversal.

Proposition 9.3.4. *For a linear topological flow Φ on a vector bundle $B \times \mathbb{R}^d$ let the corresponding time-reversed flow $\Phi^* = (\theta^*, \varphi^*)$ be defined by*

$$\Phi^*(t, b, x) = \Phi(-t, b, x), \, t \in \mathbb{R}, \, (b, x) \in B \times \mathbb{R}^d.$$

Then the Selgrade decompositions of Φ and Φ^* coincide and for every Selgrade bundle $\mathcal{V}_i$ one has $\Sigma_{Mo}(\mathcal{V}_i, \Phi^*) = -\Sigma_{Mo}(\mathcal{V}_i, \Phi)$.

Proof. It follows from Proposition 3.1.13 that the chain recurrent sets of $\mathbb{P}\Phi$ and $\mathbb{P}\Phi^*$ coincide. Hence also the chain components coincide, which determine the Selgrade bundles. This shows the first assertion. In order to show the assertion for the Morse spectrum, consider for a Selgrade bundle $\mathcal{V}_i$ the corresponding chain component $\mathcal{M}_i = \mathbb{P}\mathcal{V}_i$. For $\varepsilon, T > 0$ let ζ be a periodic (ε, T)-chain of $\mathbb{P}\Phi$ in $\mathcal{M}_i$ given by $n \in \mathbb{N}, T_0, T_1, \ldots, T_{n-1} \geq$

$T, (b_0, p_0), \ldots, (b_n, p_n) = (b_0, p_0) \in \mathbb{P}\mathcal{V}_i$. An (ε, T)-chain of $\mathbb{P}\Phi^*$ in $\mathcal{M}_i$ is obtained by going backwards: Define

$T_i^* = T_{n-i-1}$ and $(b_i^*, p_i^*) = \mathbb{P}\Phi(T_{n-i-1}, b_{n-i-1}, p_{n-i-1})$ for $i = 0, \ldots, n-1$.
With $x_i \in \mathbb{P}^{-1} p_i$ and $x_i^* \in \mathbb{P}^{-1} p_i^*$ we obtain for the chain exponents

$$\lambda(\zeta) = \frac{1}{\tau(\zeta)} \left(\sum_{i=0}^{n-1} \log \|\varphi(T_i, b_i, x_i)\| - \log \|x_i\| \right)$$

$$= \frac{1}{\tau(\zeta^*)} \left(\sum_{i=0}^{n-1} \log \|x_{n-i-1}^*\| - \log \|\varphi^*(T_{n-i-1}^*, b_{n-i-1}^*, x_{n-i-1}^*)\| \right)$$

$$= -\lambda(\zeta^*).$$

For (ε^k, T^k)-chains with $\varepsilon^k \to 0$ and $T^k \to \infty$, the assertion follows. □

Next we will show that the Morse spectrum over a Selgrade bundle is an interval. The proof is based on a 'mixing' of exponents near the extremal values of the spectrum.

Theorem 9.3.5. *For a topological linear flow Φ the Morse spectrum over a Selgrade bundle $\mathcal{V}_i$ is a compact interval,*

$$\boldsymbol{\Sigma}_{Mo}(\mathcal{V}_i) = [\kappa_i^*, \kappa_i].$$

Proof. By Lemma 9.3.1(ii), the Morse spectrum is bounded and by definition it is closed. Let $\kappa_i^* = \min \boldsymbol{\Sigma}_{Mo}(\mathcal{V}_i)$ and $\kappa_i = \max \boldsymbol{\Sigma}_{Mo}(\mathcal{V}_i)$. Then it suffices to show that for all $\lambda \in [\kappa_i^*, \kappa_i]$, all $\delta > 0$, and all $\varepsilon, T > 0$ there is a periodic (ε, T)-chain ζ in $\mathcal{M}_i = \mathbb{P}\mathcal{V}_i$ with

(9.3.3) $$|\lambda(\zeta) - \lambda| < \delta.$$

For fixed $\delta > 0$ and $\varepsilon, T > 0$, Proposition 9.3.3 shows that there are periodic (ε, T)-chains ζ^* and ζ in $\mathcal{M}$ with

$$\lambda(\zeta^*) < \kappa_i^* + \delta \text{ and } \lambda(\zeta) > \kappa_i - \delta.$$

Denote the initial points of ζ^* and ζ by (b_0^*, p_0^*) and (b_0, p_0), respectively. By chain transitivity there are (ε, T)-chains ζ_1 from (b_0^*, p_0^*) to (b_0, p_0) and ζ_2 from (b_0, p_0) to (b_0^*, p_0^*), both in $\mathcal{M}_i$. For $k \in \mathbb{N}$ let ζ^{*k} and ζ^k be the k-fold concatenation of ζ^* and of ζ, respectively. Then for $k, l \in \mathbb{N}$ the concatenation $\zeta^{k,l} = \zeta_2 \circ \zeta^k \circ \zeta_1 \circ \zeta^l$ is a periodic (ε, T)-chain in $\mathcal{M}_i$. By Lemma 9.3.2 the exponents of concatenated chains are convex combinations of the corresponding exponents. Hence for every $\lambda \in [\lambda(\zeta^*), \lambda(\zeta)]$ one finds numbers $k, l \in \mathbb{N}$ such that $|\lambda(\zeta^{k,l}) - \lambda| < \delta$. This proves (9.3.3). □

Looking at Theorem 9.1.5, we see that the objects in assertions (i), (iv), and (v) have been constructed: Selgrade's Theorem, Theorem 9.2.5, provides the subbundles related to the chain components in the projective

9.3. The Morse Spectrum

bundle and by Theorem 9.3.5 the Morse spectrum is a compact interval. Thus the Morse spectrum has a very simple structure: It consists of $\ell \leq d$ compact intervals associated to the Selgrade bundles. Before we discuss the relations to the Lyapunov exponents in Section 9.4, we will establish the assertion in (ii) that the boundary points of the Morse spectrum are strictly ordered. This needs some preparations.

The following lemma follows similarly as Fenichel's uniformity lemma, Lemma 9.1.11.

Lemma 9.3.6. *Let Φ be a linear flow Φ on a vector bundle $\mathcal{V} = B \times \mathbb{R}^d$ and let $\mathcal{V} = \mathcal{V}' \oplus \mathcal{V}''$ with invariant subbundles $\mathcal{V}^+$ and $\mathcal{V}^-$. Suppose that for all $b \in B$ and all $(b, x') \in \mathcal{V}'_b$ and $(b, x'') \in \mathcal{V}''_b$ with $x', x'' \neq 0$ one has*

$$\lim_{t \to \infty} \frac{\|\varphi(t, b, x')\|}{\|\varphi(t, b, x'')\|} = 0.$$

Then there are $c \geq 1$ and $\mu > 0$ such that

$$\|\varphi(t, b, x')\| \leq ce^{-\mu t} \|\varphi(t, b, x'')\|$$

for all $t \geq 0$ and all $(b, x') \in \mathcal{V}'_b$ and $(b, x'') \in \mathcal{V}''_b$ with $\|x'\| = \|x''\|$. Then the subbundles are called exponentially separated.

Proof. It suffices to prove the assertion for pairs in the unit sphere bundle $B \times \mathbb{S}^{d-1}$. By assumption there is for all $(b, x') \in \mathcal{V}'_b$ and $(b, x'') \in \mathcal{V}''_b$ with $\|x'\| = \|x''\| = 1$ a time $T := T((b, x'), (b, x'')) > 0$ such that

$$\frac{\|\varphi(T, b, x')\|}{\|\varphi(T, b, x'')\|} < \frac{1}{2}.$$

Let $\mathbb{S}(\mathcal{V}') := \mathcal{V}' \cap (B \times \mathbb{S}^{d-1})$ and $\mathbb{S}(\mathcal{V}'') := \mathcal{V}'' \cap (B \times \mathbb{S}^{d-1})$. Using compactness and linearity one finds a finite covering $N_1, \ldots, N_n$ of $\mathbb{S}(\mathcal{V}') \times \mathbb{S}(\mathcal{V}'')$ and times $T_1, \ldots, T_n > 0$ such that for every $i = 1, \ldots, n$,

$$\frac{\|\varphi(T_i, b, x')\|}{\|\varphi(T_i, b, x'')\|} < \frac{1}{2} \frac{\|x'\|}{\|x''\|}$$

for all $((b, x'), (b, x'')) \in \mathcal{V}' \times \mathcal{V}''$ with $\left(\left(b, \frac{x'}{\|x'\|}\right), \left(b, \frac{x''}{\|x''\|}\right)\right) \in N_i$.

Fix $(b, x') \in \mathbb{S}(\mathcal{V}'_b), (b, x'') \in \mathbb{S}(\mathcal{V}''_b), b \in B$. Choose a sequence of integers $i_j \in \{1, \ldots, n\}, j \in \mathbb{N}$, in the following way:

$i_1: \quad ((b, x'), (b, x'')) \in N_{i_1},$

$i_2: \quad \left(\frac{\varphi(T_{i_1}, b, x')}{\|\varphi(T_{i_1}, b, x')\|}, \frac{\varphi(T_{i_1}, b, x'')}{\|\varphi(T_{i_1}, b, x'')\|}\right) \in N_{i_2},$

$\vdots$

$i_j: \quad \left(\frac{\varphi(T_{i_1} + T_{i_2} \ldots + T_{i_{j-1}}, b, x')}{\|\varphi(T_{i_1} + T_{i_2} \ldots + T_{i_{j-1}}, b, x')\|}, \frac{\varphi(T_{i_1} + T_{i_2} \ldots + T_{i_{j-1}}, b, x'')}{\|\varphi(T_{i_1} + T_{i_2} \ldots + T_{i_{j-1}}, b, x'')\|}\right) \in N_{i_j}.$

Using the cocycle property and invariance of the subbundles, we obtain for $\tau_j := T_{i_1} + T_{i_2} + \ldots + T_{i_j}$,

$$\frac{\|\varphi(\tau_j, b, x')\|}{\|\varphi(\tau_j, b, x'')\|} = \frac{\|\varphi(T_{i_j}, \theta(T_{i_1} + \ldots + T_{i_{j-1}}, b), \varphi(T_{i_1} + \ldots + T_{i_{j-1}}, b, x'))\|}{\|\varphi(T_{i_j}, \theta(T_{i_1} + \ldots + T_{i_{j-1}}, b), \varphi(T_{i_1} + \ldots + T_{i_{j-1}}, b, x''))\|}$$
$$\leq \frac{1}{2} \frac{\|\varphi(T_{i_1} + \ldots + T_{i_{j-1}}, b, x')\|}{\|\varphi(T_{i_1} + \ldots + T_{i_{j-1}}, b, x'')\|} \leq \ldots \leq \left(\frac{1}{2}\right)^j \frac{\|x'\|}{\|x''\|}.$$

Now $j \geq \tau_j/T$ and $\log 2 > \frac{1}{2}$ imply $\left(\frac{1}{2}\right)^j = e^{-j \log 2} \leq e^{-\frac{\tau_j}{2T}}$. Hence with $\mu := \frac{1}{2T}$ it follows that

$$\|\varphi(\tau_j, b, x')\| \leq e^{-\mu \tau_j} \|\varphi(\tau_j, b, x'')\|.$$

This yields the assertion for all times $t = \tau_j, j \in \mathbb{N}$.

For arbitrary $t > 0$ we write $t = \tau_j + r$ for some $j \in \mathbb{N}_0$ and $0 \leq r \leq T$. Let $c_1 := \max\{\|\Phi(r, b)\| \mid r \in [0, T], b \in B\} < \infty$. Then

$$\|\varphi(t, b, x')\| \leq \|\varphi(r, \theta(\tau_j, b))\varphi(\tau_j, b, x')\| \leq c_1 \|\varphi(\tau_j, b, x')\|$$

and, with $c_2 := \min\{\|\Phi(r, b)x\| \mid \|x\| = 1\} > 0$,

$$\|\varphi(t, b, x'')\| = \|\Phi(r, \theta(\tau_j, b))\varphi(\tau_j, b, x'')\| \geq c_2 \|\varphi(\tau_j, b, x'')\|.$$

Using $t \leq \tau_j + T$ we get the following inequalities which imply the assertion with $c := \frac{c_1}{c_2} e^{-\mu T}$:

$$\|\varphi(t, b, x')\| \leq c_1 \|\varphi(\tau_j, b, x')\| \leq c_1 e^{-\mu \tau_j} \|\varphi(\tau_j, b, x'')\|$$
$$\leq \frac{c_1}{c_2} e^{-\mu T} e^{-\mu t} \|\varphi(t, b, x'')\|. \qquad \square$$

By Selgrade's Theorem we obtain the following exponentially separated subbundles.

Lemma 9.3.7. *Let Φ be a linear flow on a vector bundle $\mathcal{V} = B \times \mathbb{R}^d$ with chain transitive flow on the base space B and consider the decomposition (9.2.7) into invariant subbundles. Then for every $j \in \{1, \ldots, \ell - 1\}$ the subbundles*

$$\mathcal{V}' := \mathcal{V}_1 \oplus \ldots \oplus \mathcal{V}_j \text{ and } \mathcal{V}'' := \mathcal{V}_{j+1} \oplus \ldots \oplus \mathcal{V}_\ell.$$

are exponentially separated.

Proof. By Remark 9.2.6, the projection $A := \mathbb{P}(\mathcal{V}_1 \oplus \ldots \oplus \mathcal{V}_j)$ is an attractor with complementary repeller $A^* := \mathbb{P}(\mathcal{V}_{j+1} \oplus \ldots \oplus \mathcal{V}_\ell)$. For the time-reversed flow, A^* is an attractor with complementary repeller A; cf. Proposition 8.2.7. Hence we can apply Lemma 9.2.2(i) to the time reversed flow and obtain for (the original flow) that

$$\lim_{t \to \infty} \frac{\|\varphi(t, b, x')\|}{\|\varphi(t, b, x'')\|} = 0$$

9.3. The Morse Spectrum

for all $(b, x') \in \mathcal{V}'$ and $(b, x'') \in \mathcal{V}''$. By Lemma 9.3.6 it follows that these bundles are exponentially separated. $\square$

After these preparations we can prove that the boundary points of the Morse spectral intervals are strictly ordered. Recall that the Selgrade bundles are numbered according to the order of the Morse decomposition $\mathbb{P}\mathcal{V}_1 \preceq \ldots \preceq \mathbb{P}\mathcal{V}_\ell$ from Theorem 9.2.5.

Theorem 9.3.8. *Let $\Phi = (\theta, \varphi)$ be a topological linear flow on a vector bundle $\mathcal{V} = B \times \mathbb{R}^d$. Then the boundary points of the Morse spectral intervals $\Sigma_{Mo}(\mathcal{V}_j) = [\kappa_j^*, \kappa_j]$ are ordered according to $\kappa_j^* < \kappa_{j+1}^*$ and $\kappa_j < \kappa_{j+1}$ for all $j \in \{1, \ldots, \ell - 1\}$.*

Proof. Consider κ_j^* and κ_{j+1}^*. By Lemma 9.3.7, the vector bundles

$$\mathcal{V}' = \mathcal{V}_1 \oplus \ldots \oplus \mathcal{V}_j \text{ and } \mathcal{V}'' = \mathcal{V}_{j+1} \oplus \ldots \oplus \mathcal{V}_\ell$$

are exponentially separated. Hence there are numbers $c \geq 1$ and $\mu > 0$ such that for $(b, x') \in \mathcal{V}', (b, x'') \in \mathcal{V}''$ with $\|x'\| = \|x''\| = 1$ it follows that

$$(9.3.4) \qquad \|\varphi(t, b, x')\| \leq c e^{-\mu t} \|\varphi(t, b, x'')\|, \, t \geq 0.$$

For $\varepsilon, T > 0$ consider an (ε, T)-chain ζ'' in $\mathcal{M}_{j+1} = \mathbb{P}\mathcal{V}_{j+1}$ given by

$$n \in \mathbb{N}, T_0, \ldots, T_{n-1} > T \text{ and } (b_0, p_0''), \ldots, (b_n, p_n'') \in \mathcal{M}_{j+1}.$$

Then the growth rate $\lambda(\zeta'')$ is given by

$$\lambda(\zeta'') = \frac{1}{\tau(\zeta'')} \sum_{k=0}^{n-1} \log \|\varphi(T_k, b_k, x_k'')\|,$$

where we choose x_k'' with $\mathbb{P}x_k'' = p_k''$ and $\|x_k''\| = 1$. Using invariance of the Morse sets, one finds an (ε, T)-chain ζ' in $\mathcal{M}_j = \mathbb{P}\mathcal{V}_j$ (even without jumps in the component in $\mathbb{P}^{d-1}$) given by

$$T_0, \ldots, T_{n-1} \text{ and } (b_0, p_0'), \ldots, (b_n, p_n') \in \mathcal{M}_j.$$

Then $\tau(\zeta') = \tau(\zeta'') > nT$ and by (9.3.4) the growth rates satisfy for $\mathbb{P}x_k' = p_k'$ and $\|x_k'\| = 1$,

$$\lambda(\zeta') = \frac{1}{\tau(\zeta')} \sum_{k=0}^{n-1} \log \|\varphi(T_k, b_k, x_k')\|$$

$$\leq \frac{1}{\tau(\zeta'')} \sum_{k=0}^{n-1} \log \|\varphi(T_k, b_k, x_k'')\| + \frac{n}{\tau(\zeta')} \log c - \sum_{k=0}^{n-1} \frac{T_k}{\tau(\zeta')} \mu$$

$$\leq \lambda(\zeta'') + \frac{1}{T} \log c - \mu.$$

Now approximate κ_{i+1}^* by (ε, T)-chains ζ'' in $\mathcal{M}_{j+1}$ with $T \to \infty$. Then there are chains ζ' in $\mathcal{M}_j$ with growth rate satisfying the preceding estimate, hence
$$\kappa_j^* \leq \kappa_{j+1}^* - \mu.$$
The assertion for the κ_j follows by time reversal (or using analogous arguments). $\square$

9.4. Lyapunov Exponents and the Morse Spectrum

While we have determined the structure of the Morse spectrum and the corresponding subbundles, the relation to the Lyapunov exponents is not yet clear. We will show that the Morse spectrum contains every Lyapunov exponent of the flow Φ and hence the Morse spectrum generalizes the concept of exponential growth rates of trajectories. This is remarkable, because the Morse spectrum only considers chains in the chain recurrent set while Lyapunov exponents are defined for all initial points. Furthermore, the boundary points of every Morse spectral interval are Lyapunov exponents. This requires a characterization of the Morse spectrum based on initial pieces of trajectories, instead of chains (the resulting spectrum is called the uniform growth spectrum.) Then the proof of the main result of this chapter, Theorem 9.1.5, will be given.

First we show that the Morse spectrum contains all Lyapunov exponents.

Theorem 9.4.1. *Let $\Phi = (\theta, \varphi)$ be a topological linear flow on a vector bundle $\mathcal{V} = B \times \mathbb{R}^d$. Then for all $(b, x) \in B \times \mathbb{R}^d, x \neq 0$, the Lyapunov exponent satisfies*

$$(9.4.1) \qquad \lambda(b, x) = \limsup_{t \to \infty} \frac{1}{t} \log \|\varphi(t, b, x)\| \in \Sigma_{Mo}(\mathcal{V}_i),$$

where $\mathcal{V}_i$ is the Selgrade bundle with $\omega(b, \mathbb{P}x) \subset \mathcal{M}_i = \mathbb{P}\mathcal{V}_i$.

Proof. By Proposition 3.1.12, ω-limit sets are chain transitive and hence contained in a chain component. Thus there is a Selgrade bundle $\mathcal{V}_i$ such that $\mathcal{M}_i = \mathbb{P}\mathcal{V}_i$ contains $\omega(b, \mathbb{P}x)$. Now it suffices to show the following: For all $\delta > 0$ and all $\varepsilon > 0, T > 1$ there exists an (ε, T)-chain ζ in $\omega(b, \mathbb{P}x)$ with

$$(9.4.2) \qquad |\lambda(\zeta) - \lambda(b, x)| < \delta.$$

Fix $\delta > 0, \varepsilon > 0,$ and $T > 1$. Because $\mathbb{P}\Phi$ is uniformly continuous on the compact set $[0, 2T] \times B \times \mathbb{P}^{d-1}$, there is $\delta_0 = \delta_0(\delta, \varepsilon, T) > 0$ such that for all $(b', p'), (b'', p'') \in B \times \mathbb{P}^{d-1}$ it follows from $d((b', p'), (b'', p'')) < \delta_0$ that for all $t \in [0, 2T]$,

$$(9.4.3) \qquad d(\mathbb{P}\Phi(t, b', p'), \mathbb{P}\Phi(t, b'', p'')) < \frac{\varepsilon}{2}$$

9.4. Lyapunov Exponents and the Morse Spectrum

and

(9.4.4) $$\left|\log\|\varphi(t,b',x')\| - \log\|\varphi(t,b'',x'')\|\right| < \frac{\delta}{4},$$

where $x', x'' \in \mathbb{R}^d$ are chosen with $\mathbb{P}x' = p', \mathbb{P}x'' = p''$ and $\|x'\| = \|x''\| = 1$. There is $T_0 > 0$ such that for all $t \geq T_0$,

(9.4.5) $$d(\mathbb{P}\Phi(t,b,\mathbb{P}x), \omega(b,\mathbb{P}x)) < \delta_0.$$

The Lyapunov exponents of (b,x) and $\Phi(T_0,b,x)$ coincide, since the cocycle property implies

$$\lambda(b,x) = \limsup_{t\to\infty} \frac{1}{t-T_0} \log \|\varphi(t-T_0, \theta(T_0,b), \varphi(T_0,b,x))\| = \lambda(\Phi(T_0,b,x)).$$

There is $T_1 > 2T$ with

(9.4.6) $$\left|\lambda(b,x) - \frac{1}{T_1}\log\|\varphi(T_1, \theta(T_0,b), \varphi(T_0,b,x))\|\right| < \frac{\delta}{2}.$$

To simplify the notation, without loss of generality, we replace $\Phi(T_0,b,x) = (\theta(T_0,b), \varphi(T_0,b,x))$ by (b,x). Then (9.4.5) holds for all $t \geq 0$ and (9.4.6) yields for $t \geq 0$,

(9.4.7) $$\left|\lambda(b,x) - \frac{1}{T_1}\log\|\varphi(T_1,b,x)\|\right| < \frac{\delta}{2}.$$

We can express T_1 in the form $T_1 = (l-1)T + r$ with $l \in \mathbb{N}$ and $r \in (T, 2T]$. The relation $T > 1$ implies $T_1 > l$.

Now consider times

$$\tau_0 = \ldots = \tau_{l-2} := T, \quad \tau_{l-1} := r$$

and $x_j := \varphi(\tau_j, b, x)$ and $p_j := \mathbb{P}x_j$ for all $j = 0, \ldots, \ell-1$. Define a chain $\tilde{\zeta}$ with trivial jumps by $(b_0, p_0) = (b, \mathbb{P}x)$ and $(b_{j+1}, p_{j+1}) = \mathbb{P}\Phi(\tau_j, b_j, p_j)$ for $j = 0, \ldots, l-1$. Then the total time is $\tau(\tilde{\zeta}) = \sum_{j=0}^{l-1} \tau_j = T_1$ and

(9.4.8) $$\lambda(\tilde{\zeta}) = \frac{1}{\tau(\tilde{\zeta})} \sum_{j=0}^{l-1} [\log\|\varphi(\tau_j, b_j, x_j)\| - \log\|x_j\|]$$

$$= \frac{1}{T_1} \sum_{j=0}^{l-1} [\log\|x_{j+1}\| - \log\|x_j\|] = \frac{1}{T_1} \log\|\varphi(T_1,b,x)\|.$$

However, the chain $\tilde{\zeta}$ is not necessarily contained in $\omega(b, \mathbb{P}x)$. In order to obtain an appropriate chain ζ in $\omega(b, \mathbb{P}x)$, we use (9.4.5): For (b_j, p_j) we find points (b'_j, p'_j) in $\omega(b, \mathbb{P}x)$ with

(9.4.9) $$d((b_j, p_j), (b'_j, p'_j)) < \delta_0$$

and hence by (9.4.3) $d(\mathbb{P}\Phi(t,b_j,p_j),\mathbb{P}\Phi(t,b_j',p_j')) < \frac{\varepsilon}{2}$ for $t \in [0,2T]$. We obtain for all j,

$$d(\mathbb{P}\Phi(\tau_j,b_j',p_j'),(b_{j+1}',p_{j+1}'))$$
$$\leq d(\mathbb{P}\Phi(\tau_j,b_j',p_j'),\mathbb{P}\Phi(\tau_j,b_j,p_j)) + d(\mathbb{P}\Phi(\tau_j,b_j,p_j),(b_{j+1},p_{j+1}))$$
$$+ d((b_{j+1},p_{j+1}),(b_{j+1}',p_{j+1}'))$$
$$< \frac{\varepsilon}{2} + 0 + \frac{\varepsilon}{2} = \varepsilon.$$

Thus $\tau_1,\ldots,\tau_{l-1} \geq T$ and $(b_0',p_0'),\ldots,(b_l',p_l') \in B \times \mathbb{P}^{d-1}$ define an (ε,T)-chain ζ in $\omega(b,\mathbb{P}x)$.

In order to estimate the exponential growth rates, observe that by (9.4.9) and (9.4.4) for all j,

$$\log \|x_j\| - \log \|x_j'\| < \frac{\delta}{4} \text{ and } \log \|\varphi(\tau_j,b_j,x_j)\| - \log \|\varphi(\tau_j,b_j',x_j')\| < \frac{\delta}{4}.$$

Hence (9.4.8) yields, with $x_j' \in \mathbb{R}^d$, $\mathbb{P}x_j' = p_j'$, and appropriate sign,

$$\left|\lambda(\tilde\zeta) - \lambda(\zeta)\right|$$
$$< \frac{1}{T_1}\left|\sum_{j=0}^{l-1}[\log\|\varphi(\tau_j,b_j,x_j)\| - \log\|x_j\|] - [\log\|\varphi(\tau_j,b_j',x_j')\| - \log\|x_j'\|]\right|$$
$$< T_1 l \frac{\delta}{2} < \frac{\delta}{2}.$$

This estimate together with (9.4.7) shows (9.4.2) since

$$|\lambda(b,x) - \lambda(\zeta)| \leq \left|\lambda(b,x) - \frac{1}{T_1}\log\|\varphi(T_1,b,x)\|\right| + \left|\lambda(\tilde\zeta) - \lambda(\zeta)\right| < \delta. \quad \square$$

Remark 9.4.2. Let $\lambda^-(b,x) = \liminf_{t\to\infty} \frac{1}{t}\log\|\varphi(t,b,x)\|$ be the backward Lyapunov exponent of $(b,x) \in B \times \mathbb{R}^d$ under the flow Φ. Then $\lambda^-(b,x) \in \Sigma_{Mo}(\mathcal{V}_j,\Phi)$, where $\mathbb{P}\mathcal{V}_j = \mathcal{M}_j$ is the chain component of $\mathbb{P}\Phi$ containing $\omega^*(b,\mathbb{P}x)$. This is proved in the same way as Theorem 9.4.1 using the time-reversed flow Φ^*.

Theorem 9.4.1 shows that the Morse spectrum contains all the Lyapunov exponents. How much larger is the Morse spectrum than the set of Lyapunov exponents? We prove that the boundary points of the intervals corresponding to a Selgrade bundle, actually, are Lyapunov exponents. For the proof some preparations are necessary. In the process, an alternative characterization of the Morse spectrum will be given. We need the following technical lemma.

9.4. Lyapunov Exponents and the Morse Spectrum

Lemma 9.4.3. *Pick* $(b, x) \in B \times \mathbb{P}^{d-1}$, *fix a time* $t > 0$ *and consider* $\lambda(t, b, x) = \frac{1}{t}(\log \|\varphi(t, b, x)\| - \log \|x\|)$. *Then for any* $\varepsilon \in (0, 2M)$ *there exists a time* $t^* \leq [(2M - \varepsilon)t]/(2M)$ *such that*

$$\lambda(s, \Phi(t^*, b, x)) \leq \lambda(t, b, x) + \varepsilon \text{ for all } s \in (0, t - t^*]$$

where M *is as in Lemma 9.3.1. Furthermore,* $t - t^* \geq \frac{\varepsilon t}{2M} \to \infty$ *as* $t \to \infty$.

Proof. Abbreviate $\sigma := \lambda(t, b, x)$. Let $\varepsilon \in (0, 2M)$ and define

$$\beta := \sup_{s \in (0, t)} \lambda(s, b, x).$$

If $\beta \leq \sigma + \varepsilon$, the assertion follows with $t^* = 0$. For $\beta > \sigma + \varepsilon$, let

$$t^* := \sup\{s \in (0, t] \mid \lambda(s, b, x) \geq \sigma + \varepsilon\}.$$

Because the map $s \mapsto \lambda(s, b, x)$ is continuous, $\beta > \sigma + \varepsilon$ and $\lambda(t, b, x) = \sigma < \sigma + \varepsilon$, it follows by the intermediate value theorem, that $\lambda(t^*, b, x) = \sigma + \varepsilon$ and $0 < t^* < t$. By Lemma 9.3.1(ii), it follows with $\bar{t} := t - t^*$, that

$$\varepsilon = |\sigma - (\sigma + \varepsilon)| = |\lambda(t, b, x) - \lambda(t^*, b, x)| \leq 2M\frac{\bar{t}}{t},$$

and hence $\bar{t} \geq \varepsilon t/(2M)$ which implies $t^* \leq [(2M - \varepsilon)t]/(2M)$. For $s \in (0, t - t^*]$ it holds that $t^* + s \in [t^*, t]$ and hence $\lambda(t^* + s, b, x) < \sigma + \varepsilon$. Finally, using (9.3.2), we find the desired result:

$$\lambda(s, \Phi(t^*, b, x)) = \frac{1}{s}[(t^* + s)\lambda(t^* + s, b, x) - t^*\lambda(t^*, b, x)]$$

$$< \frac{1}{s}[(t^* + s)(\sigma + \varepsilon) - t^*(\sigma + \varepsilon)] = \sigma + \varepsilon. \quad \square$$

Next we associate to every Selgrade bundle $\mathcal{V}_i$ another set of growth rates, called the uniform growth spectrum $\Sigma_{UG}(\mathcal{V}_i)$.

Proposition 9.4.4. *For every Selgrade bundle* $\mathcal{V}_i$ *with* $\mathcal{M}_i = \mathbb{P}\mathcal{V}_i$ *the set*

$$\Sigma_{UG}(\mathcal{V}_i) := \left\{\lambda \in \mathbb{R} \ \Big| \ \begin{array}{c} \text{there are } t_k \to \infty \text{ and } (b_k, \mathbb{P}x_k) \in \mathcal{M}_i \\ \text{with } \lambda(t_k, b_k, x_k) \to \lambda \text{ for } k \to \infty \end{array}\right\}$$

is a compact interval and there are $(b^*, x^*), (b, x) \in \mathcal{M}_i$ *with*

$$\min \Sigma_{UG}(\mathcal{M}_i) = \lambda(b^*, x^*) \text{ and } \max \Sigma_{UG}(\mathcal{M}_i) = \lambda(b, x).$$

Proof. Since $\mathcal{M}_i$ is compact and connected and the map $(t, b, x) \mapsto \lambda(t, b, x)$ is continuous, the image of this map is compact and connected and hence a compact interval. Thus there are $\varepsilon_k \to 0, t_k \to \infty$ and $(b_k, x_k) \in \mathcal{V}_i$ with

$$\lambda(t_k, b_k, x_k) < \min \Sigma_{UG}(\mathcal{V}_i) + \varepsilon_k.$$

Let $\bar{\varepsilon}_k := 1/\sqrt{t_k}$, and apply Lemma 9.4.3 to x_k and t_k with $\varepsilon = \bar{\varepsilon}_k$ for each $k \in \mathbb{N}$. Thus we obtain times t_k^* such that

$$\lambda(s, \Phi(t_k^*, b_k, x_k)) \leq \min \Sigma_{UG}(\mathcal{V}_i) + \varepsilon_k + \bar{\varepsilon}_k \text{ for all } s \in (0, t_k - t_k^*],$$

where $t_k - t_k^* \geq \bar{\varepsilon}_k t_k/(2M) = \sqrt{t_k}/(2M)$. Define $(\bar{b}, \bar{x}_k) = \Phi(t_k^*, b_k, x_k)$ and $\bar{t}_k := t_k - t_k^* \to \infty$. Thus

$$\lambda(s, \bar{b}, \bar{x}_k) \leq \min \Sigma_{UG}(\mathcal{V}_i) + \varepsilon_k + \bar{\varepsilon}_k \text{ for all } s \in (0, \bar{t}_k].$$

Since $\mathcal{M}_i$ is compact, we may assume that the points $(\bar{b}, \bar{x}_k)$ converge to some $(\bar{b}, \bar{x}) \in \mathcal{M}_i$. Now fix $t > 0$ and $\varepsilon > 0$. By continuity of the map $x \mapsto \lambda(t, b, x)$, there exists a $k_0 \in \mathbb{N}$ such that $\left|\lambda(t, \bar{b}, \bar{x}) - \lambda(t, \bar{b}_k, \bar{x}_k)\right| < \varepsilon$ for all $k \geq k_0$. Hence for all $k \geq k_0$ we have

$$\lambda(t, \bar{b}, \bar{x}) < \min \Sigma_{UG}(\mathcal{V}_i) + \varepsilon_k + \bar{\varepsilon}_k + \varepsilon.$$

Since $\varepsilon > 0$ and $t > 0$ have been arbitrarily chosen and $\varepsilon_k + \bar{\varepsilon}_k \to 0$, we find

$$\lambda(t, \bar{b}, \bar{x}) \leq \min \Sigma_{UG}(\mathcal{V}_i) \text{ for all } t > 0,$$

and hence

$$\limsup_{t \to \infty} \lambda(t, \bar{b}, \bar{x}) \leq \min \Sigma_{UG}(\mathcal{V}_i).$$

Since, clearly, $\liminf_{t \to \infty} \lambda(t, \bar{b}, \bar{x}) < \min \Sigma_{UG}(\mathcal{M}_i)$ cannot occur, it follows that $\lambda(\bar{b}, \bar{x}) = \lim_{t \to \infty} \lambda(t, \bar{b}, \bar{x}) = \min \Sigma_{UG}(\mathcal{V}_i)$. For $\max \Sigma_{UG}(\mathcal{V}_i)$ one argues analogously. $\square$

Now we can prove the announced result about the boundary points of the spectral intervals.

Theorem 9.4.5. *Let Φ be a topological linear flow on a vector bundle $\mathcal{V} = B \times \mathbb{R}^d$. Then for every Selgrade bundle $\mathcal{V}_i$, the boundary points κ_i^* and κ_i of the Morse spectrum $\Sigma_{Mo}(\mathcal{V}_i) = [\kappa_i^*, \kappa_i]$ are Lyapunov exponents existing as limits.*

Proof. By Proposition 9.4.4 it suffices to prove that $\Sigma_{UG}(\mathcal{V}_i) = \Sigma_{Mo}(\mathcal{V}_i)$. The inclusion $\Sigma_{UG}(\mathcal{V}_i) \subset \Sigma_{Mo}(\mathcal{V}_i)$ is obvious. For the converse, consider an (ε, T)-chain ζ with times $T_0, \ldots, T_{n-1} \geq T$ and points $(b_0, p_0), \ldots, (b_n, p_n)$. Then

(9.4.10) $$\min_{i=0,\ldots,n-1} \lambda(T_i, b_i, x_i) \leq \lambda(\zeta) \leq \max_{i=0,\ldots,n-1} \lambda(T_i, b_i, x_i).$$

In fact, with $\tau := T_0 + \ldots + T_{n-1}$ the assertion for the maximum is seen as follows:

$$\lambda(\zeta) = \frac{1}{\tau}[T_0 \lambda(T_0, b_0, x_0) + \ldots + T_{n-1} \lambda(T_{n-1}, b_{n-1}, x_{n-1})]$$

$$\leq \frac{T_0}{\tau} \lambda(T_0, b_0, x_0) + \ldots + \frac{T_{n-1}}{\tau} \lambda(T_{n-1}, b_{n-1}, x_{n-1}) \leq \max_i \lambda(T_i, b_i, x_i).$$

Analogously, one argues for the minimum. By definition, there is a sequence of (ε^k, T^k)-chains ζ^k in $\mathcal{M}_i$ with $\varepsilon^k \to 0, T^k \to \infty$ and $\lambda(\zeta^k) \to \min \Sigma_{Mo}(\mathcal{V}_i)$. By (9.4.10) in each ζ^k there is a trajectory piece starting in a point (b_k, p_k)

with time $t_k \geq T^k$ such that $\lambda(t_k, b_k, p_k) \leq \lambda(\zeta^k)$. Hence there exists a subsequence with $\lim_{n\to\infty} \lambda(t_{k_n}, b_{k_n}, p_{k_n}) = \min \Sigma_{Mo}(\mathcal{V}_i)$. □

Note that the proof above also shows that the set $\Sigma_{UG}(\mathcal{V}_i)$ defined in Proposition 9.4.4 coincides with the Morse spectrum $\Sigma_{Mo}(\mathcal{V}_i)$.

Finally, we are in the position to complete the proof of Theorem 9.1.5.

Proof of Theorem 9.1.5. Selgrade's Theorem, Theorem 9.2.5, has constructed the subbundles $\mathcal{V}_i$ related to the chain components $\mathcal{M}_i$ in the projective bundle according to assertion (iii). They yield the decompositions in assertion (i) and assertion (iv) follows from Theorem 9.3.5 showing that the Morse spectrum over the Selgrade bundles are compact intervals. Theorem 9.4.1 shows that all Lyapunov exponents are in the Morse spectral intervals, and Theorem 9.4.5 shows that all κ_i^* and κ_i are Lyapunov exponents existing as a limit. Assertion (iv) claiming that the map associating to $b \in B$ the intersection of $\mathcal{V}_i$ with the fiber $\{b\} \times \mathbb{R}^d$ is continuous follows from the subbundle property and the definition of the Grassmannian in the beginning of Section 5.1:

Suppose $b_n \to b$ with fibers $\mathcal{V}_{i,b_n}$ of dimension d_i. For the convergence in $\mathbb{G}_{d_i}$ we have identified the subspaces with the projective space $\mathbb{P}\left(\bigwedge^{d_i} \mathbb{R}^d\right)$. Thus let $x_1^n, \ldots, x_{d_i}^n$ be an orthonormal basis of the fiber $\mathcal{V}_{i,b_n}$ which is identified with the line generated by $x_1^n \wedge \ldots \wedge x_{d_i}^n$. Then for a subsequence one obtains convergence to an orthonormal set of vectors $x_1, \ldots, x_{d_i}$ in $\mathbb{R}^d$ and hence $x_1^n \wedge \ldots \wedge x_{d_i}^n$ converges to $x_1 \wedge \ldots \wedge x_{d_i}$. Furthermore, $(b, x_j) \in \mathcal{V}_{i,b}$ for $j = 1, \ldots, d_i$, and hence $\mathcal{V}_{i,b_n}$ converges to $\mathcal{V}_{i,b}$ in $\mathbb{G}_{d_i}$.

The assertion in (ii) that $\kappa_i^* < \kappa_{i+1}^*$ and $\kappa_i < \kappa_{i+1}$ for all i follows from Theorem 9.3.8. □

9.5. Application to Robust Linear Systems and Bilinear Control Systems

Our main example of topological linear flows are families of linear differential equation of the form

$$(9.5.1) \qquad \dot{x} = A(u(t))x := A_0 x + \sum_{i=1}^m u_i(t) A_i x, u \in \mathcal{U}$$

with $A_0, \ldots, A_m \in gl(d, \mathbb{R}), \mathcal{U} = \{u : \mathbb{R} \to U \mid \text{integrable on every bounded interval}\}$ and $U \subset \mathbb{R}^m$ is compact, convex with $0 \in \text{int} U$. Unique solutions for any function $u \in \mathcal{U}$ and initial condition $x(0) = x_0 \in \mathbb{R}^d$ exist by Theorem 6.1.1.

As discussed in Section 9.1, these equations may be interpreted as robust linear systems or as bilinear control systems. It is quite obvious that they define linear skew product systems

$$\Phi = (\theta, \varphi) : \mathbb{R} \times \mathcal{U} \times \mathbb{R}^d \longrightarrow \mathcal{U} \times \mathbb{R}^d$$

with the shift $\theta : \mathbb{R} \times \mathcal{U} \longrightarrow \mathcal{U}$, $\theta(t, u(\cdot)) := u(\cdot + t)$ as base component and the skew-component consisting of the solutions $\varphi(t, u, x_0)$, $t \in \mathbb{R}$, of the differential equation. But it requires some work to construct a metric on $\mathcal{U}$ such that this defines a topological linear flow. This will be accomplished in the following, and a number of examples will be discussed.

In order to show that these systems can be considered as topological linear systems, we first endow the set $\mathcal{U}$ with a metric.

Lemma 9.5.1. *Let $\{x_n \mid n \in \mathbb{N}\}$ be a countable, dense subset of $L^1(\mathbb{R}, \mathbb{R}^m)$. Then a metric on the set $\mathcal{U} \subset L^\infty(\mathbb{R}, \mathbb{R}^m)$ is given by*

$$(9.5.2) \qquad d(u, v) = \sum_{n=1}^{\infty} \frac{1}{2^n} \frac{\left| \int_{\mathbb{R}} [u(t) - v(t)]^\top x_n(t) dt \right|}{1 + \left| \int_{\mathbb{R}} [u(t) - v(t)]^\top x_n(t) \, dt \right|}.$$

Proof. Clearly, $d(u, v) \geq 0$ for all $u, v \in \mathcal{U}$. Since $\{x_n \mid n \in \mathbb{N}\}$ is dense, it follows that $d(u, v) = 0$ if and only if $\int_{\mathbb{R}} [u(t) - v(t)]^\top x(t) dt = 0$ for all $x \in L^1(\mathbb{R}, \mathbb{R}^m)$, i.e., $u = v$.

For the proof of the triangle inequality it suffices to show it for each summand separately. The function $\rho(a) := \frac{a}{1+a}, a \geq 0$, is monotonically increasing and $\rho(a + b) \leq \rho(a) + \rho(b)$ for $a, b \geq 0$. Hence for all $a, b, c \in \mathbb{R}$ one has $\rho(|a + b|) \leq \rho(|a|) + \rho(|b|)$ and

$$\rho(|a - c|) = \rho(|a - b + b - c|) \leq \rho(|a - b|) + \rho(|b - c|).$$

For $u, v, w \in \mathcal{U}$ and $n \in \mathbb{N}$ let

$$a := \int_{\mathbb{R}} u(t)^\top x_n(t) dt, b := \int_{\mathbb{R}} v(t)^\top x_n(t) dt, c := \int_{\mathbb{R}} w(t)^\top x_n(t) dt.$$

Then the triangle inequality for n follows. □

Remark 9.5.2. An example of such a countable dense subset in the Banach space $L^1(\mathbb{R}, \mathbb{R}^m)$ is provided by all linear combinations with rational coefficients in $\mathbb{Q}^m$ of characteristic functions of intervals with rational endpoints. Alternatively, consider all polynomials $x(t) = \sum_{i=0}^{k} a_i t^i$ with $a_i \in \mathbb{Q}^m$. Then the set of all restrictions to intervals with rational endpoints is countable. Density follows by the theorem of Stone-Weierstraß, which shows that the polynomials are dense in the space of continuous functions which in turn is dense in L^1.

Lemma 9.5.1 shows that $\mathcal{U}$ is a metrizable topological space. The particular choice of the metric is irrelevant for our purposes. For notational convenience we will consider $\mathcal{U}$ as a metric space with a fixed metric given by (9.5.2). Next we discuss the associated notion of convergence.

Proposition 9.5.3. *Consider a sequence (u_k) in $\mathcal{U}$. Then $u_k \to u$ in $\mathcal{U}$ if and only for all $x \in L^1(\mathbb{R}, \mathbb{R}^m)$ one has*

$$(9.5.3) \qquad \int_{\mathbb{R}} u_k(t)^\top x(t) dt \to \int_{\mathbb{R}} u(t)^\top x(t) dt \text{ for } k \to \infty.$$

Proof. Suppose that $u_k \to u$ in $\mathcal{U}$, fix $x \in L^1(\mathbb{R}, \mathbb{R}^m)$ and $\varepsilon > 0$. There are x_N from the set in Lemma 9.5.1 and $T > 0$ large enough such that

$$\|x - x_N\|_1 = \int_{\mathbb{R}} \|x(t) - x_N(t)\| \, dt < \varepsilon \text{ and } \int_{\mathbb{R} \setminus [-T,T]} \|x_N(t)\| \, dt < \varepsilon.$$

Then for all $k \in \mathbb{N}$,

$$\left| \int_{\mathbb{R}} [u_k(t) - u(t)]^\top x_N(t) dt \right| \leq \|u - u_k\|_\infty \|x_N\|_1 \leq \operatorname{diam}\mathcal{U} \, \|x_N\|_1.$$

Furthermore, for all k large enough, one has $d(u_k, u) < \varepsilon$ and hence

$$\frac{1}{2^N} \frac{\left| \int_{\mathbb{R}} [u_k(t) - u(t)]^\top x_N(t) dt \right|}{1 + \operatorname{diam}\mathcal{U} \, \|x_N\|_1} \leq \frac{1}{2^N} \frac{\left| \int_{\mathbb{R}} [u_k(t) - u(t)]^\top x_N(t) dt \right|}{1 + \left| \int_{\mathbb{R}} [u_k(t) - u(t)]^\top x_N(t) dt \right|} < \varepsilon$$

implying

$$\left| \int_{\mathbb{R}} [u_k(t) - u(t)]^\top x_N(t) dt \right| \leq 2^N \left(1 + \operatorname{diam}\mathcal{U} \, \|x_N\|_1 \right) \varepsilon.$$

We find for all k large enough that

$$\left| \int_{\mathbb{R}} u_k(t)^\top x(t) dt - \int_{\mathbb{R}} u(t)^\top x(t) dt \right|$$

$$\leq \left| \int_{\mathbb{R}} u_k(t)^\top [x(t) - x_N(t)] \, dt \right| + \left| \int_{\mathbb{R}} [u_k(t) - u(t)]^\top x_N(t) dt \right|$$

$$+ \left| \int_{\mathbb{R}} u(t)^\top [x(t) - x_N(t)] \, dt \right|$$

$$\leq \sup_k \|u_k\|_\infty \|x - x_N\|_1 + \|u\|_\infty \|x - x_N\|_1 + \left| \int_{\mathbb{R}} [u_k(t) - u(t)]^\top x_N(t) dt \right|$$

$$\leq 2\varepsilon \max_{u \in U} \|u\| + 2^N \left(1 + \operatorname{diam}\mathcal{U} \, \|x_N\|_1 \right) \varepsilon.$$

This shows the asserted convergence. For the converse, consider a sequence in $\mathcal{U}$ such that (9.5.3) holds for all $x \in L^1(\mathbb{R}, \mathbb{R}^m)$ and fix $\varepsilon > 0$. Then there is $N \in \mathbb{N}$ such that $\sum_{n=N+1}^{\infty} 1/2^n < \varepsilon/2$. By assumption there is $K \in \mathbb{N}$ such that for all $k \geq K$ and every $n \in \{1, \ldots, N\}$,

$$\left| \int_{\mathbb{R}} u_k(t)^\top x_n(t) dt - \int_{\mathbb{R}} u(t)^\top x_n(t) dt \right| < \varepsilon/2.$$

Hence for all $k \geq K$,

$$d(u_k, u) = \sum_{n=1}^{\infty} \frac{1}{2^n} \frac{\left|\int_{\mathbb{R}} [u_k(t) - u(t)]^\top x_n(t) dt\right|}{1 + \left|\int_{\mathbb{R}} [u_k(t) - u(t)]^\top x_n(t) dt\right|}$$

$$= \sum_{n=1}^{N} \frac{1}{2^n} \frac{\left|\int_{\mathbb{R}} [u_k(t) - u(t)]^\top x_n(t) dt\right|}{1 + \left|\int_{\mathbb{R}} [u_k(t) - u(t)]^\top x_n(t) dt\right|} + \sum_{n=N+1}^{\infty} \frac{1}{2^n} < \varepsilon. \qquad \square$$

Note that $L^\infty(\mathbb{R}, \mathbb{R}^m)$ is the dual space of $L^1(\mathbb{R}, \mathbb{R}^m)$. This and Proposition 9.5.3 show that the topology generated by the metric (i.e., the family of open sets) coincides with the restriction of the weak* topology on the space $L^\infty(\mathbb{R}, \mathbb{R}^m)$; see, e.g., Dunford and Schwartz [40, Theorem 4.5.1]. Bounded convex and closed subsets are weak* compact. As a compact metric space $\mathcal{U}$ is complete and separable. These are standard facts in functional analysis; see, e.g., [40].

The following lemma analyzes the shift θ on $\mathcal{U}$. A function $u \in \mathcal{U}$ is periodic if and only if u is a periodic point of the shift flow θ.

Lemma 9.5.4. *The space $\mathcal{U}$ is a compact metric space and the shift θ defines a continuous dynamical system on $\mathcal{U}$. The periodic functions are dense in $\mathcal{U}$ and hence the shift on $\mathcal{U}$ is chain transitive (by Exercise 3.4.3).*

Proof. Compactness of $\mathcal{U}$ follows by the arguments above. Obviously $\theta_{t+s} = \theta_t \circ \theta_s$ for $t, s \in \mathbb{R}$ and $\theta_0 = \text{id}$. In order to prove continuity of θ we consider sequences $t_n \to t$ in $\mathbb{R}$ and $u_n \to u$ in $\mathcal{U}$. Then for all $x \in L^1(\mathbb{R}, \mathbb{R}^m)$,

$$\left|\int_{\mathbb{R}} u_n(t_n + \tau)^\top x(\tau) d\tau - \int_{\mathbb{R}} u(t + \tau)^\top x(\tau) d\tau\right|$$

$$\leq \left|\int_{\mathbb{R}} [u_n(t_n + \tau) - u_n(t + \tau)]^\top x(\tau) d\tau\right|$$

$$+ \left|\int_{\mathbb{R}} [u_n(t + \tau) - u(t + \tau)]^\top x(\tau) d\tau\right|$$

$$= \left|\int_{\mathbb{R}} u_n(\tau)^\top x(\tau - t_n) d\tau - \int_{\mathbb{R}} u_n(\tau)^\top x(\tau - t) d\tau\right|$$

$$+ \left|\int_{\mathbb{R}} [u_n(\tau) - u(\tau)]^\top x(\tau - t) d\tau\right|.$$

The second summand converges to zero by Proposition 9.5.3 because $u_n \to u$ in $\mathcal{U}$; the first can be estimated by

$$\leq \sup_{w \in U} \|w\| \int_{\mathbb{R}} \|x(\tau - t_n) - x(\tau - t)\| \, d\tau.$$

Here the integral converges to zero as $t_n \to t$. This well-known fact ("continuity of the norm in L^1") can be seen as follows: For a bounded interval $I \subset \mathbb{R}$ and $t \in \mathbb{R}$ define $I(t) = I + t$. Then the characteristic function $\chi_I(\tau) := 1$ for $\tau \in I$ and $\chi_I(\tau) := 0$ elsewhere, satisfies

$$\int_I |\chi_I(\tau - t) - \chi_I(\tau)|\, d\tau = \int_{I \setminus I(t)} d\tau + \int_{I(t) \setminus I} d\tau$$
$$= \lambda(I) + \lambda(I(t)) - 2\lambda(I \cap I(t)).$$

Let $|t_n| \to 0$ with $I(t_1) \cap I \subset I(t_2) \cap I \subset \ldots \subset I(t_n) \cap I \subset \ldots$ Then $\bigcup_{n=1}^\infty I(t_n) \cap I = I$ and hence we obtain for the Lebesgue measures that $\lim_{n \to \infty} \lambda(I \cap I(t_n)) = \lambda(I) = \lim_{n \to \infty} \lambda(I(t_n))$. This proves the assertion for χ_I at $t = 0$. The assertion for arbitrary $t \in \mathbb{R}$, for step functions, and then for all elements $x \in L^1$ follows in a standard way. We conclude that $\theta(t_n, u_n) = u_n(t_n + \cdot) \to \theta(t, u) = u(t + \cdot)$ in $\mathcal{U}$.

In order to see density of periodic functions pick $u^0 \in \mathcal{U}$. In view of the convergence criterion in Proposition 9.5.3 fix $x \in L^1(\mathbb{R}, \mathbb{R}^m)$. There is $T > 0$ such that

$$\int_{\mathbb{R} \setminus [-T, T]} \|x(t)\|\, dt < \frac{\varepsilon}{\operatorname{diam} U}.$$

Define a periodic function by $u_p(t) = u^0(t)$ for $t \in [-T, T]$ and extend u_p to a $2T$-periodic function on $\mathbb{R}$. Then $u_p \in \mathcal{U}$ and

$$\left| \int_\mathbb{R} [u^0(t) - u_p(t)]^\top x(t)\, dt \right| = \left| \int_{\mathbb{R} \setminus [-T, T]} [u^0(t) - u_p(t)]^\top x(t)\, dt \right|$$
$$\leq \operatorname{diam} U \int_{\mathbb{R} \setminus [-T, T]} \|x(t)\|\, dt < \varepsilon. \qquad \square$$

Next we establish the fact that perturbed linear systems or bilinear control systems define topological linear flows.

Theorem 9.5.5. *The family of equations (9.5.1) defines a continuous linear skew product flow $\Phi = (\theta, \varphi)$ on $\mathcal{U} \times \mathbb{R}^d$ where the base space $\mathcal{U}$ is a compact metric space with chain transitive shift θ.*

Proof. By Lemma 9.5.4 the shift θ on $\mathcal{U}$ is continuous and chain transitive. The group properties are clearly satisfied. It remains to prove continuity of Φ. Consider sequences $t^n \to t^0$ in $\mathbb{R}$, $u^n \to u^0$ in $\mathcal{U}$, and $x^n \to x^0$ in $\mathbb{R}^d$ and abbreviate $\varphi^n(t) = \varphi(t, x^n, u^n)$, $t \in \mathbb{R}, n = 0, 1, \ldots$ We have to show that $\varphi^n(t^n) \to \varphi^0(t^0)$ and write down the proof for $t^0 > 0$. First, observe that there is a compact set $K \subset \mathbb{R}^d$ such that for all n and $t \in [0, T], T := t^0 + 1$, one has $\varphi^n(t) \in K$. This follows, since for all $t \in [0,T]$,

$$\|\varphi^n(t)\| \le \|x^n\| + \left\|\int_0^t \left[A_0 + \sum_{i=1}^m u_i^n(s)A_i\right]\varphi^n(s)ds\right\|$$

$$\le \|x^n\| + \int_0^t \left\|A_0 + \sum_{i=1}^m u_i^n(s)A_i\right\| \|\varphi^n(s)\|\,ds$$

$$\le \|x^n\| + \beta \int_0^t \|\varphi^n(s)\|\,ds$$

with $\beta := \max_{u \in U} \|A_0 + \sum_{i=1}^m u_i A_i\|$. Now Gronwall's inequality, Lemma 6.1.3, shows that $\|\varphi^n(t)\| \le \|x^n\| e^{\beta t}$ for all n and $t \in [0,T]$ and the asserted boundedness follows.

Furthermore, the functions φ^n are equicontinuous on $[0,T]$, since for all $t_1, t_2 \in [0,T]$,

$$\|\varphi^n(t_1) - \varphi^n(t_2)\| = \left\|\int_{t_1}^{t_2}\left[A_0 + \sum_{i=1}^m u_i^n(s)A_i\right]\varphi^n(s)ds\right\|$$

$$\le \beta \sup\{\|\varphi^n(t)\| \mid t \in [0,T] \text{ and } n \in \mathbb{N}\} |t_1 - t_2|.$$

Hence the Theorem of Arzelà-Ascoli implies that a subsequence $\{\varphi^{n_k}\}$ converges uniformly to some continuous function $\psi \in C([0,T], \mathbb{R}^d)$. In order to show that $\psi = \varphi^0$ it suffices to show that ψ solves the same initial value problem as φ^0. For $t \in [0,T]$,

$$\varphi^{n_k}(t) = x^{n_k} + \int_0^t \left[A_0 + \sum_{i=1}^m u_i^{n_k}(s)A_i\right]\varphi^{n_k}(s)ds$$

$$= x^{n_k} + \int_0^t A_0 \varphi^{n_k}(s)ds + \int_0^t \sum_{i=1}^m u_i^{n_k}(s)A_i\psi(s)ds$$

$$+ \int_0^t \sum_{i=1}^m u_i^{n_k}(s)A_i\left[\varphi^{n_k}(s) - \psi(s)\right]ds.$$

For $k \to \infty$, the left-hand side converges to $\psi(t)$ and on the right-hand side

$$x^{n_k} \to x^0 \text{ and } \int_0^t A_0 \varphi^{n_k}(s)ds \to \int_0^t A_0 \psi(s)ds.$$

By weak* convergence of $u^{n_k} \to u^0$ for $k \to \infty$, one has

$$\int_0^t \sum_{i=1}^m u_i^{n_k}(s)A_i\psi(s)ds$$

$$= \int_0^T \sum_{i=1}^m u_i^{n_k}(s)\chi_{|[0,t]} A_i\psi(s)ds \to \int_0^t \sum_{i=1}^m u_i^0(s)A_i\psi(s)ds$$

and, finally,
$$\left\| \int_0^t \sum_{i=1}^m u_i^{n_k}(s) A_i \left[\varphi^{n_k}(s) - \psi(s) \right] ds \right\| \leq T\beta \sup_{s \in [0,T]} \left\| \varphi^{n_k}(s) - \psi(s) \right\| \to 0.$$

Thus $\varphi^0 = \psi$ follows. Then the entire sequence (φ^n) converges to φ^0, since otherwise there exist $\delta > 0$ and a subsequence (φ^{n_k}) with $\left\| \varphi^{n_k} - \varphi^0 \right\|_\infty \geq \delta$. This is a contradiction, since this sequence must have a subsequence converging to φ^0. $\square$

A consequence of this theorem is that all results from Sections 9.1–9.4 hold for these equations. In particular, the spectral decomposition result Theorem 9.1.5 and the characterization of stability in Theorem 9.1.12 hold.

Explicit equations for the induced system on the projective space $\mathbb{P}^{d-1}$ can be obtained as follows: The projected system on the unit sphere $\mathbb{S}^{d-1} \subset \mathbb{R}^d$ is given by
$$\dot{s}(t) = h(u(t), s(t)), \ u \in \mathcal{U}, \ s \in \mathbb{S}^{d-1};$$
here, with $h_i(s) = \left(A_i - s^\top A_i s \cdot I \right) s, i = 0, 1, \ldots, m$,
$$h(u, s) = h_0(s) + \sum_{i=1}^m u_i h_i(s).$$

Define an equivalence relation on $\mathbb{S}^{d-1}$ via $s_1 \sim s_2$ if $s_1 = -s_2$, identifying opposite points. Then the projective space can be identified as $\mathbb{P}^{d-1} = \mathbb{S}^{d-1}/\sim$. Since $h(u, s) = -h(u, -s)$, the differential equation also describes the projected system on $\mathbb{P}^{d-1}$. For the Lyapunov exponents one obtains in the same way (cf. Exercise 6.5.4)
$$\lambda(u, x) = \limsup_{t \to \infty} \frac{1}{t} \log \|\varphi(t, u, x)\| = \limsup_{t \to \infty} \frac{1}{t} \int_0^t q(u(\tau), s(\tau)) \, d\tau;$$
here, with $q_i(s) = s^\top A_i s, \ i = 0, 1, \ldots, m$,
$$q(u, s) := q_0(s) + \sum_{i=1}^m u_i q_i(s).$$

The assertions from Theorem 9.1.5 yield direct generalizations of the results in Chapter 1 for autonomous linear differential equations and in Chapter 7 for periodic linear differential equations. Here the subspaces $V_i(u)$ forming the subbundles $\mathcal{V}_i$ with $\mathbb{P}_i = \mathcal{M}_i$ depend on $u \in \mathcal{U}$. For a constant perturbation $u(t) \equiv u \in \mathbb{R}^m$ the corresponding Lyapunov exponents $\lambda(u, x)$ of the flow Φ are the real parts of the eigenvalues of the matrix $A(u)$ and the corresponding Lyapunov spaces are contained in the subspaces $V_i(u)$. Similarly, if a perturbation $u \in \mathcal{U}$ is periodic, the Floquet exponents of $\dot{x} = A(u(\cdot))x$ are part of the Lyapunov (and hence of the Morse) spectrum of the flow Φ, and the Lyapunov (i.e., Floquet) spaces are contained in the subspaces

$V_i(u)$. Note that, in general, several Lyapunov spaces may be contained in a single subspace $V_i(u)$.

Remark 9.5.6. For robust linear systems $\dot{x} = A(u(t))x$, $u \in \mathcal{U}$, of the form (9.5.1) the Morse spectrum is not much larger than the Lyapunov spectrum. Indeed, 'generically' the Lyapunov spectrum and the Morse spectrum agree; see Colonius and Kliemann [**29**, Theorem 7.3.29] for a precise definition of 'generic' in this context. This fact allows [**29**] to study stability and stabilizability of linear robust systems via the Morse spectrum.

We end this chapter with several applications illustrating the theory developed above. The first example shows that the Morse spectral intervals may overlap.

Example 9.5.7. Consider the following system in $\mathbb{R}^2$,

$$\begin{bmatrix} \dot{x} \\ \dot{y} \end{bmatrix} = A(u(t)) \begin{bmatrix} x \\ y \end{bmatrix}$$

with

$$A(u) = \begin{bmatrix} 1 & \frac{3}{4} \\ \frac{3}{4} & 1 \end{bmatrix} + u_1 \begin{bmatrix} 1 & 0 \\ 0 & 1 \end{bmatrix} + u_2 \begin{bmatrix} 0 & 1 \\ 0 & 0 \end{bmatrix} + u_3 \begin{bmatrix} 0 & 0 \\ 1 & 0 \end{bmatrix}$$

and $U = [-1,1] \times [-\frac{1}{4}, \frac{1}{4}] \times [-\frac{1}{4}, \frac{1}{4}]$. Analyzing the induced flow on $\mathcal{U} \times \mathbb{P}^1$ similar to the one-dimensional examples in Chapter 3 one finds that there are two chain components $\mathcal{M}_1, \mathcal{M}_2 \subset \mathcal{U} \times \mathbb{P}^1$. Their projections $\mathbb{P}\mathcal{M}_i$ to projective space $\mathbb{P}^1$ are given by

$$\mathbb{P}\mathcal{M}_1 = \mathbb{P}\left\{ \begin{bmatrix} x \\ y \end{bmatrix} \in \mathbb{R}^2 \mid y = \alpha x,\ \alpha \in \left[-\sqrt{2}, -\frac{1}{\sqrt{2}}\right] \right\},$$

$$\mathbb{P}\mathcal{M}_2 = \mathbb{P}\left\{ \begin{bmatrix} x \\ y \end{bmatrix} \in \mathbb{R}^2 \mid y = \alpha x,\ \alpha \in \left[\frac{1}{\sqrt{2}}, \sqrt{2}\right] \right\}.$$

Computing the eigenvalues of the constant matrices with eigenvectors in $\mathbb{P}\mathcal{M}_i$, $i = 1,2$, yields the intervals $I_1 = \left[-1, \frac{3}{2}\right]$ and $I_2 = \left[\frac{1}{2}, 3\right]$. By the definition of the Morse spectrum we know that $I_i \subset \Sigma_{Mo}(\mathcal{M}_i)$, $i = 1,2$, and hence $\left[\frac{1}{2}, \frac{3}{2}\right] \subset \Sigma_{Mo}(\mathcal{M}_1) \cap \Sigma_{Mo}(\mathcal{M}_2)$.

In general, it is not possible to compute the Morse spectrum and the associated subbundle decompositions explicitly, even for relatively simple systems, and one has to revert to numerical algorithms; compare [**29**, Appendix D].

Example 9.5.8. Let us consider the linear oscillator with uncertain restoring force

(9.5.4) $$\ddot{x} + 2b\dot{x} + [1 + u(t)]x = 0,$$

9.5. Robust Linear Systems and Bilinear Control Systems

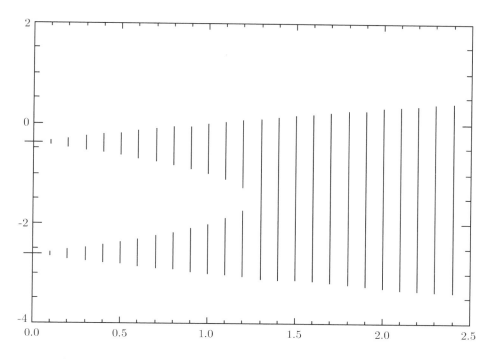

Figure 9.1. Spectral intervals of the linear oscillator (9.5.4) with uncertain restoring force

or, in state space form,

$$\begin{bmatrix} \dot{x}_1 \\ \dot{x}_2 \end{bmatrix} = \begin{bmatrix} 0 & 1 \\ -1 & -2b \end{bmatrix} \begin{bmatrix} x_1 \\ x_2 \end{bmatrix} + u(t) \begin{bmatrix} 0 & 0 \\ -1 & 0 \end{bmatrix} \begin{bmatrix} x_1 \\ x_2 \end{bmatrix}$$

with $u(t) \in [-\rho, \rho]$ and $b > 0$. Figure 9.1 shows the spectral intervals for this system depending on $\rho \in [0, 2.5]$.

Stability

Of particular interest is the upper spectral interval $\Sigma_{Mo}(\mathcal{M}_\ell) = [\kappa_\ell^*, \kappa_\ell]$, as it determines the robust stability of $\dot{x} = A(u(t))x$ (and stabilizability of the system if the set $\mathcal{U}$ is interpreted as a set of admissible control functions); compare [**29**, Section 11.3].

Definition 9.5.9. The stable, center, and unstable subbundles of $\mathcal{U} \times \mathbb{R}^d$ associated with the linear system (9.5.1) are defined as

$$L^- := \bigoplus_{j:\ \kappa_j < 0} \mathbb{P}^{-1}\mathcal{M}_j, \quad L^0 := \bigoplus_{j:\ 0 \in [\kappa_j^*, \kappa_j]} \mathbb{P}^{-1}\mathcal{M}_j, \quad L^+ := \bigoplus_{j:\ \kappa_j^* > 0} \mathbb{P}^{-1}\mathcal{M}_j.$$

As a consequence of Theorem 9.1.12 we obtain the following corollary.

Corollary 9.5.10. *The zero solution of $\dot{x} = A(u(t))x, u \in \mathcal{U}$, is exponentially stable for all perturbations $u \in \mathcal{U}$ if and only if $\kappa_\ell < 0$ if and only if $L^- = \mathcal{U} \times \mathbb{R}^d$.*

Comparing this corollary to Theorem 1.4.8 for constant matrices and Theorem 7.2.16 for periodic matrix functions, we see that this is a generalization for robust linear systems when one uses appropriate exponential growth rates and associated subspace decompositions.

More information can be obtained for robust linear systems if one considers its spectrum depending on a varying perturbation range: We introduce the family of varying perturbation ranges as $U^\rho = \rho U$ for $\rho \geq 0$. The resulting family of systems is

$$\dot{x}^\rho = A(u^\rho(t))x^\rho := A_0 x^\rho + \sum_{i=1}^m u_i^\rho(t) A_i x^\rho,$$

with $u^\rho \in \mathcal{U}^\rho = \{u : \mathbb{R} \longrightarrow U^\rho \,|\, \text{integrable on every bounded interval}\}$. As it turns out, the corresponding maximal spectral value $\kappa_\ell(\rho)$ is continuous in ρ. Hence we can define the (asymptotic-) stability radius of this family as $r = \inf\{\rho \geq 0 \,|\, \text{there exists } u_0 \in \mathcal{U}^\rho \text{ such that } \dot{x}^\rho = A(u_0(t))x^\rho \text{ is not exponentially stable}\}$. This stability radius is based on asymptotic stability under all time-varying perturbations. Similarly one can introduce stability radii based on time invariant perturbations (with values in $\mathbb{R}^m$ or $\mathbb{C}^m$) or on quadratic Lyapunov functions; compare [**29**, Chapter 11] and Hinrichsen and Pritchard [**68**, Chapter 5].

The stability radius plays a key role in the design of engineering systems if one is interested in guaranteed stability for all perturbations of a given size U^ρ. We present two simple systems to illustrate this concept.

Example 9.5.11. Consider the linear oscillator with uncertain damping

$$\ddot{y} + 2(b + u(t))\dot{y} + (1 + c)y = 0$$

with $u(t) \in [-\rho, \rho]$ and $c \in \mathbb{R}$. In equivalent first order form the system reads

$$\begin{bmatrix} \dot{x}_1 \\ \dot{x}_2 \end{bmatrix} = \begin{bmatrix} 0 & 1 \\ -1-c & -2b \end{bmatrix} \begin{bmatrix} x_1 \\ x_2 \end{bmatrix} + u(t) \begin{bmatrix} 0 & 0 \\ 0 & -2 \end{bmatrix} \begin{bmatrix} x_1 \\ x_2 \end{bmatrix}.$$

Clearly, the system is not exponentially stable for $c \leq -1$ with $\rho = 0$, and for $c > -1$ with $\rho \geq b$. It turns out that the stability radius for this system is

$$r(c) = \begin{cases} 0 & \text{for } c \leq -1, \\ b & \text{for } c > -1. \end{cases}$$

9.6. Exercises

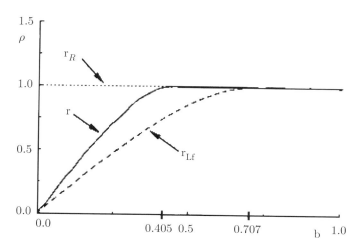

Figure 9.2. Stability radii of the linear oscillator (9.5.4) with uncertain restoring force

Example 9.5.12. Here we look again at the linear oscillator with uncertain restoring force (9.5.4) from Example 9.5.8 given by

$$\begin{bmatrix} \dot{x}_1 \\ \dot{x}_2 \end{bmatrix} = \begin{bmatrix} 0 & 1 \\ -1 & -2b \end{bmatrix} \begin{bmatrix} x_1 \\ x_2 \end{bmatrix} + u(t) \begin{bmatrix} 0 & 0 \\ -1 & 0 \end{bmatrix} \begin{bmatrix} x_1 \\ x_2 \end{bmatrix}$$

with $u(t) \in U^\rho = [-\rho, \rho]$ and $b > 0$. (For $b \leq 0$ the system is unstable even for constant perturbations.) A closed form expression of the stability radius for this system is not available and one has to use numerical methods for the computation of maximal Lyapunov exponents (or maxima of the Morse spectrum); compare [29, Appendix D]. Figure 9.2 shows the (asymptotic) stability radius r, the stability radius under constant real perturbations $r_\mathbb{R}$, and the stability radius based on quadratic Lyapunov functions r_{Lf}, all in dependence on $b > 0$; see [29, Example 11.1.12].

9.6. Exercises

Exercise 9.6.1. Let $A : \mathbb{R} \to gl(d, \mathbb{R})$ be uniformly continuous with

$$\lim_{t \to \infty} A(t) = \lim_{t \to -\infty} A(t) = A_\infty$$

for some $A_\infty \in gl(d, \mathbb{R})$. Show that the linear differential equation $\dot{x}(t) = A(t)x(t)$ defines a continuous linear skew product flow with a compact base space B in the sense of Definition 9.1.1. Show that the base flow is chain transitive.

Exercise 9.6.2. Let $A : \mathbb{R} \to gl(d, \mathbb{R})$ be continuous. Assume that there are $T_1, T_2 > 0$ such that for all i, j there are $T_{ij} \in \{T_1, T_2\}$, such that

$$a_{ij}(t) = a_{ij}(t + T_{ij}) \text{ for all } t \in \mathbb{R}.$$

Construct a compact metric base space B and a continuous linear skew product flow $(\theta, \varphi) : \mathbb{R} \times B \times \mathbb{R}^d \to B \times \mathbb{R}^d$, such that the solutions for $t \in \mathbb{R}$ of $\dot{x}(t) = A(t)x(t), x(t_0) = x_0$, are given by $\varphi(t - t_0, b, x_0)$ for some $b \in B$ depending on t_0 and such that the base flow is chain transitive.

We remark that this can also be done for more than two periods. But the general case is much harder.

Exercise 9.6.3. Verify the remaining part of Lemma 9.2.1 for the metric $\bar{d}$ on $\mathbb{P}^{d-1}$ given by

$$\bar{d}(\mathbb{P}x, \mathbb{P}y) := \sqrt{1 - \frac{\langle x, y \rangle^2}{\| x \|^2 \| y \|^2}}.$$

Let $\mathbf{e}_1, \mathbf{e}_2, \mathbf{e}_3$ be the canonical basis vectors. For unit vectors $z = \mathbf{e}_1, x = x_1\mathbf{e}_1 + x_2\mathbf{e}_2, y = y_1\mathbf{e}_1 + y_2\mathbf{e}_2 + y_3\mathbf{e}_3$ it holds that

$$\bar{d}(\mathbb{P}x, \mathbb{P}y) \leq \bar{d}(\mathbb{P}x, \mathbb{P}z) + \bar{d}(\mathbb{P}z, \mathbb{P}y).$$

9.7. Orientation, Notes and References

Orientation. The constructions in Chapter 7 for periodic systems started from the exponential growth rates and then constructed the subspace decomposition into the Lyapunov spaces. In the present chapter the converse approach has been used: Based on attractor-repeller decompositions in the projective bundle, the finest Morse decomposition is constructed. The associated decomposition into the Selgrade bundles and the Morse spectrum are appropriate generalizations of the Lyapunov spaces and the Lyapunov exponents for topological linear flows.

In Chapter 11 we will consider another class of nonautonomous systems where the matrices change randomly over time. It will turn out that the main result also gives decompositions of the state space $\mathbb{R}^d$ into subspaces corresponding to Lyapunov exponents which are nonrandom, while the decompositions depend on the considered point in the base space; again the dimensions of the subspaces will be constant. A decisive difference will be that we exclude certain points in the base space (forming a null-set). This makes it possible to get much sharper results which are very similar to the autonomous case. Before we start to analyze random dynamical systems, we will in Chapter 10 prove results from ergodic theory, which are needed in Chapter 11.

9.7. Orientation, Notes and References

Notes and references. A better understanding of subbundles introduced in Definition 9.1.2 may be gained by the fact that they have a local product structure (they are called locally trivial in a topological sense):

A closed subset $\mathcal{W}$ of the vector bundle $\mathcal{V} = B \times \mathbb{R}^d$ is a subbundle, if and only if (i) it intersects each fiber $\{b\} \times \mathbb{R}^d, b \in B$, in a linear subspace, (ii) there exist an open cover $\{U_i \mid i \in I\}$ of B and homeomorphisms $\varphi_i : \pi^{-1}(U_i) \to U_i \times \mathbb{R}^d$ such that $\pi \circ \varphi_i = \pi$, and (iii) the maps $L_{ji} : \mathbb{R}^d \to \mathbb{R}^d$ defined by $\varphi_j \circ \varphi_i^{-1}(b, w) = (b, L_{ji} w)$ are linear for every $b \in U_i \cap U_j$. See Colonius and Kliemann [**29**, Lemma B.1.13] for this characterization of subbundles. This construction can also be used to define general vector bundles, which are locally trivial.

The analysis of the boundary points κ_i^* and κ_i of the Morse spectral intervals can be approached in different ways. In Theorem 9.3.8 we have used the uniform growth spectrum introduced in Proposition 9.4.4 in order to show that these boundary points are actually Lyapunov exponents which even exist as limits. This approach is due to Grüne [**59**]; a generalization is given by Kawan and Stender [**77**]. An alternative approach in Colonius and Kliemann [**29**, Section 5.5] uses ergodic theory in the projective bundle to obtain $(b, x) \in B \times \mathbb{R}^d$ with Lyapunov exponents existing as limits and equal to κ_i^* (even forward and backward in time simultaneously); analogously for κ_i. (This result is based on the Multiplicative Ergodic Theorem discussed in Chapter 11.) More generally, one may also consider vector-valued growth rates; see Kawan and Stender [**77**] and San Martin and Seco [**122**].

One may also ask, if for robust linear systems the boundary points of the Morse spectrum can be obtained in periodic solutions in $\mathbb{P}^{d-1}$ corresponding to periodic functions $u \in \mathcal{U}$. If $d = 3$, the corresponding sphere $\mathbb{S}^2$ and projective space $\mathbb{P}^2$ are two-dimensional and the Jordan curve theorem, which is basic for dynamics on two-dimensional spaces, also holds for the two-sphere $\mathbb{S}^2$ (cf. Massey [**101**, Chapter VIII, Corollary 6.4]). The classical Poincaré-Bendixson theory yields results on periodic solutions for autonomous differential equations in the plane. There is a generalization to optimal control problems with average cost functional given by Artstein and Bright in [**10**, Theorem A]. This result together with [**10**, Proposition 6.1] can be used to show that for $d \leq 3$, the boundary points of the Morse spectral intervals are attained in points $(x_0, u) \in \mathbb{R}^d \times \mathcal{U}$ such that the projected solution $(\mathbb{P}\varphi(\cdot, x_0, u), u)$ is periodic with some period $\tau \geq 0$.

Lemma 9.1.11 is due to Fenichel [**47**]; proofs are also given in Wiggins [**139**, Lemma 3.1.4] and Salamon and Zehnder [**121**, Theorem 2.7].

See Wirth [**142**] and Shorten et al. [**127**] for a discussion of linear parameter-varying and linear switching systems, which also lead to continuous linear flows on vector bundles.

The notion of exponential dichotomies has a long tradition and is of fundamental importance for linear flows on vector bundles with many ramifications since it is a version of hyperbolicity; cf., e.g., Coppel [**36**]. The associated spectral notion is the dichotomy spectrum or Sacker-Sell spectrum going back to a series of papers by Sacker and Sell [**119**]. For extensions to infinite-dimensional spaces see, e.g., Daleckii and Krein [**39**] and Sell and You [**125**]. A general reference to the abstract linear and nonlinear theory is Bronstein and Kopanskii [**21**]. The monograph Chicone and Latushkin [**27**] develops a general theory of evolution semigroups which, based on functional analytic tools, also encompasses many results on nonautonomous dynamical systems.

The topological conjugacy problem for linear flows on vector bundles is dealt with by Ayala, Colonius and Kliemann in [**13**], under the hyperbolicity condition that the center bundle is trivial. For much more elaborate versions of topological conjugacy under weakened hyperbolicity conditions see Barreira and Pesin [**16**].

In Section 9.5 we introduced some basic ideas for robust linear systems and bilinear control systems. This approach leads to many other results in control theory (such as precise time-varying stability radii, stabilizability and feedback stabilization, periodic control, chain control sets, etc.) and in the theory of Lie semigroups. We refer to Colonius and Kliemann [**29**] for related concepts in control theory, and to San Martin and Seco [**122**] and the references therein for generalizations to Lie semigroups. For linear control theory see Hinrichsen and Pritchard [**68**] and Sontag [**129**] and for another look at bilinear control systems see Elliott [**44**].

In general, it is not possible to compute the Morse spectrum and the associated subbundle decompositions explicitly, even for relatively simple systems, and one has to revert to numerical algorithms. They can, e.g., be based on Hamilton-Jacobi equations; cf. [**29**, Appendix D].

Chapter 10

Tools from Ergodic Theory

In this short chapter, we prepare the discussion of dynamical systems from the probabilistic point of view by giving a concise introduction to the relevant parts of ergodic theory. We introduce invariant probability measures and prove some basic theorems of ergodic theory for discrete time: The maximal ergodic theorem (Theorem 10.1.5), Birkhoff's ergodic theorem (Theorem 10.2.1), and Kingman's subadditive ergodic theorem (Theorem 10.3.2).

The goal of this chapter is to present a complete proof of Kingman's subadditive ergodic theorem. It will be used in Chapter 11 for the Furstenberg-Kesten Theorem. The proof of the subadditive ergodic theorem requires some other tools from ergodic theory, mainly Birkhoff's ergodic theorem. This theorem considers an ergodic probability measure for a measurable map T and shows that for a real-valued function f the "spatial average" equals the "time average" of f along a trajectory under T. This chapter provides a condensed introduction to these parts of ergodic theory, assuming only basic knowledge of measure theory. Throughout this chapter, we use standard notation from measure theory and consider a space X endowed with a σ-algebra $\mathcal{F}$ and a probability measure μ.

10.1. Invariant Measures

Recall that a probability measure μ on a space X with σ-algebra $\mathcal{F}$ is called invariant for a measurable map $T : X \to X$ if $\mu(T^{-1}E) = \mu(E)$ for all measurable sets $E \in \mathcal{F}$ (In the probabilistic setting, probability measures are often denoted by P on a space Ω with σ-algebra $\mathcal{F}$.) The map T is called measure preserving, if it is clear from the context, which measure is meant. Analogously, the measure is called invariant, if the map T is clear from the context. An invariant measure is ergodic, if $\mu(T^{-1}E \triangle E) = 0$ implies $\mu(E) = 0$ or 1 for all $E \in \mathcal{F}$ (here $\triangle$ denotes the symmetric difference between sets). Then the map T is called ergodic as well (with respect to μ). In the following, we often omit the σ-algebra $\mathcal{F}$ in our notation.

We start with the following examples for invariant measures.

Example 10.1.1. Consider the doubling map $Tx = 2x \mod 1$ on the unit interval $X = [0,1)$ with the Lebesgue σ-algebra and Lebesgue measure μ restricted to $[0,1]$. Then, for any measurable set $E \subset [0,1]$ the preimage is $T^{-1}E = \{x \in [0,1] \mid 2x \in E \text{ or } 2x - 1 \in E\} = \frac{1}{2}E \cup \left(\frac{1}{2} + \frac{1}{2}E\right)$. Thus the preimage is the disjoint union of two sets, each having measure $\frac{1}{2}\mu(E)$. This shows that the Lebesgue measure is invariant for the doubling map on $[0,1]$.

Example 10.1.2. Consider the logistic map $Tx = \alpha x(1-x)$ with parameter $\alpha = 4$ on the unit interval $X = [0,1]$ with the Lebesgue σ-algebra. Then the probability measure μ which has density $\rho(x) = \dfrac{1}{\pi\sqrt{x(1-x)}}$ with respect to the Lebesgue measure is invariant. A computation shows that for every measurable set E,

$$\mu(E) = \int_E \rho(x)dx = \int_{T^{-1}E} \rho(x)dx = \mu(T^{-1}E).$$

The following lemma characterizes ergodicity of maps acting on probability spaces.

Lemma 10.1.3. *Let T be a measurable map on a probability space (X, μ). Then T is ergodic if and only if every measurable function $f : X \to \mathbb{R}$ satisfying $f(T(x)) = f(x)$ for almost every $x \in X$ is constant almost everywhere. A sufficient condition for ergodicity is that every characteristic function $\mathbb{I}_E$ of measurable subsets $E \subset X$ with $\mathbb{I}_E(T(x)) = \mathbb{I}_E(x)$ for almost every $x \in X$, is constant almost everywhere.*

Proof. Suppose that T is ergodic and consider an invariant function $f : X \to \mathbb{R}$, i.e., f is measurable with $f(T(x)) = f(x)$ for almost all x. For $n \in \mathbb{N}$ and $k \in \mathbb{Z}$ let

$$E_{n,k} := \{x \in X \mid 2^{-n}k \leq f(x) < 2^{-n}(k+1)\}.$$

Then for every $n \in \mathbb{N}$ the sets $E_{n,k}, k \in \mathbb{Z}$, form a partition of X. The sets $E_{n,k}$ are invariant, since by invariance of f,

$$T^{-1}(E_{n,k}) = \{x \in X \mid 2^{-n}k \leq f(T(x)) < 2^{-n}(k+1)\}$$
$$= \{x \in X \mid 2^{-n}k \leq f(x) < 2^{-n}(k+1)\} = E_{n,k}.$$

10.1. Invariant Measures

Hence, by ergodicity, each of the sets $E_{n,k}$ has measure 0 or 1. More explicitly, for each n there is a unique $k_n \in \mathbb{Z}$ such that $\mu(E_{n,k_n}) = 1$ and $\mu(E_{n,k}) = 0$ for $k \neq k_n$. Let $X_0 := \bigcap_{n=1}^{\infty} E_{n,k_n}$. Then $\mu(X_0) = 1$. Let $x \in X_0$ with $f(x) \in E_{n,k_n}$ for all n. Then, for all $n \in \mathbb{N}$ and almost all $y \in X_0$ one has $(y) \in E_{n,k_n}$, hence $|f(x) - f(y)| \leq 2^{-n}$. Then σ-additivity of μ implies $f(x) = f(y)$ for almost all $y \in X_0$.

For the converse, it suffices to prove the last assertion, since $f = \mathbb{I}_E$ are measurable functions. Suppose that $E \subset X$ is a subset satisfying $\mu(T^{-1}(E) \triangle E) = 0$. This is equivalent to $\mathbb{I}_E(T(x)) = \mathbb{I}_E(x)$ for almost every $x \in X$. Hence T is invariant and by assumption $\mathbb{I}_E(x)$ is constant almost everywhere. Thus either $\mu(E) = 1$ or $\mu(E) = 0$, and it follows that T is ergodic. $\square$

Example 10.1.4. Let us apply Lemma 10.1.3 to the doubling map in Example 10.1.1. Consider for an invariant function f the Fourier expansions of f and $f \circ T$ which, with coefficients $c_n \in \mathbb{C}$, have the form

$$f(x) = \sum_{n \in \mathbb{Z}} c_n e^{2\pi i n x} \text{ and } f(Tx) = \sum_{n \in \mathbb{Z}} c_n e^{2\pi i n 2 x} \text{ for almost all } x \in [0,1).$$

Observe that the Fourier coefficients are unique. Comparing them for f and $f \circ T$, we see that $c_n = 0$ if n is odd. Similarly, comparing the Fourier coefficients of f and $f \circ T^2$ we see that $c_n = 0$ if n is not a multiple of 4. Proceeding in this way, one finds that $c_n = 0$ if $n \neq 0$. Hence f is a constant. Thus, by Lemma 10.1.3, the doubling map is ergodic with respect to Lebesgue measure.

For the proof of Theorem 10.2.1, Birkhoff's Ergodic Theorem, we need the following result which is known as the maximal ergodic theorem.

Theorem 10.1.5. *Let $T : X \to X$ be measure preserving and consider an integrable map $f : X \to \mathbb{R}$. Define $f_0 \equiv 0$, $f_n := f + f \circ T + \ldots + f \circ T^{n-1}$, $n \geq 1$, and $F_N(x) := \max_{0 \leq n \leq N} f_n(x)$, $x \in X$. Then*

$$\int_{\{x \in X \mid F_N(x) > 0\}} f \, d\mu \geq 0 \text{ for all } N \in \mathbb{N}.$$

Proof. Observe that F_N is integrable, since the functions $f, f \circ T, \ldots$ are integrable. By definition $F_N \geq f_n$ and hence $F_N \circ T \geq f_n \circ T$ for all $0 \leq n \leq N$ implying

$$F_N \circ T + f \geq f + f_n \circ T = f_{n+1} \text{ for } n = 0, 1, \ldots, N - 1.$$

This shows $F_N(T(x)) + f(x) \geq \max_{1 \leq n \leq N} f_n(x)$ for all $x \in X$. For $x \in X$ with $F_N(x) > 0$, the right-hand side equals $\max_{0 \leq n \leq N} f_n(x) = F_N(x)$. We find

$$f(x) \geq F_N(x) - F_N(T(x)) \text{ on } \{x \in X \mid F_N(x) > 0\} =: A.$$

Integration yields
$$\int_A f\, d\mu \geq \int_A F_N\, d\mu - \int_A F_N \circ T\, d\mu = 0,$$
where the equality follows since T preserves μ. □

The next lemma is a consequence of Theorem 10.1.5.

Lemma 10.1.6. *Let $T : X \to X$ be measure preserving and consider for an integrable function $g : X \to \mathbb{R}$ and $\alpha \in \mathbb{R}$ the sets*

(10.1.1) $$B_\alpha := \left\{ x \in X \;\Big|\; \sup_{n \geq 1} \frac{1}{n} \sum_{k=0}^{n-1} g(T^k(x)) > \alpha \right\}.$$

Then for every measurable set $A \subset X$ with $\mu(T^{-1}(A) \triangle A) = 0$ it follows that
$$\int_{B_\alpha \cap A} g\, d\mu \geq \alpha \cdot \mu(B_\alpha \cap A).$$

Proof. Suppose first that $A = X$. We apply the maximal ergodic theorem, Theorem 10.1.5, to $f := g - \alpha$ implying
$$\int_{\{x : F_N(x) > 0\}} (g - \alpha)\, d\mu \geq 0 \text{ for all } N \in \mathbb{N}.$$
Note that $x \in B_\alpha$ if and only if
$$\sup_{n \geq 1} \left\{ \sum_{k=0}^{n-1} g(T^k(x)) - n\alpha \right\} > 0.$$
This shows that $B_\alpha = \bigcup_{n=0}^{\infty} \{x \in X \mid F_n(x) > 0\}$ and hence
$$\int_{B_\alpha} g\, d\mu \geq \int_{B_\alpha} \alpha\, d\mu = \alpha \cdot \mu(B_\alpha).$$
For an invariant set A as specified in the assertion, apply the arguments above to the function $f := \mathbb{I}_A \cdot (g - \alpha)$. □

10.2. Birkhoff's Ergodic Theorem

Now we are ready for a proof of the following result, Birkhoff's ergodic theorem.

Theorem 10.2.1 (Birkhoff). *Let μ be a probability measure on a measure space X and suppose that $T : X \to X$ preserves μ. If $f : X \to \mathbb{R}$ is integrable with respect to μ, i.e., $f \in L^1(X, \mu; \mathbb{R})$, then the limit*
$$f^*(x) := \lim_{n \to \infty} \frac{1}{n} \sum_{k=0}^{n-1} f(T^k(x))$$

10.2. Birkhoff's Ergodic Theorem

exists on a subset of full μ-measure, and here $f^*(T(x)) = f^*(x)$; furthermore, $f^* \in L^1(X, \mu; \mathbb{R})$ with

(10.2.2) $$\int_X f^* d\mu = \int_X f d\mu.$$

If μ is ergodic, then f^* is constant with $f^*(x) = \int_X f d\mu$ for almost all $x \in X$.

In the ergodic case, Birkhoff's ergodic theorem shows that

$$\int_X f d\mu = \lim_{n \to \infty} \frac{1}{n} \sum_{k=0}^{n-1} f(T^k(x)) \text{ for almost all } x.$$

This may be formulated by saying that "the space average equals the time average". This is an assertion which is at the origin and in the center of ergodic theory. In the special case where $f = \mathbb{I}_E$ is the characteristic function of a measurable set $E \subset X$ the limit $\lim_{n \to \infty} \frac{1}{n} \sum_{k=0}^{n-1} \mathbb{I}_E(T^k(x))$ counts how often $T^k(x)$ visits E on average. If T is ergodic, then this limit is constant on a set of full measure and equals $\mu(E)$.

Proof of Theorem 10.2.1. The consequence for an ergodic measure μ will follow from Lemma 10.1.3. First we define

$$f^*(x) := \limsup_{n \to \infty} \frac{1}{n} \sum_{k=0}^{n-1} f(T^k(x)) \text{ and } f_*(x) = \liminf_{n \to \infty} \frac{1}{n} \sum_{k=0}^{n-1} f(T^k(x)).$$

One gets

$$f^*(T(x)) = \limsup_{n \to \infty} \frac{1}{n} \sum_{k=0}^{n-1} f(T^{k+1}(x))$$

$$= \limsup_{n \to \infty} \frac{n+1}{n} \frac{1}{n+1} \sum_{k=0}^{n} f(T^k(x)) - \frac{f(x)}{n}$$

$$= \limsup_{n \to \infty} \frac{1}{n+1} \sum_{k=0}^{n} f(T^k(x)) - 0 = f^*(x).$$

It remains to prove that $f^*(x) = f_*(x)$ and that this is an integrable function for which (10.2.2) holds. In order to show that $\{x \in X \mid f_*(x) < f^*(x)\}$ has measure zero, set for rational numbers $\alpha, \beta \in \mathbb{Q}$,

$$E_{\alpha,\beta} := \{x \in X \mid f_*(x) < \beta \text{ and } f^*(x) > \alpha\}.$$

Then one obtains a countable union

$$\{x \in X \mid f_*(x) < f^*(x)\} = \bigcup_{\beta < \alpha} E_{\alpha,\beta},$$

and our strategy is to show that each $\mu(E_{\alpha,\beta}) = 0$, hence, by σ-additivity, the set on the left-hand side will have measure zero.

We find
$$T^{-1}(E_{\alpha,\beta}) = \{x \in X \mid f_*(T(x)) < \beta \text{ and } f^*(T(x)) > \alpha\}$$
$$= \{x \in X \mid f_*(x) < \beta \text{ and } f^*(x) > \alpha\} = E_{\alpha,\beta}.$$

With B_α defined as in (10.1.1), one immediately sees that $E_{\alpha,\beta} \subset B_\alpha$. Then Lemma 10.1.6 implies
$$\int_{E_{\alpha,\beta}} f \, d\mu = \int_{B_\alpha \cap E_{\alpha,\beta}} f \, d\mu \geq \alpha \cdot \mu(B_\alpha \cap E_{\alpha,\beta}) = \alpha \cdot \mu(E_{\alpha,\beta}).$$

Note that $(-f)^* = -f_*$ and $(-f)_* = -f^*$ and hence
$$E_{\alpha,\beta} = \{x \in X \mid (-f)^*(x) > -\beta \text{ and } (-f)_*(x) < -\alpha\}.$$

Replacing f, α and β by $-f, -\beta$, and $-\alpha$, respectively, one finds
$$\int_{E_{\alpha,\beta}} (-f) \, d\mu \geq (-\beta) \cdot \mu(E_{\alpha,\beta}), \text{ i.e., } \alpha \cdot \mu(E_{\alpha,\beta}) \leq \int_{E_{\alpha,\beta}} f d\mu \leq \beta \cdot \mu(E_{\alpha,\beta}).$$

Since we consider $\beta < \alpha$, it follows that $\mu(E_{\alpha,\beta}) = 0$ and hence for almost all $x \in X$,
$$f^*(x) = f_*(x) = \lim_{n \to \infty} \frac{1}{n} \sum_{k=0}^{n-1} f(T^k(x)).$$

Next we show that f^* is integrable. Let
$$g_n(x) := \left| \frac{1}{n} \sum_{k=0}^{n-1} f(T^k(x)) \right|.$$

Since T leaves μ invariant, it follows that
$$\int_X g_n d\mu \leq \frac{1}{n} \sum_{k=0}^{n-1} \int_X \left| f(T^k(x)) \right| d\mu = \frac{1}{n} \sum_{k=0}^{n-1} \int_X |f(x)| \, d\mu = \int_X |f(x)| \, d\mu < \infty.$$

Now integrability of f^* follows from Fatou's lemma and
$$\int_X |f^*| \, d\mu = \int_X \liminf_{n \to \infty} g_n d\mu \leq \liminf_{n \to \infty} \int_X g_n d\mu.$$

If μ is ergodic, then the invariance property $f^* \circ T = f^*$ and Lemma 10.1.3 imply that f^* is constant almost everywhere. Hence, in the ergodic case, (10.2.2) implies
$$\int_X f d\mu = \int_X f^* d\mu = f^*(x) \text{ almost everywhere.}$$

It only remains to prove (10.2.2). For this purpose define for $n \geq 1$ and $k \in \mathbb{Z}$ the set
$$D_{n,k} := \left\{ x \in X \mid \frac{k}{n} \leq f^*(x) \leq \frac{k+1}{n} \right\}.$$

For fixed n, the set X is the disjoint union of the sets $D_{n,k}, k \in \mathbb{Z}$. Furthermore, the sets $D_{n,k}$ are invariant, since $f^* \circ T = f^*$ implies that
$$T^{-1}(D_{n,k}) = \{x \in X \mid T(x) \in D_{n,k}\} = D_{n,k}.$$
For all $\varepsilon > 0$ one has the inclusion
$$D_{n,k} \subset \left\{ x \in X \mid \sup_{n \geq 1} \frac{1}{k} \sum_{i=0}^{k-1} f(T^i(x)) > \frac{k}{n} - \varepsilon \right\}.$$
Now Lemma 10.1.6 implies for all $\varepsilon > 0$,
$$\int_{D_{n,k}} f d\mu \geq (\frac{k}{n} - \varepsilon)\mu(D_{n,k}) \text{ and hence } \int_{D_{n,k}} f d\mu \geq \frac{k}{n}\mu(D_{n,k}).$$
Using the definition of $D_{n,k}$ this yields
$$\int_{D_{n,k}} f^* d\mu \leq \frac{k+1}{n}\mu(D_{n,k}) \leq \frac{1}{n}\mu(D_{n,k}) + \int_{D_{n,k}} f d\mu.$$
Summing over $k \in \mathbb{Z}$, one obtains $\int_X f^* d\mu \leq \frac{1}{n} + \int_X f d\mu$. For $n \to \infty$ we get the inequality
$$\int_X f^* d\mu \leq \int_X f d\mu.$$
The same procedure for $-f$ gives the complementary inequality,
$$\int_X (-f)^* d\mu \leq \int_X (-f) d\mu \text{ and hence } \int_X f^* d\mu \geq \int_X f d\mu. \qquad \square$$

10.3. Kingman's Subadditive Ergodic Theorem

Before formulating Kingman's subadditive ergodic theorem, we note the following lemma.

Lemma 10.3.1. *Let T be a measure preserving map on (X, μ) and let $f : X \to \mathbb{R} \cup \{-\infty\}$ be measurable with $f(x) \leq f(Tx)$ for almost all $x \in X$. Then $f(x) = f(Tx)$ for almost all $x \in X$.*

Proof. If, contrary to the assertion, $\mu\{x \mid f(x) < f(Tx)\} > 0$, there is $\alpha \in \mathbb{R}$ with $\mu\{x \mid f(x) \leq \alpha < f(Tx)\} > 0$. By assumption, $f(Tx) \leq \alpha$ implies $f(x) \leq \alpha$, and hence

$\mu\{x \mid f(x) \leq \alpha\}$
$= \mu\{x \mid f(x) \leq \alpha \text{ and } f(Tx) \leq \alpha\} + \mu\{x \mid f(x) \leq \alpha \text{ and } f(Tx) > \alpha\}$
$= \mu\{x \mid f(Tx) \leq \alpha\} + \mu\{x \mid f(x) \leq \alpha < f(Tx)\}.$

By invariance of μ, it follows that $\mu\{x \mid f(x) \leq \alpha < f(Tx)\} = 0$. $\qquad \square$

The following theorem due to Kingman, is called the subadditive ergodic theorem.

Theorem 10.3.2 (Kingman). *Let T be a measure preserving map on a probability space (X, μ). Suppose that $f_n : X \to \mathbb{R}$ are integrable functions satisfying the following subadditivity property for all $l, n \in \mathbb{N}$,*

(10.3.3) $$f_{l+n}(x) \leq f_l(x) + f_n(T^l x) \text{ for almost all } x \in X,$$

and $\inf_{n \in \mathbb{N}} \frac{1}{n} \int f_n \, d\mu > -\infty$. Then there are a forward invariant set $\tilde{X} \subset X$ (i.e., $T(\tilde{X}) \subset \tilde{X}$) of full measure and an integrable function $\tilde{f} : X \to \mathbb{R}$ satisfying

$$\lim_{n \to \infty} \frac{1}{n} f_n(x) = \tilde{f}(x) \text{ for all } x \in \tilde{X} \text{ and } \tilde{f} \circ T = \tilde{f} \text{ on } \tilde{X}.$$

If μ is ergodic, then $\tilde{f}$ is constant almost everywhere.

Theorem 10.3.2 generalizes Birkhoff's ergodic theorem, Theorem 10.2.1. In fact, if $f_n := \sum_{i=0}^{n-1} f \circ T^i$ for an integrable function $f : X \to \mathbb{R}$, then equality holds in assumption (10.3.3).

Proof of Theorem 10.3.2. Define for $n \geq 1$ functions $f'_n : X \to \mathbb{R}$ by

$$f'_n = f_n - \sum_{i=0}^{n-1} f_1 \circ T^i.$$

Then the sequence (f'_n) satisfies the same properties as (f_n): In fact, for $n \geq 1$ subadditivity shows that

$$f_n \leq f_1 + f_{n-1}(Tx) \leq \ldots \leq \sum_{i=0}^{n-1} f_1 \circ T^i$$

and it follows that $f'_n \leq 0$. Furthermore, the sequence (f'_n) is subadditive, since

$$f'_{l+n} = f_{l+n} - \sum_{i=0}^{l+n-1} f_1 \circ T^i \leq f_l + f_n \circ T^l - \sum_{i=0}^{l+n-1} f_1 \circ T^i$$

$$= f_l - \sum_{i=0}^{l-1} f_1 \circ T^i + f_n \circ T^l - \sum_{i=l+1}^{l+n-1} f_1 \circ T^i$$

$$= f'_l + f'_n \circ T^l.$$

It suffices to show the assertion for (f'_n), since almost sure convergence of $\left(\frac{f'_n}{n}\right)$ to a function in L^1 and Birkhoff's ergodic theorem, Theorem 10.2.1, applied to the second term in f'_n, imply almost sure convergence of $\left(\frac{f_n}{n}\right)$ to a function in L^1. This also implies $\inf_{n \in \mathbb{N}} \frac{1}{n} \int f'_n \, d\mu > -\infty$. These arguments show that we may assume, without loss of generality, $f_n \leq 0$ almost everywhere. Fatou's lemma implies

$$\int \limsup_{n\to\infty} \frac{1}{n} f_n d\mu \geq \limsup_{n\to\infty} \frac{1}{n} \int f_n d\mu =: \gamma > -\infty.$$

This shows that $\bar{f} := \limsup_{n\to\infty} \frac{1}{n} f_n \in L^1$ and $\int \bar{f} d\mu \geq \gamma$.

Now consider the measurable function

(10.3.4) $$\tilde{f}(x) := \liminf_{n\to\infty} \frac{1}{n} f_n(x), \; x \in X.$$

Subadditivity implies $\frac{f_{n+1}}{n} \leq \frac{f_1}{n} + \frac{f_n \circ T}{n}$ and hence $\tilde{f} \leq \tilde{f} \circ T$. By Lemma 10.3.1, it follows that $\tilde{f} = \tilde{f} \circ T$ almost everywhere and hence $\tilde{f} = \tilde{f} \circ T^k$ for all $k \in \mathbb{N}$.

For $\varepsilon > 0$ and $N, M \in \mathbb{N}$ let $F_M(x) := \max\{\tilde{f}(x), -M\}$ and

$$B(\varepsilon, M, N) := \{x \in X \mid \frac{f_l(x)}{l} > F_M(x) + \varepsilon \text{ for all } 1 \leq l \leq N\}.$$

The function F_M coincides with $\tilde{f}$ for the points x with $\tilde{f}(x) \leq -M$ and is invariant under T. We also observe that, for fixed ε and M, and $N \to \infty$ the characteristic function satisfies

(10.3.5) $$\mathbb{I}_{B(\varepsilon,M,N)}(x) \to 0 \text{ for almost all } x.$$

Let $x \in X$ and $n \geq N$. We will decompose the set $\{1, 2, \ldots, n\}$ into three types of (discrete) intervals. Begin with $k = 1$ and consider Tx, or suppose that $k > 1$ is the smallest integer in $\{1, \ldots, n-1\}$ not in an interval already constructed and consider $T^k x$.

(a) If $T^k x$ is in the complement $B(\varepsilon, M, N)^c$, then there is an $l = l(x) \leq N$ with

$$\frac{f_l(T^k x)}{l} \leq F_M(T^k x) + \varepsilon = F_M(x) + \varepsilon.$$

If $k + l \leq n$, take $\{k, \ldots, k + l - 1\}$ as an element in the decomposition.

(b) If there is no such l with $k + l \leq n$, take $\{k\}$.

(c) If $T^k x \in B(\varepsilon, M, N)$, again choose $\{k\}$.

We obtain P intervals of the form $\{k_i, \ldots, k_i + l_i - 1\}$, Q intervals $\{m_i\}$ of length 1 in case (c) where $\mathbb{I}_{B(\varepsilon,M,N)}(T^{m_i} x) = 1$, and R intervals of length 1 in case (b) contained in $\{n - N + 1, \ldots, n - 1\}$. Now we use subadditivity in order to show the following, beginning at the end of $\{1, \ldots, n-1\}$ and then backwards till 1:

$$f_n(x) \leq \sum_{i=1}^{P} f_{l_i}(T^{k_i}x) + \sum_{i=1}^{Q} f_1(T^{m_i}x) + \sum_{i=1}^{R} f_{l_i}(T^{n-i}x)$$

(10.3.6)
$$\leq (F_M(x) + \varepsilon) \sum_{i=1}^{P} l_i(x) \leq F_M(x) \sum_{i=1}^{P} l_i(x) + n\varepsilon.$$

Here the second inequality follows, since we may assume that $f_i \leq 0$ for all i. By construction, we also have

$$n - \sum_{k=1}^{n} \mathbb{I}_{B(\varepsilon,M,N)}(T^k x) - N \leq \sum_{i=1}^{P} l_i(x)$$

and Birkhoff's ergodic theorem, Theorem 10.2.1, yields

$$\liminf_{n\to\infty} \frac{1}{n} \sum_{i=1}^{P} l_i(x) \geq 1 - \limsup_{n\to\infty} \frac{1}{n} \sum_{k=1}^{n} \mathbb{I}_{B(\varepsilon,M,N)}(T^k x) = 1 - \int \mathbb{I}_{B(\varepsilon,M,N)} d\mu.$$

Observe that in inequality (10.3.6) one has $F_M(x) \leq 0$. Hence we obtain

$$\limsup_{n\to\infty} \frac{1}{n} f_n(x) \leq F_M(x) \left(1 - \int \mathbb{I}_{B(\varepsilon,M,N)} d\mu \right) + \varepsilon.$$

Using (10.3.5) one finds that for $N \to \infty$, the right-hand side converges for almost all x to $F_M(x) + \varepsilon$. This implies for almost all $x \in X$,

$$\limsup_{n\to\infty} \frac{1}{n} f_n(x) \leq F_M(x) + \varepsilon.$$

Using the definition of F_M, and of $\tilde{f}$ in (10.3.4), one finds for $M \to \infty$ and then $\varepsilon \to 0$, the desired convergence for almost all $x \in X$,

$$\limsup_{n\to\infty} \frac{1}{n} f_n(x) \leq \tilde{f}(x) = \liminf_{n\to\infty} \frac{1}{n} f_n(x).$$

Finally, note that, for an ergodic measure μ, Lemma 10.1.3 shows that $\tilde{f}$ is constant, if $\tilde{f} \circ T = \tilde{f}$ holds. □

10.4. Exercises

Exercise 10.4.1. In Birkhoff's Ergodic Theorem assume that the map $T : M \to M$ is bijective (invertible) and μ-invariant.

(i) Show that the limit of the negative time averages

$$\lim_{n\to\infty} \frac{1}{n} \sum_{k=0}^{n-1} f(T^{-k}x) =: \overline{f}^*(x)$$

exists on a set of full μ-measure.

(ii) Show that the limit of the two-sided time averages
$$\frac{1}{2n-1}\sum_{k=-n+1}^{n-1}f(T^k(x))$$
exists on a set of full μ-measure.

(iii) Show that $f^*(x) = \overline{f}^*(x)$ μ-almost everywhere, i.e., on a set of full μ-measure.

For the exercises below we use the following definitions: Let M be a compact metric space, and let $\mathcal{B}$ be the σ-algebra on M generated by the open sets, often called the Borel σ-algebra. Let $\mathcal{P}$ be the set of all probability measures on $(M, \mathcal{B})$. We define the weak* topology on $\mathcal{P}$ as for $\mu_n, \mu \in \mathcal{P}$: we have $\mu_n \to \mu$ if $\int_M \varphi d\mu_n \to \int_M \varphi d\mu$ for all continuous functions $\varphi : M \to \mathbb{R}$.

Exercise 10.4.2. (i) Show that $\mathcal{P}$ is compact with the weak* topology. (ii) Show that $\mathcal{P}$ is a convex space, i.e., if $\mu, \nu \in \mathcal{P}$, then $\alpha\mu + \beta\nu \in \mathcal{P}$ for all $\alpha, \beta \in [0,1]$ with $\alpha + \beta = 1$.

Exercise 10.4.3. For a measurable map $T : M \to M$ denote by $\mathcal{P}(T)$ the set of all T-invariant probability measures on $(M, \mathcal{B})$. Show that $\mathcal{P}(T)$ is a convex and closed subset of $\mathcal{P}$, hence it is compact in $\mathcal{P}$.

Exercise 10.4.4. If $\mu \in \mathcal{P}(T)$ is not an ergodic measure, then there are $\mu_1, \mu_2 \in \mathcal{P}(T)$ and $\lambda \in (0,1)$ with $\mu_1 \neq \mu_2$ and $\mu = \lambda\mu_1 + (1-\lambda)\mu_2$.

Exercise 10.4.5. Let $T : M \to M$ be a continuous map on a compact metric space M. Then M is called uniquely ergodic (with respect to T) if T has only one invariant Borel probability measure. Show that the (unique) invariant probability measure μ of a uniquely ergodic map T is ergodic.

Exercise 10.4.6. Suppose that $T : M \to M$ is uniquely ergodic. Then the time averages
$$\frac{1}{n}\sum_{k=0}^{n-1}f(T^k(x))$$
converge uniformly for $f : M \to \mathbb{R}$ continuous. Note that the converse of this statement is not true by considering, e.g., the map $T : \mathbb{S}^1 \times [0,1] \to \mathbb{S}^1 \times [0,1]$, $T(x,y) = (x+\alpha, y)$ with α irrational.

10.5. Orientation, Notes and References

Orientation. The main purpose of this chapter has been to provide a proof Kingman's subadditive ergodic theorem, Theorem 10.3.2. It is an important tool for analyzing the Lyapunov exponents for random dynamical systems in Chapter 11. It will be used for the Furstenberg-Kesten Theorem which, in particular, will allow us to verify assumption (11.4.2) of the deterministic Multiplicative Ergodic Theorem (Theorem 11.4.1).

Notes and references. An introduction to ergodic theory is given in Silva [128] including several proofs of Birkhoff's ergodic theorem. Basic references in this field are Krengel [86] and Walters [137]. The subadditive ergodic theorem, Theorem 10.3.2, is due to Kingman [80]. Our proof follows the exposition in an earlier version of Arnold's book [6] which relies, in particular, on Steele [130].

Our short introduction does not discuss many of the key ideas and results in ergodic theory. In particular, we did not comment on the existence of invariant measures, often called Krylov-Bogolyubov type results, which can be found in the references mentioned above and in Katok and Hasselblatt [75, Chapter 4]. Some ideas regarding the structure of the space of invariant measures for a given map T on a compact space M are mentioned in Exercises 10.4.3 and 10.4.4. More details in this regard are given by Choquet theory; see, e.g., [75, Theorem A.2.10]. A consequence of this theory for invariant measures is the ergodic decomposition theorem which says that there is a partition of a compact metric space M into T-invariant subsets M_α with T-invariant ergodic measures μ_α on M_α such that for a T-invariant measure μ and for any measurable function $f: M \to \mathbb{R}$ it holds that $\int f d\mu = \int \int f d\mu_\alpha d\alpha$, compare the literature mentioned above or [75, Theorem 4.1.12].

Chapter 11

Random Linear Dynamical Systems

In this chapter we consider random linear dynamical systems which are determined by time-varying matrix functions $A(\theta(\cdot,\omega))$ depending on a dynamical system $\theta(\cdot,\omega)$ with an invariant probability measure or, in other words, on an underlying stochastic process $\theta(\cdot,\omega)$. The 'linear algebra' for these matrices was developed by Oseledets [109] in 1968, using Lyapunov exponents and leading to his Multiplicative Ergodic Theorem, MET for short. The MET gives, with probability 1, a decomposition of the state space $\mathbb{R}^d$ into random subspaces, again called Lyapunov subspaces, with equal exponential growth rates in positive and negative time and these Lyapunov exponents are constants. The latter assertion (based on ergodicity of the invariant probability measure) is rather surprising. It shows that, notwithstanding the counterexamples in Chapter 6, the linear algebra from Chapter 1 is valid in this context if we take out an exceptional set of zero probability. As in the periodic case treated in Chapter 7, in the ergodic case only the Lyapunov spaces are varying, while the Lyapunov exponents are finitely many real numbers.

The notions and results on multilinear algebra from Section 5.1 are needed for an understanding of the proofs; they are further developed in Section 11.3.

We will give a proof of the Multiplicative Ergodic Theorem in discrete time for invertible systems. This is based on a deterministic MET which characterizes Lyapunov exponents and corresponding subspaces via singular values and volume growth rates. The main assumption is that the volume growth rates for time tending to infinity are limits. In order to verify this assumption, tools from ergodic theory are employed: Kingman's subadditive ergodic theorem leading to the Furstenberg-Kesten theorem. Then the proof of the MET follows from the latter theorem and the deterministic MET. Notions and facts from Multilinear Algebra are needed for the deterministic MET and the Furstenberg-Kesten Theorem.

Section 11.1 presents the framework and formulates the Multiplicative Ergodic Theorem in continuous and discrete time. Section 11.2 contains some facts about projections that are needed when defining appropriate metrics on Grassmannians and flag manifolds. Section 11.3 is devoted to singular values and their behavior under exterior powers. Furthermore, a metric is introduced which is used in Section 11.4 in the proof of the deterministic Multiplicative Ergodic Theorem. Its assumptions show which properties have to be established using ergodic theory. Section 11.5 derives the Furstenberg-Kesten Theorem and completes the proof of the Multiplicative Ergodic Theorem in discrete time. We note that the derivation of the MET for systems in continuous time (and for noninvertible systems in discrete time) will not be given here, since it requires more technicalities.

The theory in this chapter can either be formulated in the language of ergodic theory (measure theory) or in the language of random dynamical systems (probability theory). In the language of ergodic theory, we consider measurable linear flows: the coefficient matrix $A(\cdot)$ depends in a measurable way on a measurable flow on the base space where an invariant measure is given. We use the language of random dynamical systems.

11.1. The Multiplicative Ergodic Theorem (MET)

This section formulates the Multiplicative Ergodic Theorem for random linear dynamical systems in continuous and discrete time. First we define random linear dynamical systems in continuous time and show that random linear differential equations provide an example. Then the Multiplicative Ergodic Theorem in continuous time is formulated and its relation to the previously discussed cases of autonomous and periodic linear differential equations is explained. Its consequences for stability theory are discussed. Finally, we formulate the Multiplicative Ergodic Theorem in discrete time, which will be proved in the ensuing sections.

A random linear dynamical system is defined in the following way (recall Section 6.3). Let $(\Omega, \mathcal{F}, P)$ be a probability space, i.e., a set Ω with σ-algebra $\mathcal{F}$ and probability measure P. The expectation of a random variable on $(\Omega, \mathcal{F}, P)$ is denoted by $\mathbb{E}$. A measurable map $\theta : \mathbb{R} \times \Omega \longrightarrow \Omega$ is called a metric dynamical system, if (i) $\theta(0, \omega) = \omega$ for all $\omega \in \Omega$, (ii) $\theta(s+t, \omega) = \theta(s, \theta(t, \omega))$ for all $s, t \in \mathbb{R}, \omega \in \Omega$, and (iii) $\theta(t, \cdot)$ is measure preserving or P-invariant, i.e., for all $t \in \mathbb{R}$ we have $P\{\omega \in \Omega \,|\, \theta(t, \omega) \in E\} = P(E)$ for

11.1. The Multiplicative Ergodic Theorem (MET)

all measurable sets $E \in \mathcal{F}$. Where convenient, we also write $\theta_t := \theta(t) := \theta(t, \cdot) : \Omega \to \Omega$. Then condition (iii) may be formulated by saying that for every $t \in \mathbb{R}$ the measure $\theta_t P$ given by $(\theta_t P)(E) := P(\theta_t^{-1}(E)), E \in \mathcal{F}$, satisfies $\theta_t P = P$. In the following, we often omit the σ-algebra $\mathcal{F}$ in our notation.

A random linear dynamical system is a skew product flow $\Phi = (\theta, \varphi) : \mathbb{R} \times \Omega \times \mathbb{R}^d \longrightarrow \Omega \times \mathbb{R}^d$, where $(\Omega, \mathcal{F}, P, \theta)$ is a metric dynamical system, for each $(t, \omega) \in \mathbb{R} \times \Omega$ the map $\Phi(t, \omega) := \varphi(t, \omega, \cdot) : \mathbb{R}^d \longrightarrow \mathbb{R}^d$ is linear, and for each $\omega \in \Omega$ the map $\varphi(\cdot, \omega, \cdot) : \mathbb{R} \times \mathbb{R}^d \to \mathbb{R}^d$ is continuous. The flow property of Φ means that for all $\omega \in \Omega$ and all $x \in \mathbb{R}^d$ one has $\Phi(0, \omega, x) = (\omega, x)$ and for all $s, t \in \mathbb{R}$,

$$\begin{aligned}\Phi(s+t, \omega, x) &= (\theta(s+t, \omega), \varphi(s+t, \omega, x) \\ &= (\theta(s, \theta(t, \omega)), \varphi(s, \theta(t, \omega), \varphi(t, \omega, x)) \\ &= \Phi(s, \Phi(t, \omega, x)). \end{aligned}$$

In particular, the linear maps $\Phi(t, \omega) = \varphi(t, \omega, \cdot)$ on $\mathbb{R}^d$ satisfy the cocycle property

(11.1.1) $\quad \Phi(s+t, \omega) = \Phi(s, \theta(t, \omega)) \circ \Phi(t, \omega)$ for all $s, t \in \mathbb{R}, \omega \in \Omega$.

Conversely, this cocycle property of the second component together with the flow property for θ is equivalent to the flow property of Φ.

Linear differential equations with random coefficients yield an example of a random linear dynamical system. Here $L^1(\Omega, \mathcal{F}, P; gl(d, \mathbb{R}))$ is the Banach space of all (equivalence classes of) functions $f : \Omega \to gl(d, \mathbb{R})$, which are measurable with respect to the σ-algebra $\mathcal{F}$ and the Borel σ-algebra on $gl(d, \mathbb{R})$, and P-integrable.

Definition 11.1.1. Let θ be a metric dynamical system on a probability space $(\Omega, \mathcal{F}, P)$. Then, for an integrable map $A \in L^1(\Omega, \mathcal{F}, P; gl(d, \mathbb{R}))$, a random linear differential equation is given by

(11.1.2) $\quad\quad\quad\quad \dot{x}(t) = A(\theta_t \omega) x(t), t \in \mathbb{R}.$

The (ω-wise defined) solutions of (11.1.2) with $x(0) = x \in \mathbb{R}^d$ are denoted by $\varphi(t, \omega, x)$.

Proposition 11.1.2. *The solutions of (11.1.2) define a random linear dynamical system* $\Phi(t, \omega, x) := (\theta(t, \omega), \varphi(t, \omega, x))$ *for* $(t, \omega, x) \in \mathbb{R} \times \Omega \times \mathbb{R}^d$ *and* $\Phi(t, \omega) := \varphi(t, \omega, \cdot)$ *satisfies on a subset of* Ω *with full measure*

(11.1.3) $\quad\quad \Phi(t, \omega) = I + \int_0^t A(\theta(s, \omega)) \Phi(s, \omega) ds, t \in \mathbb{R}.$

Proof. For $\omega \in \Omega$ consider the matrix $A_\omega(t) := A(\theta_t \omega), t \in \mathbb{R}$. Here the map $t \mapsto A(\theta_t \omega) : \mathbb{R} \to gl(d, \mathbb{R})$ is measurable as it is the composition of the measurable maps $\theta(\cdot,\omega) : \mathbb{R} \to \Omega$ and $A(\cdot) : \Omega \to gl(d, \mathbb{R})$ The set of ω's on which $t \mapsto A(\theta(t,\omega))$ is locally integrable, is measurable (exhaust $\mathbb{R}$ by countably many bounded intervals) and it is θ-invariant, since for a compact interval $I \subset \mathbb{R}$ and $\tau \in \mathbb{R}$,

$$\int_I \|A(\theta(t, \theta(\tau,\omega)))\|\, dt = \int_{I+\tau} \|A(\theta(t,\omega))\|\, dt.$$

This set has full measure: In fact, $\mathbb{E}\left(\|A(\theta(t,\cdot))\|\right) = \int_\Omega \|A(\theta(t,\omega))\|\, P(d\omega) =: m < \infty$ with a constant m which is independent of t, and for all real $a \leq b$ Fubini's theorem implies

$$\int_a^b \int_\Omega \|A(\theta(t,\omega))\|\, P(d\omega)dt = m(b-a) = \int_\Omega \int_a^b \|A(\theta(t,\omega))\|\, dt\, P(d\omega) < \infty,$$

hence $\int_a^b \|A(\theta(t,\omega))\|\, dt < \infty$ for P-almost all $\omega \in \Omega$.

It follows that we can apply Theorem 6.1.1 to every equation $\dot{x} = A(\theta_t \omega)x$ with ω in a subset of full P-measure. The proof of this theorem shows that the principal fundamental solution $\Phi(t,\omega) = X_\omega(t,0)$ of this equation is measurable (with respect to ω), since it is the pointwise limit of measurable functions. The fundamental solution satisfies (11.1.3) which is the integrated form of the differential equation. By Theorem 6.1.4 the solution $\varphi(t,\omega,x)$ depends continuously on (t,x). The cocycle property for nonautonomous linear differential equations has been discussed in Proposition 6.1.2. Since the coefficient matrix function $A_\omega(\cdot)$ satisfies

$$A_\omega(t+s) = A_{\theta(t,\omega)}(s),\ t, s \in \mathbb{R},$$

the cocycle property (11.1.1) follows. $\square$

Remark 11.1.3. Stochastic linear differential equations (defined via Itō- or Stratonovich-integrals) are not random differential equations. However, they define random linear dynamical systems; see the notes in Section 11.8 for some more information.

As an example of a metric dynamical system we sketch Kolmogorov's construction.

Example 11.1.4. Let $(\Gamma, \mathcal{E}, Q)$ be a probability space and $\xi : \mathbb{R} \times \Gamma \longrightarrow \mathbb{R}^m$ a stochastic process with continuous trajectories, i.e., the functions $\xi(\cdot,\gamma) : \mathbb{R} \longrightarrow \mathbb{R}^m$ are continuous for all $\gamma \in \Gamma$. The process ξ can be written as a measurable dynamical system in the following way: Define $\Omega = \mathcal{C}(\mathbb{R}, \mathbb{R}^m)$, the space of continuous functions from $\mathbb{R}$ to $\mathbb{R}^m$. We denote by $\mathcal{F}$ the σ-algebra on Ω generated by the cylinder sets, i.e., by sets of the form $Z = \{\omega \in \Omega \mid \omega(t_1) \in F_1, \ldots, \omega(t_n) \in F_n,\ n \in \mathbb{N},\ F_i$ Borel sets in $\mathbb{R}^m\}$. The process ξ induces a probability measure P on $(\Omega, \mathcal{F})$ via $P(Z) := Q\{\gamma \in \Gamma \mid \xi(t_i,\gamma) \in F_i$ for $i = 1,\ldots,n\}$. Define the shift $\theta : \mathbb{R} \times \Omega \longrightarrow \Omega$ by

$\theta(t, \omega(\cdot)) = \omega(t + \cdot)$. Then θ is a measurable dynamical system on $(\Omega, \mathcal{F}, P)$. If ξ is stationary, i.e., if for all $n \in \mathbb{N}$, and $t, t_1, \ldots, t_n \in \mathbb{R}$ and all Borel sets $F_1, \ldots, F_n$ in $\mathbb{R}^m$ it holds that $Q\{\gamma \in \Gamma \mid \xi(t_i, \gamma) \in F_i \text{ for } i = 1, \ldots, n\} = Q\{\gamma \in \Gamma \mid \xi(t_i + t, \gamma) \in F_i \text{ for } i = 1, \ldots, n\}$, then the shift θ on Ω is P-invariant and hence it is a metric dynamical system on $(\Omega, \mathcal{F}, P)$.

We formulate the notion of ergodicity given in Section 10.1 for maps in the terminology of metric dynamical systems:

Definition 11.1.5. Let $\theta : \mathbb{R} \times \Omega \longrightarrow \Omega$ be a metric dynamical system on the probability space $(\Omega, \mathcal{F}, P)$. A set $E \in \mathcal{F}$ is called invariant under θ if $P(\theta_t^{-1}(E) \triangle E) = 0$ for all $t \in \mathbb{R}$. The flow θ is called ergodic, if each invariant set $E \in \mathcal{F}$ has P-measure 0 or 1.

Often, the measure P is called ergodic as well (with respect to the flow θ). If a flow is not ergodic, then the state space can be decomposed into two invariant sets with positive probability, and hence the properties of the corresponding restricted flows are independent of each other. Thus ergodicity is one of many ways to say that the state space is 'minimal'. Without such a minimality assumption on the base space, one should not expect that the behavior of a random linear dynamical system is uniform over all points in the base space.

The key result in this chapter is the following theorem, called Oseledets' Multiplicative Ergodic Theorem, that generalizes the results for constant and periodic matrix functions to the random context. It provides decompositions of $\mathbb{R}^d$ into random subspaces (i.e., subspaces depending measurably on ω) with equal exponential growth rate with P-probability 1 (i.e., over almost all points in the base space).

Theorem 11.1.6 (Multiplicative Ergodic Theorem). *Consider a random linear dynamical system $\Phi = (\theta, \varphi) : \mathbb{R} \times \Omega \times \mathbb{R}^d \longrightarrow \Omega \times \mathbb{R}^d$ and assume for $\Phi(t, \omega) := \varphi(t, \omega, \cdot)$ the integrability condition*

$$(11.1.4) \qquad \sup_{0 \le t \le 1} \log^+ \left\| \Phi(t, \omega)^{\pm 1} \right\| \in L^1 := L^1(\Omega, \mathcal{F}, P; \mathbb{R}),$$

where $\log^+$ denotes the positive part of $\log$. Assume that the base flow θ is ergodic. Then there exists a set $\widehat{\Omega} \subset \Omega$ of full P-measure, invariant under the flow $\theta : \mathbb{R} \times \Omega \longrightarrow \Omega$, such that for each $\omega \in \widehat{\Omega}$ the following assertions hold:

(i) There is a decomposition

$$\mathbb{R}^d = L_1(\omega) \oplus \ldots \oplus L_\ell(\omega)$$

of $\mathbb{R}^d$ into random linear subspaces $L_j(\omega)$ which are invariant under the flow Φ, i.e., $\Phi(t, \omega) L_j(\omega) = L_j(\theta_t \omega)$ for all $j = 1, \ldots, \ell$ and their dimensions d_j are constant. These subspaces have the following properties:

(ii) There are real numbers $\lambda_1 > \ldots > \lambda_\ell$, such that for each $x \in \mathbb{R}^d \setminus \{0\}$ the Lyapunov exponent $\lambda(\omega, x) \in \{\lambda_1, \ldots, \lambda_\ell\}$ exists as a limit with

$$\lambda(\omega, x) = \lim_{t \to \pm\infty} \frac{1}{t} \log \|\varphi(t, \omega, x)\| = \lambda_j \text{ if and only if } x \in L_j(\omega) \setminus \{0\}.$$

(iii) For each $j = 1, \ldots, \ell$ the maps $L_j : \Omega \longrightarrow \mathbb{G}_{d_j}$ with the Borel σ-algebra on the Grassmannian $\mathbb{G}_{d_j}$ are measurable.

(iv) The limit $\lim_{t \to \infty} \left(\Phi(t, \omega)^\top \Phi(t, \omega) \right)^{1/2t} := \Psi(\omega)$ exists and is a positive definite random matrix. The different eigenvalues of $\Psi(\omega)$ are constants and can be written as $e^{\lambda_1} > \ldots > e^{\lambda_\ell}$; the corresponding random eigenspaces are given by $L_1(\omega), \ldots, L_\ell(\omega)$. Furthermore, the Lyapunov exponents are obtained as limits from the singular values δ_k of $\Phi(t, \omega)$: The set of indices $\{1, \ldots, d\}$ can be decomposed into subsets $\Sigma_j, j = 1, \ldots, \ell$, such that for all $k \in \Sigma_j$,

$$\lambda_j = \lim_{t \to \infty} \frac{1}{t} \log \delta_k(\Phi(t, \omega)).$$

The set of numbers $\{\lambda_1, \ldots, \lambda_\ell\}$ is called the Lyapunov spectrum and the subspaces $L_j(\omega)$ are called the Lyapunov (or sometimes the Oseledets) spaces of the flow Φ. They form measurable subbundles of $\Omega \times \mathbb{R}^d$. A proof of this theorem is given in Arnold [6, Theorem 3.4.11]. If the underlying metric dynamical system, the base flow, is not ergodic, the Multiplicative Ergodic Theorem holds in a weaker form.

Remark 11.1.7. If the base flow $\theta : \mathbb{R} \times \Omega \longrightarrow \Omega$ is not ergodic but just a metric dynamical system, then the Lyapunov exponents, the number of Lyapunov spaces, and their dimensions depend on $\omega \in \widehat{\Omega}$, hence $\ell(\omega), \lambda_j(\omega)$ and $d_j(\omega)$ are random numbers, and they are invariant under θ, e.g., $\lambda(\theta_t \omega) = \lambda(\omega)$ for all $t \in \mathbb{R}$.

To understand the relation of Oseledets' theorem to our previous results, we apply it to the cases of constant and periodic matrix functions.

Example 11.1.8. Let $A \in gl(d, \mathbb{R})$ and consider the dynamical system $\varphi : \mathbb{R} \times \mathbb{R}^d \longrightarrow \mathbb{R}^d$ generated by the solutions of the autonomous linear differential equation $\dot{x} = Ax$. The flow φ can be considered as the skew-component of a random linear dynamical system over the trivial base flow given by $\Omega = \{0\}$, $\mathcal{F}$ the trivial σ-algebra, P the Dirac measure at $\{0\}$, and $\theta : \mathbb{R} \times \Omega \longrightarrow \Omega$ defined as the constant map $\theta_t \omega = \omega$ for all $t \in \mathbb{R}$. Since the flow is ergodic and satisfies the integrability condition, Theorem 11.1.6 yields the decomposition of $\mathbb{R}^d$ into Lyapunov spaces and the results from Theorem 1.4.3 on Lyapunov exponents are recovered. Note that here no use is made of explicit solution formulas.

Floquet theory follows in a weakened form.

Example 11.1.9. Let $A : \mathbb{R} \longrightarrow gl(d, \mathbb{R})$ be a continuous, T-periodic matrix function. Define the base flow as follows: Let $\Omega = \mathbb{S}^1$ parametrized by $\tau \in [0, T)$, take $\mathcal{B}$ as the Borel σ-algebra on $\mathbb{S}^1$, let P be defined by the uniform distribution on $\mathbb{S}^1$, and let θ be the shift $\theta(t, \tau) = t + \tau$. Then θ is an ergodic metric dynamical system (Exercise 11.7.3 asks for a proof of this assertion.) The solutions $\varphi(\cdot, \tau, x)$ of $\dot{x}(t) = A(t)x(t), x(\tau) = x$, define a random linear dynamical system $\Phi : \mathbb{R} \times \Omega \times \mathbb{R}^d \longrightarrow \Omega \times \mathbb{R}^d$ via $\Phi(t, \tau, x) = (\theta_t \tau, \varphi(t, \tau, x))$. With this set-up, the Multiplicative Ergodic Theorem recovers the results of Floquet Theory in Theorem 7.2.9 with P-probability 1.

For random linear differential equations the Multiplicative Ergodic Theorem yields the following.

Example 11.1.10. Consider a random linear differential equation of the form (11.1.2) and assume that θ is ergodic. If the generated random linear dynamical system Φ satisfies the integrability assumption (11.1.4), the assertions of the Multiplicative Ergodic Theorem, Theorem 11.1.6, then holds. Hence for random linear differential equations of the form (11.1.2) the Lyapunov exponents and the associated Oseledets spaces replace the real parts of eigenvalues and the Lyapunov spaces of constant matrices $A \in gl(d, \mathbb{R})$.

Stability

The problem of stability of the zero solution of random linear differential equations can now be analyzed in analogy to the case of a constant matrix or a periodic matrix function. We consider the following notion of stability.

Definition 11.1.11. The zero solution $\varphi(t, \omega, 0) \equiv 0$ of the random linear differential equation (11.1.2) is P-almost surely exponentially stable, if there exists a set $\widehat{\Omega} \subset \Omega$ of full P-measure, invariant under the flow $\theta : \mathbb{R} \times \Omega \longrightarrow \Omega$ and a constant $\alpha > 0$ such that for each $\omega \in \widehat{\Omega}$ the following assertions hold: for any $\varepsilon : \widehat{\Omega} \to (0, \infty)$ there is $\delta : \widehat{\Omega} \to (0, \infty)$ such that $\|x\| < \delta(\omega)$ implies

$$\|\varphi(t, \omega, x)\| \leq \varepsilon(\omega)$$

and

$$\|\varphi(t, \omega, x)\| \leq e^{-\alpha t} \text{ for } t \geq T(\omega, x).$$

The stable, center, and unstable subspaces are defined for $\omega \in \widehat{\Omega}$ using the Multiplicative Ergodic Theorem, Theorem 11.1.6, as

$$L^-(\omega) := \bigoplus_{\lambda_j < 0} L_j(\omega), L^0(\omega) := \bigoplus_{\lambda_j = 0} L_j(\omega), \text{ and } L^+(\omega) := \bigoplus_{\lambda_j > 0} L_j(\omega),$$

respectively. These subspaces form measurable stable, center, and unstable subbundles. We obtain the following characterization of exponential stability.

Corollary 11.1.12. *The zero solution $x(t, \omega, 0) \equiv 0$ of the random linear differential equation (11.1.2) is P-almost surely exponentially stable, if and only if all Lyapunov exponents are negative if and only if $P\{\omega \in \Omega \mid L^-(\omega) = \mathbb{R}^d\} = 1$.*

Proof. Consider for $\omega \in \Omega$ the decomposition of $\mathbb{R}^d$ into the subspaces determining the Lyapunov exponents. If there is a positive Lyapunov exponent λ_j, then a corresponding solution satisfies for some sequence $t_n \to \infty$,

$$\lim_{n \to \infty} \frac{1}{t_n} \log \|\varphi(t_n, \omega, x)\| = \lambda_j > 0.$$

Then for $\lambda \in (0, \lambda_j)$ and some constant $c > 0$,

$$\|\varphi(t, \omega, x)\| > ce^{t\lambda}, t \geq 0,$$

contradicting exponential stability. In fact, if there is $\lambda \in (0, \lambda_j)$ such that for all $c > 0$ and all $k \in \mathbb{N}$ there is $t_n > k$ with

$$\|\varphi(t_n, \omega, x)\| \leq ce^{t_n \lambda},$$

one obtains the contradiction

$$\lim_{n \to \infty} \frac{1}{t_n} \log \|\varphi(t_n, \omega, x)\| \leq \lambda < \lambda_j.$$

Conversely, suppose that all Lyapunov exponents are negative. Then write $x \in \mathbb{R}^d$ as $x = \sum_{i=1}^\ell x_j$ with $x_j \in L_j(\omega)$ for all j. Lemma 6.2.2 shows that $\lambda(\omega, x) \leq \max \lambda_j < 0$. $\square$

In general, Lyapunov exponents for random linear systems are difficult to compute explicitly—numerical methods are usually the way to go. An example will be presented in Section 11.6.

Discrete Time Random Dynamical Systems

Linear (invertible) random dynamical systems in discrete time are defined in the following way. For a probability space $(\Omega, \mathcal{F}, P)$ let $\theta : \mathbb{Z} \times \Omega \longrightarrow \Omega$ be a measurable map with the dynamical system properties $\theta(0, \cdot) = \mathrm{id}_\Omega$ and $\theta(n + m, \omega) = \theta(n, \theta(m, \omega))$ for all $n, m \in \mathbb{Z}$ and $\omega \in \Omega$ such that the probability measure P is invariant under θ, i.e., $\theta(n, \cdot)P = P$ for all $n \in \mathbb{Z}$. For conciseness, we also write $\theta := \theta(1, \cdot)$ for the time-one map and hence

$$\theta^n \omega = \theta(n, \omega) \text{ for all } n \in \mathbb{Z} \text{ and } \omega \in \Omega.$$

We call θ a metric dynamical system in discrete time and consider it fixed in the following. A random linear dynamical system in discrete time is a map $\Phi = (\theta, \varphi) : \mathbb{Z} \times \Omega \times \mathbb{R}^d \longrightarrow \Omega \times \mathbb{R}^d$ of the form

$$\Phi(n, \omega, x) = (\theta^n \omega, \varphi(n, \omega, x))$$

11.1. The Multiplicative Ergodic Theorem (MET)

such that the maps $\Phi(n,\omega) := \varphi(n,\omega,\cdot)$ are linear and

$$\Phi(0,\cdot,\cdot) = \mathrm{id}_{\Omega \times \mathbb{R}^d} \text{ and } \Phi(n+m,\cdot,\cdot) = \Phi(n,\Phi(m,\cdot,\cdot)) \text{ for all } m,n \in \mathbb{Z}.$$

Thus Φ is a linear skew product dynamical system in discrete time. The second component satisfies a cocycle property in the form

$$\varphi(0,\omega,x) = x, \ \varphi(n+m,\omega,x) = \varphi(n,\theta(m,\omega),\varphi(m,\omega,x))$$

for all $\omega \in \Omega$, $x \in \mathbb{R}^d$ and $n,m \in \mathbb{Z}$.

Equivalently, consider a measurable map A defined on Ω with values in the group $Gl(d,\mathbb{R})$ of invertible $d \times d$ matrices. Then, for $(\omega,x) \in \Omega \times \mathbb{R}^d$, the solutions $(\varphi(n,\omega,x))_{n \in \mathbb{Z}}$ of the nonautonomous linear difference equation

(11.1.5) $\qquad x_0 = x, \ x_{n+1} = A(\theta^n \omega) x_n, \ n \in \mathbb{Z},$

define a random dynamical system $\Phi = (\theta, \varphi)$ in discrete time. We say that the map A is the generator of Φ. The map $\varphi : \mathbb{Z} \times \Omega \times \mathbb{R}^d \longrightarrow \mathbb{R}^d$ is linear in its $\mathbb{R}^d$-component, i.e., the maps $\Phi(n,\omega) := \varphi(n,\omega,\cdot)$ are linear on $\mathbb{R}^d$ and the cocycle property can be written as

(11.1.6) $\qquad \Phi(0,\omega) = I, \ \Phi(n+m,\omega) = \Phi(n,\theta^m \omega) \Phi(m,\omega)$

for all $\omega \in \Omega$ and $n,m \in \mathbb{Z}$. One also sees that for all $\omega \in \Omega$,

(11.1.7) $\qquad \Phi(n,\omega) = A(\theta^{n-1}\omega) \cdots A(\omega) \text{ for } n \in \mathbb{Z}.$

Conversely, a random linear dynamical system in discrete time defines a nonautonomous linear difference equation of the form (11.1.5) with $A(\omega) := \Phi(1,\omega), \omega \in \Omega$.

The following theorem is the Multiplicative Ergodic Theorem in discrete time $\mathbb{Z}$.

Theorem 11.1.13 (Multiplicative Ergodic Theorem). *Consider a random linear dynamical system* $\Phi = (\theta, \varphi) : \mathbb{Z} \times \Omega \times \mathbb{R}^d \longrightarrow \Omega \times \mathbb{R}^d$ *in discrete time and assume the integrability condition*

(11.1.8) $\qquad \log^+ \|A\|, \log^+ \|A^{-1}\| \in L^1 := L^1(\Omega, \mathcal{F}, P; \mathbb{R}).$

Assume that the probability measure P is ergodic. Then there exists a set $\widehat{\Omega} \subset \Omega$ of full P-measure which is invariant, i.e., $\theta(\widehat{\Omega}) = \widehat{\Omega}$, such that for each $\omega \in \widehat{\Omega}$ the following assertions hold:

(i) There is a decomposition

$$\mathbb{R}^d = L_1(\omega) \oplus \ldots \oplus L_\ell(\omega)$$

of $\mathbb{R}^d$ into invariant random linear subspaces $L_j(\omega)$, i.e., $A(\omega) L_j(\omega) = L_j(\theta \omega)$ for all $j = 1, \ldots, \ell$ and their dimensions d_j are constants.

(ii) There are real numbers $\lambda_1 > \ldots > \lambda_\ell$ such that for each $x \in \mathbb{R}^d \setminus \{0\}$ the Lyapunov exponent $\lambda(\omega, x) \in \{\lambda_1, \ldots, \lambda_\ell\}$ exists as a limit and

$$\lambda(\omega, x) = \lim_{n \to \pm \infty} \frac{1}{n} \log \|\varphi(n, \omega, x)\| = \lambda_j \text{ if and only if } x \in L_j(\omega) \setminus \{0\}.$$

(iii) For each $j = 1, \ldots, \ell$ the maps $L_j : \Omega \longrightarrow \mathbb{G}_{d_j}$ with the Borel σ-algebra on the Grassmannian $\mathbb{G}_{d_j}$ are measurable.

(iv) The limit $\lim_{n \to \infty} \left(\Phi(n, \omega)^\top \Phi(n, \omega) \right)^{1/2n} := \Psi(\omega)$ exists and is a positive definite random matrix. The different eigenvalues of $\Psi(\omega)$ are constants and can be written as $e^{\lambda_1} > \ldots > e^{\lambda_\ell}$; the corresponding random eigenspaces are $L_1(\omega), \ldots, L_\ell(\omega)$. Furthermore, the Lyapunov exponents are obtained as limits from the singular values δ_k of $\Phi(n, \omega)$: The set of indices $\{1, \ldots, d\}$ can be decomposed into subsets $\Sigma_j, j = 1, \ldots, \ell$, such that for all $k \in \Sigma_j$,

$$\lambda_j = \lim_{n \to \infty} \frac{1}{n} \log \delta_k(\Phi(n, \omega)).$$

Again, the subspaces $L_j(\omega)$ are called the Lyapunov (or sometimes the Oseledets) spaces of Φ. Comparing this Multiplicative Ergodic Theorem to Theorem 11.1.6 in continuous time, we see that assertions (i)–(iv) are completely parallel. In the discrete time case considered here, analogous applications to stability theory can be given and one obtains stable, center, and unstable subspaces forming measurable stable, center, and unstable subbundles.

The rest of this chapter is devoted to a proof of the Multiplicative Ergodic Theorem in discrete time. A brief sketch of the proof is as follows: Section 11.4 presents a deterministic MET which, for fixed ω, essentially yields the assertions of the MET in positive time (in the autonomous case, Theorem 1.5.8 shows that this only leads us to a flag of subspaces). The verification of its assumptions needs a considerable amount of ergodic theory: Kingman's subadditive ergodic theorem, when applied to random linear dynamical systems, yields the Furstenberg-Kesten Theorem, which is presented in Section 11.5. This theorem establishes the limit property in assertion (iv) and verifies the assumptions of the deterministic MET. Then the construction of the Lyapunov spaces follows by applying the deterministic MET in forward and in backward time.

The proof is based on an analysis of the eigenspaces of the matrices in assertion (iv) of Theorem 11.1.13. They are related to the singular values of $\Phi(n, \omega)$. Hence, to begin with, the next sections present relevant notions and facts on projections, on singular values, and from multilinear algebra complementing Section 5.2. Furthermore, a metric is introduced which will allow us to analyze convergence properties of sets of subspaces in Section 11.4.

11.2. Some Background on Projections

This section contains some facts about projections that are needed when defining appropriate metrics on Grassmannians and flag manifolds in the next section. We begin with a result on the convergence of matrix power series.

Lemma 11.2.1. *Consider the scalar power series $\sum_{n=0}^{\infty} a_n x^n$ with r its radius of convergence. Given a matrix $A \in gl(d, \mathbb{R})$, the matrix power series converges if $\rho(A) \leq \|A\| < r$, where ρ denotes the spectral radius of A.*

Proof. Let $\|\cdot\|$ the operator norm obtained from the Euclidean inner product on $\mathbb{R}^d$. Then it holds for any $k \in \mathbb{N}$ that

$$\left\| \sum_{n=0}^{k} a_n A^n \right\| \leq \sum_{n=0}^{k} \|a_n A^n\| \leq \sum_{n=0}^{k} |a_n| \|A\|^n$$

and hence $\sum_{n=0}^{k} a_n A^n$ converges if $\|A\| < r$. Noting that $\rho(A) \leq \|A\|$ for any matrix norm finishes the proof. □

Next we discuss some facts on projections. Let $\mathbb{R}^d = M \oplus N$ and define the projection operator $P : \mathbb{R}^d \to M \oplus N$ onto M along N as $P(x) = u$ for $x = u \oplus v$ with $u \in M$ and $v \in N$. Denoting by I the identity, we note that $I - P : \mathbb{R}^d \to \mathbb{R}^d$ is the projection onto N along M. Projections are idempotent, i.e., $P \circ P = P^2 = P$ and it holds that $\text{tr} P = \dim M$. This is seen by noticing that the trace of a linear operator $T : \mathbb{R}^d \to \mathbb{R}^d$ does not depend on the basis chosen for its representation.

If we are given a vector space with inner product, such as $(\mathbb{R}^d, \langle \cdot, \cdot \rangle)$, we can consider orthogonal projections: $P : \mathbb{R}^d \to \mathbb{R}^d = M \oplus M^\perp$, where $M^\perp$ is the orthogonal complement of M. Obviously, such a projection is idempotent, symmetric and positive semidefinite, i.e., $\langle Px, x \rangle = \langle u, u \rangle \geq 0$, where $x = u \oplus v$ with $u \in M$ and $v \in M^\perp$. Vice versa it holds that any symmetric, idempotent operator P on $\mathbb{R}^d$ is an orthogonal projection onto $P[\mathbb{R}^d]$. If P is an orthogonal projection, then $\|P\| = 1$ if and only if $P \neq 0$, and $I - P : \mathbb{R}^d \to \mathbb{R}^d$ is an orthogonal projection along with P. Furthermore we have for the spectral radius $\rho(P) = \|P\|$, which follows from the fact that P is symmetric.

Remark 11.2.2. Note that for orthogonal projections P, Q on $\mathbb{R}^d$ we always have $\|PQ\| = \|QP\|$ and $\|(I - P)Q\| = \|(I - Q)P\|$. This follows from the property of the operator norm $\|A\| = \max\{|\langle Ax, y\rangle| \mid \|x\| = 1, \|y\| = 1\}$ for any matrix $A \in gl(d, \mathbb{R})$: With $\langle PQx, y\rangle = \langle Qx, Py\rangle$ one finds

$$\|PQ\| = \max\{|\langle PQx, y\rangle| \mid \|x\| = 1, \|y\| = 1\}$$
$$= \max\{|\langle x, y\rangle| \mid \|x\| = 1, \|y\| = 1, x \in \mathrm{im}\, Q, y \in \mathrm{im}\, P\} = \|QP\|.$$

Next we prove an important result on the dimension of the images of orthogonal projections:

Lemma 11.2.3. *Let $P, Q : \mathbb{R}^d \to \mathbb{R}^d$ be two orthogonal projections. Then one obtains:*

(i) $\|P - Q\| \leq 1$.

(ii) If $\|P - Q\| < 1$, then P and Q are similar, in particular, we have $\dim P[\mathbb{R}^d] = \dim Q[\mathbb{R}^d]$.

Proof. The proof of (i) is left as Exercise 11.7.2. Statement (ii) can be seen as follows:

Denote $P[\mathbb{R}^d] =: M$ and $Q[\mathbb{R}^d] =: N$. We compare the action of P and Q by considering the maps

$$(11.2.1) \qquad U := QP + (I - Q)(I - P), V := PQ + (I - P)(I - Q).$$

Note that U maps M into N, and $M^\perp$ into $N^\perp$, while V maps N into M, and $N^\perp$ into $M^\perp$. However, these two maps are not inverses since

$$UV = VU = I - (P - Q)^2 = I - R \text{ with } R := (P - Q)^2.$$

Note that R commutes with P, Q, U, and V. To construct inverse maps with the same behavior on range and kernel as U and V, we can define U_0 and V_0 via

$$U_0(I - R)^{\frac{1}{2}} = U, V_0(I - R)^{\frac{1}{2}} = V.$$

These matrices exist if $(I - R)^{-\frac{1}{2}}$ exists. To this end, recall that

$$\begin{pmatrix} -\frac{1}{2} \\ 0 \end{pmatrix} := 1 \text{ and } \begin{pmatrix} -\frac{1}{2} \\ n \end{pmatrix} := \frac{(-\frac{1}{2})(-\frac{1}{2} - 1)\ldots(-\frac{1}{2} - (n-1))}{n!} \text{ for } n \in \mathbb{N}$$

and consider the scalar negative binomial series

$$(1 - x)^{-\frac{1}{2}} = \sum_{n=0}^{\infty} (-1)^n \begin{pmatrix} -\frac{1}{2} \\ n \end{pmatrix} x^n$$

with radius of convergence $r = 1$. By Lemma 11.2.1 the corresponding matrix series converges if $\|R\| = \|(P - Q)^2\| \leq \|P - Q\|^2 < 1 = r$, which is satisfied by assumption. One shows by direct calculation that the sum S of the matrix series

$$S := (I - R)^{-\frac{1}{2}} = \sum_{n=0}^{\infty} (-1)^n \begin{pmatrix} -\frac{1}{2} \\ n \end{pmatrix} R^n$$

11.2. Some Background on Projections

satisfies $S^2 = (I - R)^{-1}$ (just like in the scalar case) and it commutes with the operators P, Q, U, V, U_0 and V_0. We obtain as desired

$$V_0 U_0 = U_0 V_0 = I, \text{ hence } V_0 = U_0^{-1}, U_0 = V_0^{-1}.$$

For the projections P and Q this means the following: From equation (11.2.1) we obtain $UP = QP = QU$ and $PV = PQ = VQ$, and therefore, using the commutation properties of R, we have $U_0 P = QU_0$ and $PV_0 = V_0 Q$. This means that $U_0 P U_0^{-1} = Q$ and $P = V_0^{-1} Q V_0$. In particular, the ranges $P[\mathbb{R}^d] = M$ and $Q[\mathbb{R}^d] = N$ are isomorphic, since they are mapped onto each other by U_0 and V_0. $\square$

Remark 11.2.4. The proof of Lemma 11.2.3(ii) goes through even if P and Q are just projections. In the case of orthogonal projections one can easily see that the operators U_0 and V_0 are actually orthogonal.

The following result refines the second part of the previous lemma.

Proposition 11.2.5. *Let P, Q be two orthogonal projections on $\mathbb{R}^d$ with $P[\mathbb{R}^d] =: M$ and $Q[\mathbb{R}^d] =: N$. Assume that $\|(I - Q)P\| =: \delta < 1$. Then there are two possible alternatives:*

(i) *Q maps M onto N bijectively, in which case we have*

$$\|P - Q\| = \|(I - P)Q\| = \|(I - Q)P\| = \delta.$$

(ii) *Q maps M onto $N_0 \subsetneq N$ bijectively, in which case $\|P - Q\| = \|(I - P)Q\| = 1$ and the orthogonal projection Q_0 onto N_0 satisfies*

$$\|P - Q_0\| = \|(I - P)Q_0\| = \|(I - Q_0)P\| = \|(I - Q)P\| = \delta.$$

Proof. For $x \in M$ it holds that $\|x - Qx\| = \|(I - Q)Px\| \leq \delta \|x\|$, which implies $(1 - \delta)\|x\| \leq \|Qx\|$. Hence the linear operator $Q : M \to N$ is injective. Define $Q[M] := N_0$ and let $Q_0 : \mathbb{R}^d \to N_0$ be the orthogonal projection onto N_0.

Note that for any $x \in \mathbb{R}^d$ there exists $y \in M$ with $Q_0 x = Qy$, and $y \neq 0$ if $Q_0 x \neq 0$. Since $(I - P)y = 0$, we have for any $x \in \mathbb{R}^d$ with $Q_0 x \neq 0$,

$$\|(I - P)Q_0 x\| = \|(I - P)Qy\| = \left\|(I - P)\left(Qy - \frac{\|Qy\|^2}{\|y\|^2} y\right)\right\|$$

$$\leq \left\|Qy - \frac{\|Qy\|^2}{\|y\|^2} y\right\|.$$

This implies for $\|x\| = 1$,

$$\|(I-P)Q_0 x\|^2 \le \|Qy\|^2 - \|Qy\|^4 \frac{1}{\|y\|^2} = \|Qy\|^2 \frac{1}{\|y\|^2}(\|y\|^2 - \|Qy\|^2)$$
$$= \|Q_0 x\|^2 \frac{1}{\|y\|^2}\|(I-Q)y\|^2 \le \frac{1}{\|y\|^2}\|(I-Q)Py\|^2 \le \delta^2.$$

Hence we obtain

(11.2.2) $\quad\quad\quad \|(I-P)Q_0\| \le \delta = \|(I-Q)P\|.$

To estimate $\|P - Q_0\|$ we note that the ranges of Q_0 and $I - Q_0$ are orthogonal, which gives for $x \in \mathbb{R}^d$ the equalities

$$\|(P-Q_0)x\|^2 = \|(I-Q_0)Px - Q_0(I-P)x\|^2$$
$$= \|(I-Q_0)Px\|^2 + \|Q_0(I-P)x\|^2.$$

Projections are idempotent, and hence we obtain

$$\|(P-Q_0)x\|^2 \le \|(I-Q_0)P\|^2 \|Px\|^2 + \|Q_0(I-P)\|^2 \|(I-P)x\|^2.$$

Now by definition $Q_0 P = Q_0 Q P = QP$ and therefore $\delta = \|(I-Q)P\| = \|(I-Q_0)P\|$, which together with

$$\|Q_0(I-P)\| = \left\|(Q_0(I-P))^\top\right\| = \|(I-P)Q_0\| \le \delta$$

yields

$$\|(P-Q_0)x\|^2 \le \delta^2(\|Px\|^2 + \|(I-P)x\|^2) = \delta^2 \|x\|^2.$$

Note that we have shown

(11.2.3) $\quad \delta = \|(I-Q)P\| = \|(I-Q_0)P\| = \|(P-Q_0)P\| \le \|P-Q_0\| \le \delta$

and hence equality holds here. By Lemma 11.2.3(ii) $\|P-Q_0\| = \delta < 1$ means that $P[N_0] = P[Q_0[\mathbb{R}^d]] = M$. Hence in equation (11.2.2) we can replace P, Q by Q_0, P to obtain $\|(I-Q_0)P\| \le \|(I-P)Q_0\|$, and with equation (11.2.3) we then have

$$\|(I-P)Q_0\| = \|(I-Q)P\| = \delta.$$

In other words, if $Q[M] = N$, i.e., in case (i), we have shown the desired equalities. In case (ii), i.e., if $Q[M] = N_0 \subsetneq N$, what is left to show is $\|P - Q\| = \|(I-P)Q\| = 1$.

To this end, let $x \in N \setminus N_0$. Then by $P[N_0] = P[Q_0[\mathbb{R}^d]] = M$ there is $x_0 \in N_0$ with $Px_0 = Px$, and we have $y := x - x_0 \in N$, $y \ne 0$ but $Py = 0$. We compute $(P-Q)y = -y$ and $Q(I-P)y = Qy = y$, which implies $\|P-Q\| \ge 1$ and $\|(I-P)Q\| = \|Q(I-P)\| \ge 1$. Now Lemma 11.2.3(i) provides the reverse inequalities to finish the proof. $\square$

11.3. Singular Values, Exterior Powers, and the Goldsheid-Margulis Metric

This section prepares tools from multilinear algebra which will be needed in the ensuing sections. First basic properties of singular values of matrices are proved and their relevance for exterior powers is explained. Then the Goldsheid-Margulis metric on flags is introduced which will be needed for the proof of the deterministic multiplicative ergodic theorem in the next section.

In order to introduce singular values and associated decompositions let $\mathbb{R}^d$ be endowed with the Euclidean scalar product and consider the group of orthogonal matrices, $O(d, \mathbb{R}) := \{R \in Gl(d, \mathbb{R}) \mid R^\top R = I\}$. For a matrix $A \in gl(d, \mathbb{R})$ we say that

$$A = RDQ^\top$$

is a singular value decomposition if $R, Q \in O(d, \mathbb{R})$ and D is a diagonal matrix, $D = \text{diag}(\delta_1, \ldots, \delta_d)$ with $\delta_1 \geq \ldots \geq \delta_d \geq 0$. Then the δ_i are called the singular values of A. Such singular value decompositions always exist and the δ_i are unique. We will need this only in the following special case, where A is an invertible square matrix.

Proposition 11.3.1. *Any matrix $A \in Gl(d, \mathbb{R})$ has a singular value decomposition with singular values $\delta_1 \geq \ldots \geq \delta_d > 0$. Then the δ_i^2 are the eigenvalues of $A^\top A$, hence they are uniquely determined by A. In particular, $\|A\| = \delta_1$ where $\|\cdot\|$ is the operator norm associated with the Euclidean scalar product in $\mathbb{R}^d$, and $|\det A| = \delta_1 \cdots \delta_d$. Furthermore, let x be an element in an eigenspace of $A^\top A$ corresponding to δ_j^2. Then $\|Ax\| = \delta_j \|x\|$.*

Proof. The matrix $A^\top A$ is symmetric and positive definite, since for all $0 \neq x \in \mathbb{R}^d$,

$$x^\top A^\top A x = \langle Ax, Ax \rangle = \|Ax\|^2 > 0.$$

Hence there exists $Q \in O(d, \mathbb{R})$ such that $Q^\top A^\top A Q = \text{diag}(\delta_1^2, \ldots, \delta_d^2) = D^2$. Here we may order the numbers δ_i that $\delta_1 \geq \ldots \geq \delta_d > 0$. Writing this as $A^\top A Q = \text{diag}(\delta_1^2, \ldots, \delta_d^2) Q$, one sees that the columns q_j of Q are (right) eigenvectors of $A^\top A$ for the eigenvalues δ_j. Thus they form a basis of $\mathbb{R}^d$. Let $R := [r_1, \ldots, r_d] \in gl(d, \mathbb{R})$ be the matrix with columns $r_j := \delta_j^{-1} A q_j$, $j = 1, \ldots, d$. Then

$$R^\top R = \text{diag}(\delta_1^{-2}, \ldots, \delta_d^{-2})[q_1, \ldots, q_d]^\top A^\top A [q_1, \ldots, q_d]$$
$$= D^{-2} Q^\top A^\top A Q = I_d,$$

hence $R \in O(d, \mathbb{R})$. Furthermore, with the standard basis vectors $\mathbf{e}_j$ one has $RDQ^\top q_j = RD\mathbf{e}_j = \delta_j r_j = Aq_j$ for $j = 1, \ldots, d$. Since the q_j form a basis of $\mathbb{R}^d$, it follows that $RDQ^\top = A$, as desired. The proof also shows that the δ_j

are unique. Furthermore, $(\det A)^2 = \det A^\top \det A = \det(A^\top A) = \delta_1^2 \ldots \delta_d^2$, and

$$\|A\|^2 = \left(\sup_{\|x\|=1} \|Ax\|\right)^2 = \sup_{\|x\|=1} \langle Ax, Ax \rangle = \sup_{\|x\|=1} \langle A^\top Ax, x \rangle = \delta_1^2.$$

Finally, for x in the eigenspace of $A^\top A$ for δ_j^2 it follows that $\|Ax\|^2 = x^\top A^\top Ax = \delta_j^2 \|x\|^2$. $\square$

In the decomposition $A = RDQ$ the columns of R are called the left-singular vectors of A: they are the (normalized) eigenvectors of $AA^\top$. Similarly, the columns of Q are called the right-singular eigenvectors of A: they are the normalized eigenvectors of $A^\top A$.

Now we come back to exterior powers of linear spaces as introduced in Section 5.1 and show that one can in a similar way define exterior powers of linear operators.

Proposition 11.3.2. *If $A : \mathbb{R}^d \to \mathbb{R}^d$ is a linear operator, then the linear extension of*

$$\left(\bigwedge^k A\right)(u_1 \wedge \ldots \wedge u_k) := Au_1 \wedge \ldots \wedge Au_k$$

defines a linear operator $\bigwedge^k A$ on $\bigwedge^k \mathbb{R}^d$ with $\bigwedge^1 A = A$, $\bigwedge^d A = \det A$, and $\bigwedge^k I = I$. For a linear operator $B : \mathbb{R}^d \to \mathbb{R}^d$ one has

$$\bigwedge^k(AB) = \left(\bigwedge^k A\right)\left(\bigwedge^k B\right) \text{ and } \left\|\bigwedge^k(AB)\right\| \leq \left\|\left(\bigwedge^k A\right)\right\|\left\|\left(\bigwedge^k B\right)\right\|.$$

Proof. This follows from the definitions and properties given in Section 5.1: On the simple vectors, the map $\bigwedge^k A$ is well defined. Then it can be extended to $\bigwedge^k \mathbb{R}^d$, since there is a basis consisting of simple vectors. Similarly, one verifies the other assertions. $\square$

The next proposition shows that the singular values determine the norm of exterior powers.

Proposition 11.3.3. *Let $A = RDQ^\top$ be a singular value decomposition of $A \in Gl(d, \mathbb{R})$ and let $\delta_1 \geq \ldots \geq \delta_d$ be the singular values of A. Then the following assertions hold.*

(i) A singular value decomposition of $\bigwedge^k A$ is

$$\bigwedge^k A = \left(\bigwedge^k R\right)\left(\bigwedge^k D\right)\left(\bigwedge^k Q^\top\right).$$

(ii) The matrix $\bigwedge^k D$ is diagonal with entries $(\delta_{i_1} \cdots \delta_{i_k}), 1 \leq i_1 \leq \ldots \leq i_k \leq d$. In particular, the top singular value of $\bigwedge^k A$ is $\delta_1 \cdots \delta_k$ and the smallest is $\delta_{d-k+1} \cdots \delta_d$. If A is positive definite, then $\bigwedge^k A$ is positive definite.

11.3. Singular Values and the Goldsheid-Margulis Metric

(iii) The operator norm is given by $\|\bigwedge^k A\| = \delta_1 \cdots \delta_k$. In particular, $\|\bigwedge^d A\| = \delta_1 \cdots \delta_d = |\det A|$, and $\|\bigwedge^{k+l} A\| \le \|\bigwedge^k A\| \|\bigwedge^l A\|$, $1 \le k, l$ and $k+l \le d$. In particular, $\|\bigwedge^k A\| \le \|A\|^k$.

Proof. The assertions in (iii) follow from (i), (ii), and Propositions 11.3.1 and 11.3.2. The equality in assertion (i) follows from Proposition 11.3.2. Hence it remains to show that the exterior power of an orthogonal matrix is orthogonal and that $\bigwedge^k D$ has the form indicated in assertion (ii). For $Q \in O(d)$ the scalar product (5.1.4) yields

$$\left\langle \left(\bigwedge\nolimits^k Q\right)(x_1 \wedge \ldots \wedge x_k), y_1 \wedge \ldots \wedge y_k \right\rangle = \langle Qx_1 \wedge \ldots \wedge Qx_k, y_1 \wedge \ldots \wedge y_k \rangle$$
$$= \det(\langle Qx_i, y_j \rangle) = \det\left(\left\langle x_i, Q^\top y_j \right\rangle\right)$$
$$= \left\langle x_1 \wedge \ldots \wedge x_k, \left(\bigwedge\nolimits^k Q^\top\right)(y_1 \wedge \ldots \wedge y_k) \right\rangle.$$

Hence $(\bigwedge^k Q)^\top = \bigwedge^k Q^\top$ and this matrix is orthogonal, since

$$I = \bigwedge\nolimits^k (Q^\top Q) = \left(\bigwedge\nolimits^k Q^\top\right)\left(\bigwedge\nolimits^k Q\right).$$

It is immediate from the definition that the exterior power of a diagonal matrix is a diagonal matrix. Then it is also clear that the singular values of $\bigwedge^k A$ are as given in (ii). $\square$

Geometrically, the equality $\|\bigwedge^k A\| = \delta_1 \cdots \delta_k$ means that the largest growth of a parallelepiped is achieved when its sides are in the direction of the basis vectors which are eigenvectors of $A^\top A$ corresponding to the k largest singular values.

Next we define convenient metrics on the Grassmannian $\mathbb{G}_k(d)$ and on flag manifolds.

Lemma 11.3.4. *Let P and $\hat{P}$ be the orthogonal projections onto U and $\hat{U} \in \mathbb{G}_k(d)$, respectively. Then $d(U, \hat{U}) := \|(I-P)\hat{P}\|$ defines a complete metric on the Grassmannian $\mathbb{G}_k(d)$. Here $\|\cdot\|$ denotes again the operator norm relative to the Euclidean inner product on $\mathbb{R}^d$.*

Proof. First note that by Remark 11.2.2 $\|(I-P)\hat{P}\| = \|(I-\hat{P})P\|$. Hence it follows that $d(U, \hat{U}) = d(\hat{U}, U)$. Furthermore, $\|(I-P)\hat{P}\| = 0$ if and only if $\hat{U} \subset U$ and, similarly, $\|(I-\hat{P})P\| = 0$ means that $U \subset \hat{U}$. Thus $d(U, \hat{U}) = 0$ if and only if $U = \hat{U}$.

Let $\tilde{P}$ denote the orthogonal projection onto $\tilde{U} \in \mathbb{G}_k(d)$, then the triangle inequality follows from

$$\left\| (I-P)\hat{P} \right\| = \left\| (I-P)\hat{P}(I-\tilde{P}+\tilde{P}) \right\|$$
$$\leq \left\| (I-P)\hat{P}(I-\tilde{P}) \right\| + \left\| (I-P)\hat{P}\tilde{P} \right\|$$
$$\leq \left\| \hat{P}(I-\tilde{P}) \right\| + \left\| (I-P)\tilde{P} \right\|,$$

hence $d(U,\hat{U}) \leq d(U,\tilde{U}) + d(\tilde{U},\hat{U})$.

Completeness of the metric $d(U,\hat{U})$ on $\mathbb{G}_k(d)$ can be seen directly using the fact that $\left\| (I-P)\hat{P} \right\| = \left\| P - \hat{P} \right\|$ in $\mathbb{G}_k(d)$ if $\left\| (I-P)\hat{P} \right\| < 1$; see Proposition 11.2.5. Hence for a Cauchy sequence the projections converge. □

We now turn to the Goldsheid-Margulis metric on flag manifolds. (Actually, we do not need that the flags form manifolds; the metric space structure is sufficient for our purposes.) Consider finite sequences of increasing subspaces, i.e., flags: For a multi-index $\tau = (\tau_1, \ldots, \tau_\ell)$ with $\tau_\ell < \tau_{\ell-1} < \ldots < \tau_1 = d$ let $\mathbb{F}_\tau$ be the set of flags F in $\mathbb{R}^d$ given by increasing sequences of subspaces V_i with $\dim V_i = \tau_i, i = 1, \ldots, \ell$,

$$F = (V_\ell, \ldots, V_1).$$

Define $U_\ell := V_\ell$ and U_i as the unique orthogonal complement of V_{i+1} in $V_i, i = \ell - 1, \ldots, 1$, thus

$$V_i = U_\ell \oplus \ldots \oplus U_i, \ i = \ell, \ldots, 1.$$

Lemma 11.3.5. *Let $\lambda_1 > \ldots > \lambda_\ell > 0, \ell \geq 2$, be pairwise different and abbreviate $\Delta := \min_i(\lambda_i - \lambda_{i+1})$ and $h := \Delta/(\ell - 1)$.*

(i) Then a metric, called Goldsheid-Margulis metric, on the space $\mathbb{F}_\tau$ is defined by

$$(11.3.1) \qquad d_{GM}(\hat{F}, F) := \max_{i,j=1,\ldots,\ell, i \neq j} \left\| \hat{P}_i P_j \right\|^{h/|\lambda_i - \lambda_j|},$$

where for $F, \hat{F} \in \mathbb{F}_\tau$ the orthogonal projections to the subspaces U_j and $\hat{U}_i$ are denoted by P_j and $\hat{P}_i$, respectively.

(ii) Furthermore,

$$(11.3.2) \qquad d_{GM}(\hat{F}, F) = \max_{i,j=1,\ldots,\ell, i \neq j} \max_{x,y} |\langle x, y \rangle|^{h/|\lambda_i - \lambda_j|} \leq 1,$$

where the inner maximum is taken over all unit vectors $x \in \hat{U}_i, y \in U_j$.

(iii) If (F_n) is a Cauchy sequence in $\mathbb{F}_\tau$, then the corresponding subspaces $U_i(n)$ form Cauchy sequences in the Grassmannians $\mathbb{G}_{\tau_i}(d)$. Hence, with this metric, the space $\mathbb{F}_\tau$ becomes a complete metric space.

11.3. Singular Values and the Goldsheid-Margulis Metric

Proof. First we observe that one always has $d_{GM}(\hat{F}, F) \leq 1$, since $\|\hat{P}_i P_j\| \leq 1$ and $0 < h/|\lambda_i - \lambda_j| < 1/(\ell - 1) \leq 1$. Remark 11.2.2 shows that

$$\|\hat{P}_i P_j\| = \|P_j \hat{P}_i\| = \max |\langle x, y \rangle|,$$

where the maximum is taken over all $x \in \mathrm{im}\hat{P}_i, y \in \mathrm{im}P_j$ with $\|x\| = \|y\| = 1$. This proves equality (11.3.2).

In order to show that d_{GM} defines a metric note that $d_{GM}(\hat{F}, F) = d_{GM}(F, \hat{F})$, and $d_{GM}(\hat{F}, F) = 0$ if and only if $\|\hat{P}_i P_j\| = \|P_j \hat{P}_i\| = 0$ for all $i \neq j$. Thus $\hat{U}_i$ is contained in the orthogonal complement of every U_j with $j \neq i$, hence in U_i and conversely, U_i is contained in the orthogonal complement of every $\hat{U}_j$ with $j \neq i$, hence in $\hat{U}_i$. It follows that $\hat{U}_i = U_i$ for every i (the distance in the Grassmannian vanishes) and $F = \hat{F}$. It remains to show the triangle inequality

$$d_{GM}(\hat{F}, F) \leq d_{GM}(\hat{F}, \tilde{F}) + d_{GM}(\tilde{F}, F)$$

for a flag $\tilde{F} = (\tilde{V}_\ell \subset \ldots \subset \tilde{V}_1) \in \mathbb{F}_\tau$ with orthogonal projections $\tilde{P}_k$ to $\tilde{U}_k$.

Let $x \in \hat{U}_i$ and $y \in U_j$ with $\|x\| = \|y\| = 1$ and $d_{GM}(\hat{F}, F) = |\langle x, y \rangle|^{h/|\lambda_i - \lambda_j|}$ for certain indices i, j. Write

$$x = x_1 + \ldots + x_\ell \text{ and } y = y_1 + \ldots + y_\ell,$$

where $x_k = \tilde{P}_k \hat{P}_i x$, $y_k = \tilde{P}_k P_j y$ are the projections to $\tilde{U}_k$. By orthogonality

$$|\langle x, y \rangle| = \left| \sum_{k=1}^{\ell} \langle x_k, y_k \rangle \right| \leq \sum_{k=1}^{\ell} \|x_k\| \|y_k\|.$$

We set $\delta_1 := d_{GM}(\tilde{F}, F)$ and $\delta_2 := d_{GM}(\hat{F}, \tilde{F})$. Then, by the definitions,

$$\|x_k\| \leq \delta_2^{|\lambda_i - \lambda_k|/h} \text{ and } \|y_k\| \leq \delta_1^{|\lambda_k - \lambda_j|/h}.$$

Together this yields

$$d_{GM}(\hat{F}, F) = |\langle x, y \rangle|^{h/|\lambda_i - \lambda_j|} \leq \left(\sum_{k=1}^{\ell} \|x_k\| \|y_k\| \right)^{h/|\lambda_i - \lambda_j|}$$

$$\leq \left(\sum_{k=1}^{\ell} \delta_2^{|\lambda_i - \lambda_k|/h} \delta_1^{|\lambda_k - \lambda_j|/h} \right)^{h/|\lambda_i - \lambda_j|}.$$

Hence we have to verify

$$\sum_{k=1}^{\ell} \delta_2^{|\lambda_i - \lambda_k|/h} \delta_1^{|\lambda_k - \lambda_j|/h} \leq (\delta_1 + \delta_2)^{|\lambda_i - \lambda_j|/h}.$$

We can suppose without loss of generality that $\delta_2 = \alpha\delta_1$ with $\alpha \in [0,1]$. Then we rewrite this inequality in the form

$$(11.3.3) \qquad \sum_{k=1}^{\ell} \delta_1^{(|\lambda_i-\lambda_k|+|\lambda_k-\lambda_j|)/h} \alpha^{|\lambda_k-\lambda_j|/h} \leq (1+\alpha)^{|\lambda_i-\lambda_j|/h} \delta_1^{|\lambda_i-\lambda_j|/h}.$$

Since the distance $\delta_1 \leq 1$, one has for every k that

$$\delta_1^{(|\lambda_i-\lambda_k|+|\lambda_k-\lambda_j|)/h} \leq \delta_1^{|\lambda_i-\lambda_j|/h}$$

and hence (11.3.3) follows from the inequality

$$(11.3.4) \qquad \sum_{k=1}^{\ell} \alpha^{|\lambda_k-\lambda_j|/h} \leq (1+\alpha)^{|\lambda_i-\lambda_j|/h}.$$

Finally, we verify (11.3.4) by using the definition of $h = \Delta/(\ell-1)$,

$$\sum_{k=1}^{\ell} \alpha^{|\lambda_k-\lambda_j|/h} \leq 1 + (\ell-1)\alpha^{\ell-1} \leq (1+\alpha)^{\ell-1} \leq (1+\alpha)^{|\lambda_i-\lambda_j|/h}.$$

This proves that d_{GM} is a metric.

For assertion (iii) recall the metric d on the Grassmannian $\mathbb{G}_{d_i}(d)$ from Lemma 11.3.4. With the orthogonal projections $P_i(n)$ onto $U_i(n)$ and using the identity $I = \sum_{j=1}^{\ell} P_j(n+m)$, one finds from $d_{GM}(F(n+m), F(n))) < \varepsilon$

$$d(U_i(n+m), U_i(n)) = \max\left(\|(I - P_i(n+m))P_i(n)\|, \|P_i(n+m)P_i(n)\|\right)$$

$$\leq \sum_{j=1,\ldots,\ell, j\neq i} \|P_j(n+m)P_i\| < \varepsilon.$$

Hence $F(n) \to F$ in $\mathbb{F}_\tau$ implies $U_i(n) \to U$ in $\mathbb{G}_{k_i}(d)$ for all $i = 1,\ldots,\ell$. Vice versa, from the proof of Lemma 11.3.4 we have $\|(I-P)\hat{P}\| = \|P - \hat{P}\|$ in $\mathbb{G}_k(d)$ if $\|(I-P)\hat{P}\| < 1$. Hence $d(U_i(n), U_i) \to 0$ for all $i = 1,\ldots,\ell$ implies $P_i(n) \to P_i$ for all i, which in turn implies $d_{GM}(F(n), F) \to 0$. Since the metric $d(\cdot,\cdot)$ from Lemma 11.3.4 is complete, so is $d_{GM}(\cdot,\cdot)$. □

11.4. The Deterministic Multiplicative Ergodic Theorem

In this section we formulate and prove the deterministic MET, explain its meaning in the autonomous case, and discuss the difficulties in verifying its assumptions in the nonautonomous case. For a sequence (A_n) of matrices, the MET studies, in the limit for $n \to \infty$,

- the singular values of $\Phi_n := A_n A_{n-1} \cdots A_1$ and the exponential growth rates of the exterior powers of Φ_n,
- the associated eigenspaces,
- and the corresponding Lyapunov exponents.

11.4. The Deterministic Multiplicative Ergodic Theorem

Theorem 11.4.1 (Deterministic MET). *Let $A_n \in Gl(d, \mathbb{R}), n \in \mathbb{N}$, be a sequence of invertible $d \times d$-matrices which satisfies the following conditions:*

$$\text{(11.4.1)} \qquad \limsup_{n \to \infty} \frac{1}{n} \log \|A_n^{\pm 1}\| \leq 0,$$

and with $\Phi_n := A_n \cdots A_1$ the limits

$$\text{(11.4.2)} \qquad \lim_{n \to \infty} \frac{1}{n} \log \|\wedge^i \Phi_n\| = \gamma^{(i)}$$

exist for $i = 1, \ldots, d$.

(i) Then the limit $\lim_{n \to \infty} (\Phi_n^\top \Phi_n)^{1/2n} =: \Psi$ exists and the eigenvalues of Ψ coincide with the e^{Λ_i}, where the Λ_i's are successively defined by $\Lambda_1 + \ldots + \Lambda_i := \gamma^{(i)}, i = 1, \ldots, d$. For the ith singular value $\delta_i(\Phi_n)$ of Φ_n one has

$$\Lambda_i = \lim_{n \to \infty} \log \delta_i(\Phi_n), \ i = 1, \ldots, d.$$

(ii) Let $e^{\lambda_1} > \ldots > e^{\lambda_\ell}$ be the different eigenvalues of Ψ and $U_1, \ldots, U_\ell$ their corresponding eigenspaces with multiplicities $d_i = \dim U_i$. Let

$$V_i := U_\ell \oplus \ldots \oplus U_i, \ i = 1, \ldots, \ell,$$

so that one obtains the flag

$$\text{(11.4.3)} \qquad V_\ell \subset \ldots \subset V_1 = \mathbb{R}^d.$$

Then for each $0 \neq x \in \mathbb{R}^d$ the Lyapunov exponent

$$\text{(11.4.4)} \qquad \lambda(x) = \lim_{n \to \infty} \frac{1}{n} \log \|\Phi_n x\|$$

exists as a limit, and $\lambda(x) = \lambda_i$ if and only if $x \in V_i \setminus V_{i+1}$.

Before we start the rather nontrivial proof, involving what Ludwig Arnold calls "the hard work in Linear Algebra" associated with the MET, we discuss the meaning of this result in the autonomous case, where $\Phi_n = A^n, n \in \mathbb{N}$. In this case, assumption (11.4.1) trivially holds by

$$\lim_{n \to \infty} \frac{1}{n} \log \|A^{\pm 1}\| = 0.$$

In order to verify assumption (11.4.2) we apply Proposition 11.3.2 to $\Phi_n = A^n$ and obtain the following subadditivity property for $m, n \in \mathbb{N}$

$$\log \|\wedge^i A^{m+n}\| \leq \log \left(\|\wedge^i A^m\| \cdot \|\wedge^i A^n\| \right) = \log \|\wedge^i A^m\| + \log \|\wedge^i A^n\|.$$

Write $a_n := \log \|\wedge^i A^n\|, n \in \mathbb{N}$. Then (11.4.2) is an immediate consequence of the following lemma, which is often used in ergodic theory.

Lemma 11.4.2. *Let $a_n \geq 0, n \in \mathbb{N}$, be a subadditive sequence, i.e., $a_{m+n} \leq a_m + a_n$ for all $m, n \in \mathbb{N}$. Then $\frac{a_n}{n}$ converges, and*

$$\lim_{n \to \infty} \frac{a_n}{n} = \inf_{n \in \mathbb{N}} \frac{a_n}{n} =: \gamma.$$

Proof. Fix $N \in \mathbb{N}$ and write $n = k(n)N + r(n)$ with $k(n) \in \mathbb{N}_0$ and $0 \leq r(n) < N$, hence $k(n)/n \to 1/N$ for $n \to \infty$. Clearly, a_k is bounded for $0 \leq k \leq N$ for any N. By subadditivity, for any $n, N \in \mathbb{N}$,

$$\gamma \leq \frac{a_n}{n} \leq \frac{1}{n}\left[k(n)a_N + a_{r(n)}\right].$$

Hence, for $\varepsilon > 0$ there exists an $N_0(\varepsilon, N) \in \mathbb{N}$ such that for all $n > N_0(\varepsilon.N)$,

$$\gamma \leq \frac{a_n}{n} \leq \frac{a_N}{N} + \varepsilon.$$

Since ε and N are arbitrary, the result follows. □

In the autonomous case, the assertion of the deterministic MET is that

$$\lim_{n \to \infty} (A^{\top n} A^n)^{1/2n} =: \Psi$$

exists; the eigenvalues of Ψ coincide with the e^{Λ_i}, where the Λ_i's are the limits of the singular values of A^n. At the same time, the Λ_i are the Lyapunov exponents of A, i.e., the real parts of the corresponding eigenvalues. Note that the deterministic MET shows that also the associated subspaces converge nicely. This is a maybe unexpected relation between singular values and eigenvalues. From a geometric point of view, it may not be so surprising, since the real parts of the eigenvalues determine the volume growth of parallelepipeds (Theorem 5.2.5) and the singular values determine the maximal volume growth of a parallelepiped, see Proposition 11.3.3(iii). (They also determine the volume growth of hyperellipsoids.)

Concerning the subspaces V_i in Theorem 11.4.1, note that the Lyapunov spaces L_j in Chapter 1 are characterized by the property that $\lambda(x) = \lim_{n \to \pm\infty} \frac{1}{n} \log \|A^n x\| = \lambda_j$ for $x \in L_j$. Thus the Lyapunov spaces cannot solely be characterized by limits for time tending to $+\infty$ (see Theorem 1.4.4 for continuous time and Theorem 1.5.8 for discrete time). One only obtains a flag, i.e., a sequence of increasing subspaces as in (11.4.3). Hence, for a proof of the MET, we will have to apply the deterministic MET in forward and in backward time.

Our discussion shows that, for a single matrix A, assumption (11.4.2) is easily verified, based on the simple subadditivity property exploited in Lemma 11.4.2. For a general matrix function $\Phi_n = A_n \cdots A_1$ subadditivity of $n \mapsto \log \left\|\bigwedge^i (A_n \cdots A_1)\right\|$ and hence the limit property (11.4.2) need not hold.

11.4. The Deterministic Multiplicative Ergodic Theorem

Proof of Theorem 11.4.1. Let $\Phi_n = A_n \cdots A_1 = R_n D_n Q_n^\top$ be the singular value decomposition of Φ_n, with $D_n = \operatorname{diag}(\delta_1(\Phi_n), \ldots, \delta_d(\Phi_n))$ and orthogonal matrices R_n and Q_n. By Proposition 11.3.3(iii)

$$\log \left\| \wedge^i \Phi_n \right\| = \log [\delta_1(\Phi_n) \cdots \delta_i(\Phi_n)] = \log \delta_1(\Phi_n) + \ldots + \log \delta_i(\Phi_n).$$

By assumption (11.4.2) for all i the limits

$$\lim_{n \to \infty} \frac{1}{n} \log \left\| \wedge^i \Phi_n \right\| = \lim_{n \to \infty} \frac{1}{n} [\log \delta_1(\Phi_n) + \ldots + \log \delta_i(\Phi_n)]$$

exist. Thus there is $\Lambda_1 \in \mathbb{R}$ with

$$\lim_{n \to \infty} \frac{1}{n} \log \delta_1(\Phi_n) = \Lambda_1, \text{ hence } \lim_{n \to \infty} (\delta_1(\Phi_n))^{1/n} = e^{\Lambda_1},$$

and there is $\Lambda_2 \in \mathbb{R}$ with

$$\lim_{n \to \infty} \frac{1}{n} [\log \delta_1(\Phi_n) + \log \delta_2(\Phi_n))] = \Lambda_1 + \Lambda_2, \text{ hence } \lim_{n \to \infty} (\delta_2(\Phi_n))^{1/n} = e^{\Lambda_2}.$$

Proceeding in this way, one finds $\Lambda_1, \ldots, \Lambda_d \in \mathbb{R}$ with

(11.4.5) $$\lim_{n \to \infty} D_n^{1/n} = \operatorname{diag}(e^{\Lambda_1}, \ldots, e^{\Lambda_d}).$$

Denote by $\lambda_1 > \ldots > \lambda_\ell$ the different numbers among the Λ_i. Since

$$\Phi_n^\top \Phi_n = Q_n D_n R_n^\top R_n D_n Q_n^\top = Q_n D_n^2 Q_n^\top,$$

one obtains the symmetric positive definite matrix

$$\left(\Phi_n^\top \Phi_n \right)^{1/2n} = Q_n D_n^{1/n} Q_n^\top$$

with eigenvalues $\delta_i(\Phi_n)^{1/n}$.

The case $\ell = 1$: Here (cf. (11.4.5)), one has for the diagonal matrices $D_n^{1/n} \to \operatorname{diag}(e^{\Lambda_1}, \ldots, e^{\Lambda_d}) = \operatorname{diag}(e^{\lambda_1}, \ldots, e^{\lambda_1}) = e^{\lambda_1} I$. Thus

$$\left(\Phi_n^\top \Phi_n \right)^{1/2n} = Q_n D_n^{1/n} Q_n^\top \to e^{\lambda_1} Q_n Q_n^\top = e^{\lambda_1} I$$

and assertion (i) of Theorem 11.4.1 follows. For assertion (ii), observe that the flag (11.4.3) is trivial, $V_1 = \mathbb{R}^d$. We only need to prove that for each $0 \neq x \in \mathbb{R}^d$,

(11.4.6) $$\lambda(x) = \lim_{n \to \infty} \frac{1}{n} \log \| \Phi_n x \| = \lambda_1.$$

With $x_n = [x_n^1, \ldots, x_n^1]^\top := Q_n^\top x$,

(11.4.7) $$\| \Phi_n x \|^2 = \| D_n x \|^2 = \delta_1(\Phi_n)^2 \left(x_n^1 \right)^2 + \ldots + \delta_d(\Phi_n)^2 (x_n^d)^2.$$

For every $\varepsilon > 0$ there is $C_\varepsilon > 0$ such that for all $i = 1, \ldots, d$ and all n,

$$\frac{1}{C_\varepsilon} e^{n(\lambda_1 - \varepsilon)} \leq \delta_i(\Phi_n) \leq C_\varepsilon e^{n(\lambda_1 + \varepsilon)},$$

and hence (11.4.7) shows
$$\frac{1}{C_\varepsilon} e^{n(\lambda_1-\varepsilon)} \|x\| \le \|\Phi_n x\| \le C_\varepsilon e^{n(\lambda_1+\varepsilon)} \|x\| \text{ for every } x.$$
Now assertion (11.4.6) follows for $n \to \infty$ from
$$\frac{1}{n}\left[\log\|x\| - \log C_\varepsilon + n(\lambda_1 - \varepsilon)\right]$$
$$\le \frac{1}{n} \log \|\Phi_n x\| \le \frac{1}{n}\left[\log\|x\| + \log C_\varepsilon + n(\lambda_1 + \varepsilon)\right].$$
This concludes the proof for $\ell = 1$.

The case $\ell \ge 2$: We will not be able to prove that the matrices Q_n converge. However, using the Goldsheid-Margulis metric introduced in Proposition 11.3.5 we will prove convergence of the spaces spanned by eigenvectors.

Let Σ_i be the set of d_i indices k with $\lim_{n\to\infty} \delta_k(\Phi_n)^{1/n} = e^{\lambda_i}$. In fact, Σ_i is well defined by (11.4.5) showing that for $\varepsilon > 0$ there is $N_\varepsilon \in \mathbb{N}$ such that for all $i = 1, \ldots, \ell$ and $k \in \Sigma_i$
$$\left|\frac{1}{n} \log \delta_k(\Phi_n) - \lambda_i\right| < \varepsilon \text{ for all } n \ge N_\varepsilon.$$
Consider for $n \ge N_\varepsilon$ the subspace $U_i(n)$ with $d_i := \dim U_i(n)$ spanned by the eigenvectors of $\Phi_n^\top \Phi_n$ corresponding to the eigenvalues $\delta_k(\Phi_n)^{1/n}$ with $k \in \Sigma_i$. Using the canonical basis $\mathbf{e}_1, \ldots, \mathbf{e}_d$ of $\mathbb{R}^d$ it may be written explicitly as
$$U_i(n) = \text{span}(Q_n \mathbf{e}_{d_1+\ldots+d_{i-1}+1}, \ldots, Q_n \mathbf{e}_{d_1+\ldots+d_i}), \ i = 1, \ldots, \ell.$$
These subspaces are pairwise orthogonal and we may consider the corresponding orthogonal projections $P_i(n)$. Define
$$V_i(n) := U_\ell(n) \oplus \ldots \oplus U_i(n), \ i = 1, \ldots, \ell.$$
The proof will show that $\left(\Phi_n^\top \Phi_n\right)^{1/2n}$ converges to a matrix Ψ with the desired properties by analyzing the convergence of the subspaces $V_i(n)$ simultaneously. This can conveniently be done by considering the flags
$$F(n) := (V_\ell(n) \subset \ldots \subset V_1(n)) \text{ with } V_1(n) := \mathbb{R}^d.$$
Here the dimensions of the subspaces are given by the multi-index
$$\tau = (d_\ell, d_\ell + d_{\ell-1}, \ldots, d_\ell + \ldots + d_1 = d).$$
This multi-index remains fixed in the following and we denote the corresponding space of flags by $\mathbb{F}_\tau$. For the study of convergence we will use the Goldsheid-Margulis metric d_{GM} (see Proposition 11.3.5), on the set $\mathbb{F}_\tau$ of flags. A moment of reflection shows that the speed of convergence of the flags $F(n)$ will be influenced by the maximal difference $\Delta := \min_{i=1,\ldots,\ell-1}(\lambda_i - \lambda_{i+1})$ between consecutive numbers λ_i.

11.4. The Deterministic Multiplicative Ergodic Theorem

Recall the construction of d_{GM} in Lemma 11.3.5: Take a flag $F = (V_\ell \subset \ldots \subset V_1) \in \mathbb{F}_\tau$ and define $U_\ell := V_\ell$ and U_i as the unique orthogonal complement of V_{i+1} in V_i, $i = \ell-1, \ldots, 1$, thus

$$V_i = U_\ell \oplus \ldots \oplus U_i, \ i = 1, \ldots, \ell.$$

Then for any two flags $F, \hat{F} \in \mathbb{F}_\tau$ with associated orthogonal projections P_j and $\hat{P}_i$ onto U_i and $\hat{U}_i$, respectively,

$$(11.4.8) \qquad d_{GM}(\hat{F}, F) = \max_{i,j=1,\ldots,\ell, i \neq j} \left\| \hat{P}_i P_j \right\|^{h/|\lambda_i - \lambda_j|} \text{ for } F, \hat{F} \in \mathbb{F}_\tau.$$

The subspaces $V_i(n)$ and $U_i(n)$ constructed above via eigenvectors of $\Phi_{n\Phi_n}^\top$ fit into this construction: $U_\ell(n) := V_\ell(n)$ and $U_i(n)$ is the orthogonal complement of $V_{i+1}(n)$ in $V_i(n)$ for $i = \ell-1, \ldots, 1$. We denote the corresponding orthogonal projections onto $U_i(n)$ by $P_i(n)$.

Before proceeding with the proof of the deterministic multiplicative ergodic theorem, Theorem 11.4.1, we show the following lemma.

Lemma 11.4.3. *Under the assumptions of Theorem* 11.4.1 *the Goldsheid-Margulis metric satisfies*

$$(11.4.9) \qquad \limsup_{n \to \infty} \frac{1}{n} \log d_{GM}(F(n), F(n+1)) \leq -h.$$

Hence the sequence $(F(n))_{n \in \mathbb{N}}$ *is a Cauchy sequence and thus converges in $\mathbb{F}_\tau$ to a flag F. Furthermore, for every $\varepsilon > 0$ there is $C_\varepsilon > 0$ with*

$$(11.4.10) \qquad d_{GM}(F(n), F) \leq C_\varepsilon e^{-n(h-\varepsilon)} \text{ for all } n \in \mathbb{N}.$$

Proof. First we show that assertion (11.4.10) is a consequence of (11.4.9): For any summable series $f(n)_{n \in \mathbb{N}}$ we have

$$\limsup_{n \to \infty} \frac{1}{n} \log \left\| \sum_{k=n}^\infty f(k) \right\| \leq \limsup_{n \to \infty} \frac{1}{n} \log \|f(n)\| \text{ if } \limsup_{n \to \infty} \frac{1}{n} \log \|f(n)\| < 0.$$

This follows as Lemma 6.2.2 by an elementary computation with exponential growth rates. Now define $f(n) := d_{GM}(F(n), F(n+1))$. Then we can assume from (11.4.9) that $0 < d_{GM}(F(n), F(n+1)) < 1$ for all n sufficiently large. Hence with

$$d_{GM}(F(n), F) \leq \sum_{k=n}^\infty d_{GM}(F(k), F(k+1))$$

we obtain, using also monotonicity of log,

$$\limsup_{n \to \infty} \frac{1}{n} \log d_{GM}(F(n), F) \leq \limsup_{n \to \infty} \frac{1}{n} \log \left[\sum_{k=n}^\infty d_{GM}(F(k), F(k+1)) \right]$$

$$\leq \limsup_{n \to \infty} \frac{1}{n} \log d_{GM}(F(n), F(n+1)) \leq -h.$$

This implies that for every $\varepsilon > 0$, arbitrarily small, there is $C_\varepsilon > 0$ with
$$d_{GM}(F(n), F(n+1)) \leq C_\varepsilon e^{-n(h-\varepsilon)} \text{ for all } n \in \mathbb{N}.$$
Hence it suffices to prove (11.4.9).

With the orthogonal projections onto $U_i(n)$ and $U_j(n+1)$ let (compare Proposition 11.2.5 and Lemma 11.3.5)
$$\Delta_{ij}(n) := \|P_i(n)P_j(n+1)\| = \|P_j(n+1)P_i(n)\|.$$
Then by the definition of the metric d_{GM}
$$\log d_{GM}(F(n), F(n+1)) = \max_{i \neq j} \frac{h \log \|P_i(n) P_j(n+1)\|}{|\lambda_i - \lambda_j|} = h \max_{i \neq j} \frac{\log \Delta_{ij}(n)}{|\lambda_i - \lambda_j|}.$$
Hence (11.4.9) holds if
$$(11.4.11) \qquad \limsup_{n \to \infty} \frac{1}{n} \log \Delta_{ij}(n) \leq -|\lambda_i - \lambda_j| \text{ for } i \neq j.$$
Case 1: $i > j$, hence $\lambda_i < \lambda_j$. With
$$(11.4.12) \qquad \underline{\delta}_j(\Phi_n) := \min_{k \in \Sigma_i} \delta_k(\Phi_n), \; \bar{\delta}_j(\Phi_n) := \max_{k \in \Sigma_i} \delta_k(\Phi_n)$$
and using the formula (11.4.7) we can estimate for every $x \in \mathbb{R}^d$,
$$(11.4.13) \qquad \|P_j(n)x\| \underline{\delta}_j(\Phi_n) \leq \|\Phi_n P_j(n)x\| \leq \|P_j(n)x\| \bar{\delta}_j(\Phi_n).$$
Observe that
$$(11.4.14) \qquad \lim_{n \to \infty} \frac{1}{n} \log \underline{\delta}_j(\Phi_n) = \lim_{n \to \infty} \frac{1}{n} \log \bar{\delta}_j(\Phi_n) = \lambda_j.$$
For a unit vector $x \in U_i(n)$ and $y = P_j(n+1)x \in U_j(n+1)$ we have
$$(11.4.15) \qquad \|\Phi_{n+1} x\| = \|A_{n+1} \Phi_n x\| \leq \|A_{n+1}\| \|\Phi_n x\| \leq \|A_{n+1}\| \bar{\delta}_i(\Phi_n).$$
Using orthogonality in $\Phi_{n+1} x = \Phi_{n+1} y + \Phi_{n+1}(x-y)$ one finds
$$\|\Phi_{n+1} x\|^2 = \langle \Phi_{n+1} x, \Phi_{n+1} x \rangle$$
$$= \langle \Phi_{n+1} y, \Phi_{n+1} y \rangle + \|\Phi_{n+1}(x-y)\|^2$$
$$\geq \langle \Phi_{n+1}^\top \Phi_{n+1} y, \Phi_{n+1} y \rangle.$$
Hence by (11.4.13)
$$\|\Phi_{n+1} x\| \geq \underline{\delta}_j(\Phi_{n+1}) \|P_j(n+1)x\|.$$
Together with (11.4.15) this implies
$$\Delta_{ij}(n) = \|P_j(n+1)P_i(n)\| = \sup_{\|x\|=1, x \in U_i(n)} \|P_j(n+1)x\|$$
$$\leq \sup_{\|x\|=1, x \in U_i(n)} \frac{\|\Phi_{n+1} x\|}{\underline{\delta}_j(\Phi_{n+1})} \leq \|A_{n+1}\| \frac{\bar{\delta}_i(\Phi_n)}{\underline{\delta}_j(\Phi_{n+1})},$$

11.4. The Deterministic Multiplicative Ergodic Theorem

and assertion (11.4.11) follows from assumption (11.4.1) and

$$\limsup_{n\to\infty} \frac{1}{n} \log \Delta_{ij}(n) \leq \limsup_{n\to\infty} \frac{1}{n} \left[\log \|A_{n+1}\| + \log \bar{\delta}_i(\Phi_n) - \log \underline{\delta}_j(\Phi_{n+1}) \right]$$
$$\leq \lambda_i - \lambda_j.$$

Case 2: $i < j$, hence $\lambda_i > \lambda_j$. For a unit vector $x \in U_j(n+1)$ and $y = P_i(n)x \in U_i(n)$,

$$\|\Phi_n x\| = \|A_{n+1}^{-1} \Phi_{n+1} x\| \leq \|A_{n+1}^{-1}\| \|\Phi_{n+1} x\| \leq \|A_{n+1}^{-1}\| \bar{\delta}_j(\Phi_{n+1})$$

and, similarly to Case 1,

$$\|\Phi_n x\| = \left(\|\Phi_n y\|^2 + \|\Phi_n (x-y)\|^2 \right)^{1/2} \geq \|\Phi_n y\| = \underline{\delta}_i(\Phi_n) \|P_i(n)x\|.$$

Hence it follows that

$$\Delta_{ij}(n) = \|P_i(n) P_j(n+1)\| \leq \|A_{n+1}^{-1}\| \frac{\bar{\delta}_j(\Phi_{n+1})}{\underline{\delta}_i(\Phi_n)}$$

and assertion (11.4.11) holds by

$$\limsup_{n\to\infty} \frac{1}{n} \log \Delta_{ij}(n) \leq \limsup_{n\to\infty} \frac{1}{n} \left[\log \|A_{n+1}^{-1}\| + \log \bar{\delta}_j(\Phi_{n+1}) - \log \underline{\delta}_i(\Phi_n) \right]$$
$$\leq \lambda_j - \lambda_i = -|\lambda_i - \lambda_j|. \qquad \square$$

Completion of the proof of Theorem 11.4.1: Lemma 11.4.3 shows that $F(n)$ converges to the limit $F = (V_\ell \subset \ldots \subset V_1)$. By Lemma 11.3.5(iii), the convergence $F(n) \to F$ implies that for each i one has $U_i(n) \to U_i$ in the Grassmannian $\mathbb{G}_{d_i}(d)$. We will consider the orthogonal projections P_i onto the subspaces U_i as $d \times d$-matrices.

In order to prove part (i) of Theorem 11.4.1 we claim

$$(11.4.16) \qquad \left(\Phi_n^\top \Phi_n \right)^{1/2n} \to \Psi := \sum_{i=1}^{\ell} e^{\lambda_i} P_i.$$

Since $\left(\Phi_n^\top \Phi_n \right)^{1/2n}$ is a symmetric matrix we can write

$$\left(\Phi_n^\top \Phi_n \right)^{1/2n} = \sum_{k=1}^{d} \lambda_k(n) P_{kn},$$

where the P_{kn} are the orthogonal projections to the eigenspaces corresponding to the eigenvalue $\lambda_k(n)$ of $\left(\Phi_n^\top \Phi_n \right)^{1/2n}$. Recall that for $k \in \Sigma_i$ the eigenvalues $\lambda_k(n)$ of $\left(\Phi_n^\top \Phi_n \right)^{1/2n}$ converge to e^{λ_i} which is an eigenvalue of Ψ. The space $U_i(n)$ is the sum of the eigenspaces of $\left(\Phi_n^\top \Phi_n \right)^{1/2n}$ for $k \in \Sigma_i$ and we have shown that $U_i(n)$ converges to the eigenspace U_i of $\Psi = \sum_{i=1}^{\ell} e^{\lambda_i} P_i$, hence $\sum_{k \in \Sigma_i} P_{kn} = P_i(n) \to P_i$. We find that

$$\left(\Phi_n^\top \Phi_n\right)^{1/2n} - \Psi = \sum_{k=1}^{d} \lambda_k(n) P_{kn} - \sum_{i=1}^{\ell} e^{\lambda_i} P_i$$

$$(11.4.17) \qquad = \sum_{i=1}^{\ell} \left[\sum_{k \in \Sigma_i} (\lambda_k(n) - e^{\lambda_i}) P_{kn} + e^{\lambda_i} \left(\sum_{k \in \Sigma_i} P_{kn} - P_i \right) \right] \xrightarrow[n \to \infty]{} 0.$$

This proves (11.4.16) and hence part (i) of Theorem 11.4.1.

For part (ii), it remains to show formula (11.4.4). Note first that every $x \in \mathbb{R}^d$ lies in exactly one set $V_i \setminus V_{i+1}$, hence it suffices to prove the limit property (11.4.4) for an arbitrary unit vector $x \in V_i \setminus V_{i+1}$. With the orthogonal projections for Φ_n and Ψ we can write

$$x = \sum_{j=1}^{\ell} P_j(n) x = \sum_{j=i}^{\ell} P_j x.$$

Then $P_i x \neq 0$ by the assumption on x. For $j \neq i$ the vectors $\Phi_n P_j(n) x$ are orthogonal to $\Phi_n P_i(n) x$, since

$$\left(\Phi_n P_j(n) x\right)^\top \Phi_n P_i(n) x = (P_j(n) x)^\top \Phi_n^\top \Phi_n P_i(n) x$$

where $P_i(n) x \in U_i(n)$ and $P_j(n) x$ is in the orthogonal subspace $U_j(n)$ which is invariant under $\Phi_n^\top \Phi_n$. Recall the definitions of $\underline{\delta}_j(\Phi_n)$ and $\bar{\delta}_j(\Phi_n)$ from (11.4.12) and estimate (11.4.13). Then

$$\sum_{j=1}^{\ell} \|P_j(n) x\|^2 \underline{\delta}_j(\Phi_n)^2 \leq \|\Phi_n x\|^2 = \sum_{j=1}^{\ell} \|\Phi_n P_j(n) x\|^2$$

$$(11.4.18) \qquad \leq \sum_{j=1}^{\ell} \|P_j(n) x\|^2 \bar{\delta}_j(\Phi_n)^2.$$

In particular, since $\|P_i(n) x\| \underline{\delta}_i(\Phi_n) \leq \|\Phi_n x\|$ and $\|P_i(n) x\| \to \|P_i x\| > 0$ one obtains from the limit property given in (11.4.14) that

$$(11.4.19) \qquad \lambda_i = \lim_{n \to \infty} \frac{1}{n} \left[\log \|P_i(n) x\| + \log \underline{\delta}_i(\Phi_n) \right] \leq \liminf_{n \to \infty} \frac{1}{n} \log \|\Phi_n x\|.$$

It remains to get the reverse inequality. We estimate the summands on the right-hand side of (11.4.18) from above:

For $j = i, \ldots, \ell$ we use $\|P_j(n) x\| \leq 1$ and (11.4.14) to obtain

$$(11.4.20) \qquad \limsup_{n \to \infty} \frac{1}{n} \log \left[\|P_j(n) x\| \bar{\delta}_j(\Phi_n) \right] = \lambda_j \leq \lambda_i.$$

11.4. The Deterministic Multiplicative Ergodic Theorem

Now consider the other indices $j = 1, \ldots, i-1$. Here we use

$$(11.4.21) \qquad \|P_j(n)x\| = \left\| P_j(n) \sum_{k=i}^{\ell} P_k x \right\| \leq \sum_{k=i}^{\ell} \|P_j(n)P_k\|.$$

Then (11.4.10) in Lemma 11.4.3 shows for all $k \geq i$,

$$\limsup_{n \to \infty} \frac{1}{n} \log \|P_j(n)P_k\|^{h/|\lambda_j - \lambda_k|} \leq \limsup_{n \to \infty} \frac{1}{n} \log d_{GM}(F(n), F) \leq -h,$$

and hence

$$\limsup_{n \to \infty} \frac{1}{n} \log \|P_j(n)P_k\| \leq -|\lambda_j - \lambda_k| \leq \max_{k=i,\ldots,\ell} (-|\lambda_j - \lambda_k|) = \lambda_i - \lambda_j.$$

Using also (11.4.21) one finds for all $j \leq i-1$,

$$\limsup_{n \to \infty} \frac{1}{n} \log \|P_j(n)x\| \leq \limsup_{n \to \infty} \frac{1}{n} \left[\log \ell + \log \max_{k=i,\ldots,\ell} \|P_j(n)P_k\| \right]$$

$$(11.4.22) \qquad = \limsup_{n \to \infty} \frac{1}{n} \log \max_{k=i,\ldots,\ell} \|P_j(n)P_k\| \leq \lambda_i - \lambda_j.$$

Estimate (11.4.18) yields

$$\frac{2}{n} \log \|\Phi_n x\| \leq \frac{1}{n} \log \left[\ell \max_{j=1,\ldots,\ell} \|P_j(n)x\|^2 \; \bar{\delta}_j(\Phi_n)^2 \right]$$

$$= \frac{1}{n} \log \ell + \frac{2}{n} \max_{j=1,\ldots,\ell} \log \left[\|P_j(n)x\| \bar{\delta}_j(\Phi_n) \right].$$

Taking the limit for $n \to \infty$ one obtains

$$\limsup_{n \to \infty} \frac{1}{n} \log \|\Phi_n x\| \leq \limsup_{n \to \infty} \frac{1}{n} \max_{j=1,\ldots,\ell} \log \left[\|P_j(n)x\| \bar{\delta}_j(\Phi_n) \right]$$

$$= \max_{j=1,\ldots,\ell} \limsup_{n \to \infty} \frac{1}{n} \log \left[\|P_j(n)x\| \bar{\delta}_j(\Phi_n) \right].$$

For $j \geq i$ estimate (11.4.20) shows that

$$(11.4.23) \qquad \limsup_{n \to \infty} \frac{1}{n} \log \left[\|P_j(n)x\| \bar{\delta}_j(\Phi_n) \right] \leq \lambda_i$$

and for $j < i$ estimate (11.4.22) shows that

$$\limsup_{n \to \infty} \frac{1}{n} \log \left[\|P_j(n)x\| \bar{\delta}_j(\Phi_n) \right]$$

$$\leq \limsup_{n \to \infty} \frac{1}{n} \log \|P_j(n)x\| + \limsup_{n \to \infty} \frac{1}{n} \log \bar{\delta}_j(\Phi_n)$$

$$\leq \lambda_i - \lambda_j + \lambda_j = \lambda_i.$$

This estimate and (11.4.23) applied to the right-hand side of (11.4.18) yield the desired converse inequality, and together with (11.4.19) it follows that

$$\limsup_{n\to\infty} \frac{1}{n} \log \|\Phi_n x\| = \lambda_i,$$

hence formula (11.4.4) holds and the proof the deterministic MET, Theorem 11.4.1 is complete. □

11.5. The Furstenberg-Kesten Theorem and Proof of the MET in Discrete Time

In this section we derive the Furstenberg-Kesten theorem using Kingman's subadditive ergodic theorem and prove the Multiplicative Ergodic Theorem in discrete time, Theorem 11.1.13.

Our discussion of the assumptions in the deterministic MET, Theorem 11.4.1, shows that, for a single matrix A, assumption (11.4.2) is easily verified, based on the subadditivity property exploited in Lemma 11.4.2. For a general matrix function $\Phi_n = A_n \cdots A_1$ subadditivity of the sequence $n \mapsto \log \|\bigwedge^i (A_n \cdots A_1)\|$ and hence the limit property (11.4.2) cannot be verified. This is a crucial point, where the additional structure provided by the metric dynamical system in the base is needed. We will show that the cocycle property implies a weakened version of subadditivity taking into account the flow in the base space. Then Kingman's subadditive ergodic theorem implies existence of limits in such a situation. This will be used in the Furstenberg-Kesten Theorem to establish existence of the limits required in (11.4.2) and the MET will follow.

Consider a random linear system Φ in discrete time. The deterministic MET, Theorem 11.4.1 shows that the singular values of $\Phi(n,\omega)$ play a decisive role. Hence, in order to prepare the proof of the MET, we study the asymptotic behavior of $\Phi(n,\omega)$ and its singular values $\delta_k(\Phi(n,\omega))$ for $n \to \infty$. By Proposition 11.3.2 the cocycle property (11.1.6) lifts to the exterior products

(11.5.1) $\qquad \bigwedge^k \Phi(n+m,\omega) = \bigwedge^k \Phi(n, \theta^m \omega) \cdot \bigwedge^k \Phi(m,\omega).$

The following result is the Furstenberg-Kesten Theorem.

Theorem 11.5.1 (Furstenberg-Kesten). *Let $\Phi = (\theta, \varphi) : \mathbb{Z} \times \Omega \times \mathbb{R}^d \longrightarrow \Omega \times \mathbb{R}^d$ be a random linear dynamical system in discrete time over the ergodic metric dynamical system θ and write $\Phi(n,\omega)$ for the corresponding linear cocycle. Assume that the generator $A : \Omega \to Gl(d, \mathbb{R})$ satisfies*

(11.5.2) $\qquad \log^+ \|A\|, \log^+ \|A^{-1}\| \in L^1 = L^1(\Omega, \mathcal{F}, P; \mathbb{R}).$

Then the following statements hold on an invariant set $\tilde{\Omega} \in \mathcal{F}$ of full measure, i.e., $\theta \tilde{\Omega} = \tilde{\Omega}$ and $P(\tilde{\Omega}) = 1$:

11.5. Furstenberg-Kesten Theorem and MET in discrete time

(i) For each $k = 1, \ldots, d$ the functions
$$f_n^{(k)}(\omega) := \log \left\| \wedge^k \Phi(n, \omega) \right\|, \ n \geq 0,$$
are subadditive, i.e., $f_{m+n}^{(k)}(\omega) \leq f_m^{(k)}(\omega) + f_n^{(k)}(\theta^m \omega)$, and $f_n^{(k)} \in L^1$.

(ii) There are constants $\gamma^{(k)} \in \mathbb{R}$ such that
$$\gamma^{(k)} = \lim_{n \to \infty} \frac{1}{n} \log \left\| \wedge^k \Phi(n, \omega) \right\| \ \text{and} \ \gamma^{(k+l)} \leq \gamma^{(k)} + \gamma^{(l)}.$$

(iii) The constants $\Lambda_k \in \mathbb{R}$ successively defined by
$$\Lambda_1 + \ldots + \Lambda_k = \gamma^{(k)}, \ k = 1, \ldots, d,$$
satisfy
$$\Lambda_k = \lim_{n \to \infty} \frac{1}{n} \log \delta_k(\Phi(n, \omega)),$$
where $\delta_k(\Phi(n, \omega))$ are the singular values of $\Phi(n, \omega)$, and $\Lambda_1 \geq \ldots \geq \Lambda_d$.

(iv) Furthermore, in backward time we have for $k = 1, \ldots, d$,
$$\gamma^{(k)-} := \lim_{n \to \infty} \frac{1}{n} \log \left\| \wedge^k \Phi(-n, \omega) \right\| = \gamma^{(d-k)} - \gamma^{(d)} = -\Lambda_d - \ldots - \Lambda_{d+1-k}$$
and
$$\Lambda_k^- := \lim_{n \to \infty} \frac{1}{n} \log \delta_k(\Phi(-n, \omega)) = -\Lambda_{d+1-k}.$$

Proof. Note that $A(\omega) := \Phi(1, \omega)$ and $A(\theta\omega) := \Phi(-1, \omega)$ define discrete-time linear cocycles $\Phi(n, \omega)$ over θ with $\theta^{\pm 1}\omega = \theta(\pm 1, \omega), \omega \in \Omega, n \geq 1$. We want to apply Kingman's subadditive ergodic theorem, Theorem 10.3.2, with $(X, \mu) = (\Omega, P)$ and $T = \theta$. First we verify that for all k,

(11.5.3) $\qquad f_n^{(k)}(\cdot) = \log \left\| \wedge^k \Phi(n, \cdot) \right\| \in L^1(\Omega, \mathcal{F}, P; \mathbb{R})$

for all $n \geq 1$, and

(11.5.4) $\qquad \inf_{n \in \mathbb{N}} \frac{1}{n} \mathbb{E}(f_n^{(k)}) = \inf_{n \in \mathbb{N}} \frac{1}{n} \int f_n^{(k)}(\omega)) P(d\omega) > -\infty.$

The cocycle properties
$$\left\| \wedge^k \Phi(n+1, \omega) \right\| \leq \left\| \wedge^k \Phi(n, \theta\omega) \right\| \left\| \wedge^k A(\omega) \right\|,$$
$$\left\| \wedge^k \Phi(n, \theta\omega) \right\| \leq \left\| \wedge^k \Phi(n+1, \omega) \right\| \left\| \wedge^k A(\omega)^{-1} \right\|,$$

and (see Proposition 11.3.3(iii)) $\log \left\| \wedge^k A(\omega) \right\| \leq k \log^+ \|A(\omega)\|$ result in

(11.5.5)
$$f_n^{(k)} \circ \theta(\omega) - k \log^+ \left\| A(\omega)^{-1} \right\| \leq f_{n+1}^{(k)}(\omega) \leq f_n^{(k)} \circ \theta(\omega) + k \log^+ \|A(\omega)\|.$$

Since P is an invariant measure, $f_n^{(k)} \circ \theta \in L^1$ if $f_n^{(k)} \in L^1$. Hence (11.5.5), together with assumptions (11.5.2), implies that $f_n^{(k)} \in L^1$ if $f_1^{(k)} \in L^1$. In order to show $f_1^{(k)} \in L^1$, observe that $1 \leq \|AA^{-1}\| \leq \|A\| \, \|A^{-1}\|$ implies

$$-\log^+ \|A^{-1}\| \leq \log \|A\| \leq \log^+ \|A\|$$

and hence

$$|\log \|A\|| \leq \log^+ \|A\| + \log^+ \|A^{-1}\|.$$

In the same way one shows that

$$\left|\log \left\|\bigwedge^k A\right\|\right| \leq k \left(\log^+ \|A\| + \log^+ \|A^{-1}\|\right).$$

Therefore $f_1^{(k)} = \log \left\|\bigwedge^k \Phi(1, \cdot)\right\| = \log \left\|\bigwedge^k A\right\| \in L^1$, by assumption (11.5.2). This verifies (11.5.3).

Using the definition of Φ_n in (11.1.7), invariance of P, and Theorem 11.3.3(iii) one finds

$$\frac{1}{n}\left|\mathbb{E}(f_n^{(k)})\right| = \frac{1}{n}\left|\mathbb{E}\log\left\|\bigwedge^k \Phi(n,\cdot)\right\|\right| \leq \frac{1}{n}\sum_{i=0}^{n-1} \mathbb{E}\left|\log\left\|\bigwedge^k A(\theta^i \cdot)\right\|\right|$$

$$\leq k\frac{1}{n}\sum_{i=0}^{n-1} \mathbb{E}\left|\log^+\|A(\theta^i \cdot)\|\right| = k\mathbb{E}\left|\log^+\|A\|\right| < \infty.$$

Thus (11.5.4) also holds.

Next we verify properties (i)–(iv) for each k.

(i) The cocycle property (11.5.1) for $\bigwedge^k \Phi(n, \omega)$ and Proposition 11.3.2 imply that $f_n^{(k)}(\omega) = \log \|\bigwedge^k \Phi(n, \omega)\|$ is subadditive. Thus, together with (11.5.3) and (11.5.4) we have verified the assertions in (i) and also all assumptions of the subadditive ergodic theorem, Theorem 10.3.2.

(ii) The limit property is an immediate consequence of Theorem 10.3.2 in the ergodic case. The forward invariant set $\tilde{\Omega}$ can be taken to be the intersection of the d forward invariant sets $\tilde{\Omega}_k, k = 1, \ldots, d$. The inequality $\|\bigwedge^{k+l} \Phi\| \leq \|\bigwedge^k \Phi\| \, \|\bigwedge^l \Phi\|$ and Proposition 11.3.3(iii)) imply $\gamma^{(k+l)} \leq \gamma^{(k)} + \gamma^{(l)}$.

(iii) By Proposition 11.3.3 the norm of the exterior power is determined by the singular values, hence for $k = 1, \ldots, d$,

$$\frac{1}{n}\log\left\|\bigwedge^k \Phi(n,\omega)\right\| = \frac{1}{n}\sum_{i=1}^{k} \log \delta_i(\Phi(n,\omega)).$$

We proceed inductively as follows: Choose $\omega \in \tilde{\Omega}$. For $k=1$,

$$\gamma^{(1)} = \lim_{n\to\infty} \frac{1}{n} \log \delta_1(\Phi(n,\omega)) =: \Lambda_1$$

exists. Then

$$\gamma^{(2)} - \gamma^{(1)} = \lim_{n\to\infty} \frac{1}{n} \log \delta_2(\Phi(n,\omega)) =: \Lambda_2$$

exists, etc. Finally,

$$\gamma^{(d)} - \gamma^{(d-1)} = \lim_{n\to\infty} \frac{1}{n} \log \delta_d(\Phi(n,\omega)) =: \Lambda_d$$

exists. The inequalities $\Lambda_1 \geq \ldots \geq \Lambda_d$ follow, since the singular values are ordered according to $\delta_1 \geq \ldots \geq \delta_d$.

(iv) The cocycle $\Phi(-n,\omega), n \geq 0$, over θ^{-1} satisfies the integrability conditions (11.5.2) and the cocycle property relates positive and negative times via $\Phi(-n,\omega) = \Phi(n,\theta^{-n}\omega)^{-1}$. If $\delta_1 \geq \ldots \geq \delta_d$ are the singular values of a matrix $G \in Gl(d,\mathbb{R})$, then $1/\delta_d \geq \ldots \geq 1/\delta_1$ are the singular values of G^{-1}. Hence

(11.5.6) $\quad \delta_k(\Phi(-n,\omega)) = \delta_k(\Phi(n,\theta^{-n}\omega)^{-1}) = \delta_{d+1-k}^{-1}(\Phi(n,\theta^{-n}\omega)).$

Taking logarithms and dividing by n one finds for $\omega \in \tilde{\Omega}$,

$$\frac{1}{n} \log \delta_k(\Phi(-n,\omega)) = -\frac{1}{n} \log \delta_{d+1-k}(\Phi(n,\theta^{-n}\omega))) \to -\Lambda_{d+1-k}.$$

Hence the statements in (i), (ii), and (iii) hold in backward time and also (iv) follows. $\square$

If we compare the Furstenberg-Kesten Theorem, Theorem 11.5.1, with the statement of the MET, Theorem 11.1.13, we see that the former is concerned with the asymptotic behavior of the singular values of $\Phi(n,\omega)$, i.e., with part (iv) of Theorem 11.1.13. What remains to be done is the analysis of the exponential limit behavior of trajectories and of the subspaces on which the Lyapunov exponents are attained. For this purpose, we will invoke the deterministic Multiplicative Ergodic Theorem, Theorem 11.4.1. In fact, part (ii) of the Furstenberg-Kesten Theorem (which essentially is a consequence of the subadditive ergodic theorem, Theorem 10.3.2) assures us that assumption (11.4.2) in Theorem 11.4.1 is satisfied.

Before we turn to the proof of the Multiplicative Ergodic Theorem in discrete time, we note the following elementary lemma.

Lemma 11.5.2. *Let $(\Omega, \mathcal{F}, P, \theta)$ be a metric dynamical system in discrete time and $f : \Omega \to \mathbb{R}$ a random variable.*

(i) If $f \geq 0$, then for all $\varepsilon > 0$,

$$\sum_{n=1}^{\infty} P\{\omega \mid f(\omega) > \varepsilon n\} \leq \frac{1}{\varepsilon} \mathbb{E} f \leq \sum_{n=0}^{\infty} P\{\omega \mid f(\omega) > \varepsilon n\}.$$

(ii) If $f^+ \in L^1(\Omega, P; \mathbb{R})$, then the invariant sets

$$\Omega_1 := \{\omega \in \Omega \mid \limsup_{n \to \infty} \frac{1}{n} f(\theta^{n-1}\omega) \leq 0\},$$

$$\Omega_2 := \{\omega \in \Omega \mid \liminf_{n \to \infty} \frac{1}{n} f(\theta^{n-1}\omega) \leq 0 \leq \limsup_{n \to \infty} \frac{1}{n} f(\theta^{n-1}\omega)\}$$

have full measure.

Proof. (i) This is the well-known approximation of

$$\mathbb{E} f = \int_{\Omega} f \, dP = \int_0^{\infty} P\{\omega \mid f(\omega) > x\} dx = \varepsilon \int_0^{\infty} P\{\omega \mid f(\omega) > \varepsilon x\} dx$$

by sums from above and below: We first define $F := \{(\omega, x) \mid 0 < x < f(\omega)\} \subset \Omega \times [0, \infty)$ and note that $F = \bigcup_{r \in \mathbb{Q}^+} (\{\omega \in \Omega \mid r < f(\omega)\} \times (0, r))$. Using the product measure $P \times \nu$ on $\Omega \times [0, \infty)$, where ν is the Lebesgue measure, we obtain

$$\mathbb{E} f = \int_{\Omega} f \, dP = \int_{\Omega} \nu(F_{\omega}) P(d\omega) = P \times \nu(F)$$

$$= \int_{[0,\infty)} P(F_x) \nu(dx) = \int_{[0,\infty)} P\{\omega \mid f(\omega) > x\} \nu(dx).$$

Hence we obtain the estimates

$$\sum_{n \geq 1} P\{\omega \mid f(\omega) > n\} = \sum_{n \geq 1} \int_{(n-1,n]} P\{\omega \mid f(\omega) > n\} \, dx$$

$$\leq \sum_{n \geq 1} \int_{(n-1,n]} P\{\omega \mid f(\omega) > x\} \, dx$$

$$= \int_{[0,\infty)} P\{\omega \mid f(\omega) > x\} \, dx = \mathbb{E}(f)$$

$$\leq \sum_{n \geq 1} \int_{(n-1,n]} P\{\omega \mid f(\omega) > n - 1\} \, dx$$

$$= \sum_{n \geq 0} P\{\omega \mid f(\omega) > n\}.$$

(ii) The invariance of these sets follows from their definitions. The Borel-Cantelli Lemma (see, e.g., Durrett [41, Section 2.3]), states that for any

11.5. Furstenberg-Kesten Theorem and MET in discrete time

sequence of measurable sets B_n one has

$$P\left(\limsup_{n \to \infty} B_n\right) = P\left(\bigcap_{m \geq 1} \bigcup_{n \geq m} B_n\right) = 0 \text{ if } \sum_{n=1}^{\infty} P(B_n) < \infty.$$

Let $B_n := \{\omega \mid \frac{1}{n} f(\theta^{n-1}\omega) > \varepsilon\}$. Then by (i),

$$\sum_{n=1}^{\infty} P(B_n) = \sum_{n=1}^{\infty} P\{\omega \mid \frac{1}{n} f(\theta^{n-1}\omega) > \varepsilon\} = \sum_{n=1}^{\infty} P\{\omega \mid f > \varepsilon n\}$$

$$= \sum_{n=1}^{\infty} P\{\omega \mid f^+ > \varepsilon n\}$$

$$\leq \frac{1}{\varepsilon} \mathbb{E} f^+ < \infty,$$

and hence $P\left(\bigcap_{m \geq 1} \bigcup_{n \geq m} B_n\right) = P\{\omega \mid \limsup_{n \to \infty} \frac{1}{n} f(\theta^{n-1}\omega) \geq \varepsilon\} = 0$. Thus for every $\varepsilon > 0$,

$$P\{\omega \mid \limsup_{n \to \infty} \frac{1}{n} f(\theta^{n-1}\omega) < \varepsilon\} = 1$$

and it follows that Ω_1 has full measure. Similarly, for any $\varepsilon > 0$,

$$\lim_{n \to \infty} P\{\omega \mid \frac{1}{n} |f(\theta^{n-1}\omega)| > \varepsilon\} = \lim_{n \to \infty} P\{\omega \mid |f(\theta^{n-1}\omega)| > \varepsilon n\} = 0,$$

hence

$$P\{\omega \in \Omega \mid \liminf_{n \to \infty} \frac{1}{n} f(\theta^{n-1}\omega) \leq -\varepsilon \text{ and } \varepsilon < \limsup_{n \to \infty} \frac{1}{n} f(\theta^{n-1}\omega)\} = 0.$$

Then it follows that Ω_2 has full measure. $\square$

The following lemma already contains essential information concerning positive time and negative time considered separately.

Lemma 11.5.3. *Consider a random linear dynamical system $\Phi = (\theta, \varphi) : \mathbb{Z} \times \Omega \times \mathbb{R}^d \longrightarrow \Omega \times \mathbb{R}^d$ in discrete time, assume that the integrability conditions (11.5.2) hold and that the probability measure P is ergodic.*

Then there exists a set $\widehat{\Omega} \subset \Omega$ of full P-measure which is invariant, i.e., $\theta \widehat{\Omega} = \widehat{\Omega}$, such that for each $\omega \in \widehat{\Omega}$ the following holds:

(i) The limit $\lim_{n \to \infty} \left(\Phi(n, \omega)^\top \Phi(n, \omega)\right)^{1/2n} := \Psi(\omega)$ exists and is a positive definite random matrix with eigenvalues $\Lambda_1 \geq \ldots \geq \Lambda_\ell$ which are constants.

(ii) Let the different eigenvalues of $\Psi(\omega)$ be written as $e^{\lambda_1} > \ldots > e^{\lambda_\ell}$; the corresponding random eigenspaces $U_1(\omega), \ldots, U_\ell(\omega)$ are invariant, i.e., $A(\omega) U_j(\omega) = U_j(\theta \omega)$ for all $j = 1, \ldots, \ell$ and their dimensions $d_j = \dim U_j(\omega)$ are constants.

(iii) For $i = 1, \ldots, \ell$ let $V_i(\omega) := U_\ell(\omega) \oplus \ldots \oplus U_i(\omega)$, which yields the flag

$$V_\ell(\omega) \subset \ldots \subset V_1(\omega) = \mathbb{R}^d.$$

Then for each $0 \neq x \in \mathbb{R}^d$ the Lyapunov exponent

$$\lambda(\omega, x) = \lim_{n \to \infty} \frac{1}{n} \log \| \Phi(n, \omega) x \|$$

exists as a limit, and, with $V_{\ell+1} := \{0\}$,

$$\lambda(\omega, x) = \lambda_i \text{ if and only if } x \in V_i(\omega) \setminus V_{i+1}(\omega).$$

(iv) Similarly, for backward time $\lim_{n \to \infty} \left(\Phi(-n, \omega)^\top \Phi(-n, \omega) \right)^{1/2n} := \Psi^-(\omega)$ exists and is a positive definite random matrix with eigenvalues $\Lambda_1^- \geq \ldots \geq \Lambda_\ell^-$ satisfying

$$\Lambda_k^- = \lim_{n \to \infty} \frac{1}{n} \log \delta_k(\Phi(-n, \omega)) = -\Lambda_{d+1-k}.$$

Hence the different eigenvalues of $\Psi^-(\omega)$ can be written as $\mathrm{e}^{\lambda_1^-} > \ldots > \mathrm{e}^{\lambda_\ell^-}$; the corresponding random eigenspaces $U_1^-(\omega), \ldots, U_\ell^-(\omega)$ are invariant, i.e., $A(\omega) U_j(\omega) = U_j(\theta^{-1} \omega)$ for all $j = 1, \ldots, \ell$ and $\dim U_j^-(\omega) = d_{\ell+1-j}$. For $i = 1, \ldots, \ell$ let $V_i^-(\omega) := U_\ell^-(\omega) \oplus \ldots \oplus U_i^-(\omega)$, which yields the backward flag

$$V_\ell^-(\omega) \subset \ldots \subset V_1^-(\omega) = \mathbb{R}^d.$$

Then for each $0 \neq x \in \mathbb{R}^d$ the backward Lyapunov exponent

$$\lambda^-(\omega, x) := \lim_{n \to \infty} \frac{1}{n} \log \| \Phi(-n, \omega) x \|$$

exists as a limit, $\lambda_i = -\lambda_{\ell+1-i}^-$ and, with $V_{\ell+1}^- := \{0\}$,

$$\lambda(\omega, x) = \lambda_i \text{ if and only if } x \in V_i^-(\omega) \setminus V_{i+1}^-(\omega).$$

Proof. By the Furstenberg-Kesten theorem, Theorem 11.5.1, the equality

(11.5.7) $$\lim_{n \to \infty} \frac{1}{n} \log \left\| \bigwedge^k \Phi(n, \omega) \right\| = \gamma^{(k)} \in \mathbb{R}$$

holds on an invariant set $\tilde{\Omega}$ of full measure. Furthermore, the constants $\Lambda_k \in \mathbb{R}$ successively defined by

$$\Lambda_1 + \ldots + \Lambda_k = \gamma^{(k)}, \ k = 1, \ldots, d,$$

have the following properties on $\tilde{\Omega}$:

$$\Lambda_k = \lim_{n \to \infty} \frac{1}{n} \log \delta_k(\Phi(n, \omega)),$$

where $\delta_k(\Phi(n,\omega))$ are the singular values of $\Phi(n,\omega)$, and $\Lambda_1 \geq \ldots \geq \Lambda_d$. Furthermore, on $\tilde{\Omega}$ we have for $k = 1, \ldots, d$,

$$\gamma^{(k)-} := \lim_{n \to \infty} \frac{1}{n} \log \left\| \bigwedge^k \Phi(-n,\omega) \right\|$$
$$= \gamma^{(d-k)} - \gamma^{(d)} = -\Lambda_d - \Lambda_{d-1} - \ldots - \Lambda_{d+1-k}$$

and

$$\Lambda_k^- := \lim_{n \to \infty} \frac{1}{n} \log \delta_k(\Phi(-n,\omega)) = -\Lambda_{d+1-k}.$$

In order to use the deterministic MET, Theorem 11.4.1, forward in time define

$$A_n(\omega) := A(\theta^{n-1}\omega), \ n \in \mathbb{N}.$$

The integrability assumption (11.5.2) implies $f := \log^+ \|A\| \in L^1(\Omega, \mathcal{F}, P; \mathbb{R})$. Hence Lemma 11.5.2(ii) shows that for all ω in an invariant set Ω_1 of full measure

$$\limsup_{n \to \infty} \frac{1}{n} A(\theta^{n-1}\omega) \leq 0$$

verifying assumption (11.4.1), and assumption (11.4.2) holds by (11.5.7). Hence Theorem 11.4.1 implies that the limit

$$\lim_{t \to \infty} (\Phi(n,\omega)^\top \Phi(n,\omega))^{1/2n} =: \Psi(\omega)$$

exists and the eigenvalues of $\Psi(\omega)$ coincide with the e^{Λ_k}, where the Λ_k's are successively defined by $\Lambda_1 + \ldots + \Lambda_k := \gamma^{(k)}$, $k = 1, \ldots, d$. For the kth singular value $\delta_k(\Phi(n,\omega))$ of $\Phi(n,\omega)$ one has

$$\Lambda_k = \lim_{n \to \infty} \log \delta_k(\Phi(n,\omega)), \ k = 1, \ldots, d.$$

Let $e^{\lambda_1} > \ldots > e^{\lambda_\ell}$ be the different eigenvalues of $\Psi(\omega)$ and $U_1(\omega), \ldots, U_\ell(\omega)$ their corresponding eigenspaces with multiplicities $d_i = \dim U_i(\omega)$. Write $V_{\ell+1} := \{0\}$ and let

$$V_i(\omega) = U_\ell(\omega) \oplus \ldots \oplus U_i(\omega), \ i = 1, \ldots, \ell,$$

so that one obtains the flag

$$V_\ell(\omega) \subset \ldots \subset V_i(\omega) \subset \ldots \subset V_1 = \mathbb{R}^d.$$

Then for each $x \in \mathbb{R}^d \setminus \{0\}$ the Lyapunov exponent

$$\lambda(\omega, x) = \lim_{n \to \infty} \frac{1}{n} \log \|\Phi(n,\omega)x\|$$

exists as a limit, and $\lambda(\omega, x) = \lambda_i$ if and only if $x \in V_i(\omega) \setminus V_{i+1}(\omega)$. Similarly, we can apply the deterministic MET, Theorem 11.4.1, backward in time by defining

$$A_n^-(\omega) := A(\theta^{-n+1}\omega), \ n \in \mathbb{N}.$$

This proves the claims in (iv). □

Now we are in the position to complete the proof of the Multiplicative Ergodic Theorem in discrete time, Theorem 11.1.13. What remains to be done is to combine the forward and backward flags and to relate them to the Lyapunov exponents forward and backward in time, respectively.

Proof of the MET, Theorem 11.1.13. We will construct the Lyapunov spaces and determine the exponential growth rates on them, forward and backward in time. By Lemma 11.5.3 there is an invariant set $\hat{\Omega}$ of full measure with the forward flag

$$V_\ell(\omega) \subset \ldots \subset V_1(\omega) = \mathbb{R}^d, \ A(\omega)V_i(\omega) = V_i(\theta\omega)$$

and numbers λ_i, d_i. The same lemma yields on an invariant set of full measure the backward flag

$$V_{\ell^-}^-(\omega) \subset \ldots \subset V_1^-(\omega) = \mathbb{R}^d, \ A^{-1}(\omega)V_i^-(\omega) = V_i^-(\theta^{-1}\omega)$$

and corresponding numbers λ_i^-, d_i^-. The Furstenberg-Kesten theorem, Theorem 11.5.1, has assured us that on an invariant set $\tilde{\Omega}$ of full measure statement (iv) holds, and together with $\ell = \ell^-$ for $i = 1, \ldots, \ell$ we have

(11.5.8) $$d_i = d_{\ell+1-i}^-, \ \lambda_i = -\lambda_{\ell+1-i}^-.$$

We now construct the Lyapunov spaces $L_i(\omega)$ as intersections of certain spaces from the forward and backward flags. We set

$$L_i(\omega) := V_i(\omega) \cap V_{\ell+1-i}^-(\omega), \ 1 \leq i \leq \ell.$$

In particular, one has $L_1(\omega) = L_\ell^-(\omega)$ and $L_\ell(\omega) = V_\ell(\omega)$.

We interrupt the proof of Theorem 11.1.13 in order to show the following lemma which gives information on these subspaces.

Lemma 11.5.4. *On an invariant set $\tilde{\Omega}_1 \subset \Omega$ of full measure and for $1 \leq i \leq \ell$*

$$V_{i+1}(\omega) \cap V_{\ell+1-i}^-(\omega) = \{0\} \ and \ V_{i+1}(\omega) \oplus V_{\ell+1-i}^-(\omega) = \mathbb{R}^d.$$

Proof. First we prove that the set

$$B := \{\omega \in \Omega \mid V_{i+1}(\omega) \cap V_{\ell+1-i}^-(\omega) \neq \{0\}\}$$

has measure 0. The idea of the proof is as follows: For an $\omega \in B$ we move a vector $x \in V_{i+1}(\omega) \cap V_{\ell+1-i}^-(\omega)$ from time 0 back to time $-N$ by $\Phi(-N, \omega)$ with exponential rate at most $-\lambda_i$. Then we move the same point from time $-N$ forward to time 0 by $\Phi(N, \theta^{-N}\omega)$ with exponential rate at most λ_{i+1}. Since $\lambda_{i+1} < \lambda_i$ the result will be a small quantity. But this would contradict the cocycle property which gives $x = \Phi(N, \theta^{-N}\omega)\Phi(-N, \omega)x$.

11.5. Furstenberg-Kesten Theorem and MET in discrete time

Put $\delta = \frac{1}{2}(\lambda_i - \lambda_{i+1})$. For each sufficiently small $\varepsilon > 0$ we can choose an integer N_ε so big that $P(C_\varepsilon) > 1 - \frac{\varepsilon}{2}$ and $P(D_\varepsilon) = P(\theta^{N_\varepsilon} D_\varepsilon) > 1 - \frac{\varepsilon}{2}$, where

$$C_\varepsilon := \{\omega \in \Omega \mid \|\Phi(-N_\varepsilon, \omega)x\| < e^{N_\varepsilon(-\lambda_i + \delta)} \|x\| \text{ for all } 0 \neq x \in V^-_{\ell+1-i}(\omega)\}$$

and

$$D_\varepsilon := \{\omega \in \Omega \mid \|\Phi(N_\varepsilon, \omega)x\| < e^{N_\varepsilon(\lambda_{i+1} + \delta)} \|x\| \text{ for all } 0 \neq x \in V_{i+1}(\omega)\}.$$

Now

$$P(B) = P(B \cap C_\varepsilon \cap \theta^{N_\varepsilon} D_\varepsilon) + P(B \cap (C_\varepsilon \cap \theta^{N_\varepsilon} D_\varepsilon)^c) \leq P(B \cap C_\varepsilon \cap \theta^{N_\varepsilon} D_\varepsilon) + \varepsilon.$$

We prove that $B \cap C_\varepsilon \cap \theta^{N_\varepsilon} D_\varepsilon = \emptyset$ by the argument explained above.

Suppose $\omega \in B$. Then there exists $0 \neq x \in V_{i+1}(\omega) \cap V^-_{\ell+1-i}(\omega)$ such that by the cocycle property

$$(11.5.9) \qquad 1 = \|x\| = \left\| \Phi(N, \theta^{-N_\varepsilon} \omega) \frac{\Phi(-N_\varepsilon, \omega)x}{\|\Phi(-N_\varepsilon, \omega)x\|} \right\| \|\Phi(-N_\varepsilon, \omega)x\|.$$

Since $\omega \in C_\varepsilon$,

$$(11.5.10) \qquad \|\Phi(-N_\varepsilon, \omega)x\| < e^{N_\varepsilon(-\lambda_i + \delta)} \|x\|,$$

and since $\omega \in \theta^{N_\varepsilon} D_\varepsilon$, for any $0 \neq y \in V_{i+1}(\theta^{-N_\varepsilon} \omega)$,

$$(11.5.11) \qquad \|\Phi(N_\varepsilon, \theta^{-N_\varepsilon} \omega)y\| < e^{N_\varepsilon(\lambda_{i+1} + \delta)} \|y\|.$$

Since $\Phi(-N_\varepsilon, \omega) V_{i+1}(\omega) = V_{i+1}(\theta^{-N_\varepsilon} \omega)$ on $\tilde{\Omega}$,

$$y := \frac{\Phi(-N_\varepsilon, \omega)x}{\|\Phi(-N_\varepsilon, \omega)x\|} \in V_{i+1}(\theta^{-N_\varepsilon} \omega).$$

Estimating (11.5.9) with (11.5.10) and (11.5.11), one obtains the contradiction

$$1 \leq \|\Phi(N, \theta^{-N_\varepsilon} \omega)y\| \|\Phi(-N_\varepsilon, \omega)x\| < e^{N_\varepsilon(\lambda_{i+1} + \delta)} e^{N_\varepsilon(-\lambda_i + \delta)} = 1.$$

Concerning the second equality in the lemma, we already know that the sum is direct. We have $\dim V_{i+1}(\omega) = \sum_{k=i+1}^{\ell} d_k$ and by (11.5.8)

$$\dim V^-_{\ell+1-i} = \sum_{k=1}^{i} d^-_{\ell+1-k} = \sum_{k=1}^{i} d_k$$

giving $\dim V_{i+1}(\omega) + \dim V^-_{\ell+1-i}(\omega) = d$ and thus the assertion. $\square$

Continuation of the proof of Theorem 11.1.13: Recall that for all subspaces $U, V \subset \mathbb{R}^d$,

$$\dim(U \cap V) = \dim U + \dim V - \dim(U + V).$$

Using the second equality in Lemma 11.5.4, we have $V_i(\omega) + V_{\ell+1-i}^-(\omega) \supset V_{i+1}(\omega) + V_{\ell+1-i}^-(\omega) = \mathbb{R}^d$. For brevity, we omit the argument ω in the following and find

$$\dim L_i = \dim V_i + \dim V_{\ell+1-i}^- - \dim\left(V_i + V_{\ell+1-i}^-\right) = \sum_{k=i}^{\ell} d_k + \sum_{k=1}^{i} d_k - d = d_i.$$

We now prove that the sum of the L_i is direct. This is the case if and only if $L_i \cap F_i = \{0\}$ for all i, where $F_i := \sum_{j \neq i} L_j$. By definition

$$L_i \cap F_i = L_i \cap (U + V), \text{ where } U := \sum_{j=1}^{i-1} L_j \text{ and } V := \sum_{j=i+1}^{\ell} L_j.$$

Note that for $i = 1$, the subspace U is trivial, and for $i = \ell$, the subspace V is trivial. By definition of L_j and monotonicity of the $V_{\ell+1-j}^-$ and V_j, the inclusions

$$U = \sum_{j=1}^{i-1}\left(V_j \cap V_{\ell+1-j}^-\right) \subset V_{\ell+1-(i-1)}^-, \quad V = \sum_{j=i+1}^{\ell}\left(V_j \cap V_{\ell+1-j}^-\right) \subset V_{i+1}$$

hold. Suppose that $x \in L_i \cap F_i$. Then $x \in V_i$ and $x \in V_{\ell+1-i}^-$ and $x = u + v$ with $u \in U$ and $v \in V$. Since $x \in V_i$ and $v \in V_{i+1} \subset V_i$, we have $x - v = u \in V_i \cap V_{\ell+1-(i-1)}^- = \{0\}$ by Lemma 11.5.4, i.e., $u = 0$. (Note that this trivially holds for $i = 0$, since here $U = \{0\}$.) Since $x \in V_{\ell+1-i}^-$ and $x = v \in V_{i+1}$, we have $v \in V_{i+1} \cap V_{\ell+1-i}^- = \{0\}$, again by Lemma 11.5.4. Together, we have shown that $x = 0$ and hence $\sum L_i$ is direct. In particular, $\dim \sum L_i = \sum \dim L_i = d$, and the $L_i(\omega)$ form a splitting of $\mathbb{R}^d$ for each $\omega \in \Omega$.

By the invariance of the forward and backward flags,

$$A(\omega)V_i(\omega) = V_i(\theta\omega) \text{ and } V_{\ell+1-i}^-(\omega) = A^{-1}(\omega)V_{\ell+1-i}^-(\theta\omega)$$

and $A(U \cap V) = AU \cap AV$, we obtain the invariance property for L_i,

$$A(\omega)L_i(\omega) = A(\omega)V_i(\omega) \cap A(\omega)V_{\ell+1-i}^-(\omega) = V_i(\theta\omega) \cap V_{\ell+1-i}^-(\theta\omega) = L_i(\theta\omega).$$

This proves part (i) of the MET.

Next we prove part (ii) of the MET. By Lemma 11.5.3 applied in forward and backward time,

$$\lim_{n\to\pm\infty} \frac{1}{n}\log\|\Phi(n,\omega)x\| = \lambda_i$$

if and only if (here ^{c} denotes the complement in $\mathbb{R}^d$)

$$x \in [V_i(\omega) \setminus V_{i+1}(\omega)] \cap \left[V_{\ell+1-i}^-(\omega) \setminus V_{\ell+1-(i-1)}^-(\omega)\right]$$
$$= V_i(\omega) \cap [V_{i+1}(\omega)]^c \cap V_{\ell+1-i}^-(\omega) \cap \left[V_{\ell+1-(i-1)}^-(\omega)\right]^c =: W_i(\omega).$$

Since by Lemma 11.5.4 $V_i \cap V^-_{\ell+1-(i-1)} = \{0\}$ it follows that

$$V_i \cap \left[V^-_{\ell+1-(i-1)}\right]^c = V_i \setminus \{0\}.$$

Similarly, $V^c_{i+1} \cap V^-_{\ell+1-i} = \{0\}$ and it follows that $V_{i+1} \cap [V^-_{\ell+1-i}]^c = V_{i+1} \setminus \{0\}$. Together

$$W_i(\omega) = [V_i(\omega) \setminus \{0\}] \cap [V_{i+1}(\omega) \setminus \{0\}] = L_i(\omega) \setminus \{0\}.$$

Thus assertion (ii) follows.

Measurability $\omega \mapsto L_i(\omega)$ claimed in assertion (iii) follows from the proof of the deterministic MET, Theorem 11.4.1, since the subspaces $U_i(\omega)$ are the limits of measurable functions $U_i(n, \omega)$ with values in a Grassmannian and hence measurable themselves. Part (iv) follows directly from Lemma 11.5.3. □

Remark 11.5.5. The image of the unit sphere under the linear map $\Phi(n, \omega)$ is a hyperellipsoid, whose ith principal axis has length given by the ith singular value $\delta_i(\Phi(n, \omega))$. The Furstenberg-Kesten Theorem, Theorem 11.5.1, says that the exponential growth rate of the length of the ith principal axis is given by $\Lambda_i(\omega)$ almost surely. The Multiplicative Ergodic Theorem additionally asserts that the span of the principal axes corresponding to the Lyapunov exponent $\lambda_i(\omega)$ also converges to a limit $L_i(\omega)$. Moreover, each point $x \neq 0$ has a Lyapunov exponent $\lambda(x)$, which is a random variable taking values in the finite set $\{\lambda_i \,|\, i = 1, \ldots, \ell\}$.

11.6. The Random Linear Oscillator

In this section, we come back to the linear oscillator. Section 7.3 has analyzed in detail its stability properties when the coefficients are periodic. In Section 9.5 the case of uncertain (or controlled) coefficients has been discussed as an application of the topological theory. Here the case of randomly perturbed coefficients will be discussed. The average Lyapunov exponent is determined analytically and numerical results for a specific base (or background) process θ_t are presented.

Consider the linear oscillator with random restoring force

(11.6.1) $$\ddot{x}(t) + 2b\dot{x}(t) + (1 + \rho f(\theta_t \omega))x(t) = 0,$$

where $b, \rho \in \mathbb{R}$ are positive constants and $f : \Omega \to \mathbb{R}$ is in $L^1(\Omega, \mathbb{R})$; compare (7.3.3) and (7.3.4) for the periodic case and Example 9.5.8 for the uncertain case. We assume that the background process θ on Ω is ergodic. Using the notation $x_1 = x$ and $x_2 = \dot{x}$ we can write this equation as

$$\begin{bmatrix} \dot{x}_1(t) \\ \dot{x}_2(t) \end{bmatrix} = A(\theta_t \omega) \begin{bmatrix} x_1(t) \\ x_2(t) \end{bmatrix}, \text{ where } A(\theta_t \omega) := \begin{bmatrix} 0 & 1 \\ -1 - \rho f(\theta_t \omega) & -2b \end{bmatrix}.$$

It is not possible to compute the Lyapunov exponents of such a system analytically, except in trivial cases. This is different for average Lyapunov exponents defined as

$$\overline{\lambda} := \frac{1}{d} \sum_{j=1}^{\ell} d_j \lambda_j,$$

under the assumptions of the Multiplicative Ergodic Theorem, Theorem 11.1.6. For a system with $d = 2$ one has

$$\overline{\lambda} = \frac{1}{2} (\lambda_1 + \lambda_2) \text{ or } \overline{\lambda} = \lambda_1,$$

if there are two or one Lyapunov exponents, respectively. We will show that $\overline{\lambda} = -b$ for the random linear oscillator (11.6.1).

For the proof, we first eliminate the damping $2b$ by introducing the new state variable

$$y(t) = e^{bt} x(t).$$

This is analogous to the construction in Exercise 7.4.2 and in the analysis of the pendulum equation in Remark 7.3.1. With $\dot{y}(t) = b e^{bt} x(t) + e^{bt} \dot{x}(t)$ one computes (omitting the argument t)

$$\ddot{y} = b^2 e^{bt} x + 2b e^{bt} \dot{x} + e^{bt} \ddot{x} = b^2 e^{bt} x - [1 + \rho f(\theta_t \omega)] e^{bt} x = b^2 y - [1 + \rho f(\theta_t \omega)] y,$$

hence

$$\ddot{y}(t) + [1 - b^2 + \rho f(\theta_t \omega)] y(t) = 0$$

or, with $y_1 = y, y_2 = \dot{y}$,

$$\begin{bmatrix} \dot{y}_1(t) \\ \dot{y}_2(t) \end{bmatrix} = B(\theta_t \omega) \begin{bmatrix} y_1(t) \\ y_2(t) \end{bmatrix}, \text{ where } B(\theta_t \omega) := \begin{bmatrix} 0 & 1 \\ -1 + b^2 - \rho f(\theta_t \omega) & 0 \end{bmatrix}.$$

Note that the trace of B satisfies $\operatorname{tr} B(\theta_t \omega) = 0$ for all t and all ω. The Lyapunov exponents $\hat{\lambda}_i$ of this transformed equation are given by

(11.6.2) $$\hat{\lambda}_i = b + \lambda_i.$$

In fact, one computes (we use the max-norm)

$$\|(y_1(t), y_2(t))\|_\infty = \max_{i=1,2} |y_i(t)| = \max \left(\left| e^{bt} x_1(t) \right|, \left| b e^{bt} x_1(t) + e^{bt} x_2(t) \right| \right)$$

$$= e^{bt} \max \left(|x_1(t)|, |b x_1(t) + x_2(t)| \right),$$

and hence one finds

$$\hat{\lambda}_i = \lim_{t \to \infty} \frac{1}{t} \log \|(y_1(t), y_2(t))\|_\infty = b + \lim_{t \to \infty} \frac{1}{t} \log \|(x_1(t), x_2(t))\|_\infty = b + \lambda_i.$$

For the fundamental solution $\Phi(t, \omega)$ given by

$$\frac{d}{dt} \Phi(t, \omega) = B(\theta_t \omega) \Phi(t, \omega) \text{ with } \Phi(0, \omega) = I,$$

11.6. The Random Linear Oscillator

Liouville's trace formula (see Theorem 6.1.5) shows that
$$\det \Phi(t,\omega) = e^{\int_0^t \operatorname{tr} B(\theta_s \omega) ds} = 1.$$
Proposition 11.3.1 implies that the singular values of $\Phi(t,\omega)$ satisfy
$$1 = \det \Phi(t,\omega) = \delta_1(\Phi(t,\omega))\delta_2(\Phi(t,\omega)).$$
Now the Multiplicative Ergodic Theorem, Theorem 11.1.6, yields
$$\lambda_i = \lim_{t\to\infty} \frac{1}{t} \log \delta_i(\Phi(t,\omega)),$$
and hence, if there are two Lyapunov exponents $\hat\lambda_i$, they satisfy
$$\begin{aligned}\hat\lambda_1 + \hat\lambda_2 &= \lim_{t\to\infty} \frac{1}{t} \log \delta_1(\Phi(t,\omega)) + \lim_{t\to\infty} \frac{1}{t} \log \delta_2(\Phi(t,\omega)) \\ &= \lim_{t\to\infty} \frac{1}{t} \log\left(\delta_1(\Phi(t,\omega))\delta_2(\Phi(t,\omega))\right) \\ &= 0.\end{aligned}$$
Hence
$$\bar\lambda = \frac{1}{2}(\lambda_1 + \lambda_2) = \frac{1}{2}(\hat\lambda_1 - b + \hat\lambda_2 - b) = -b$$
as claimed. If there is only one Lyapunov exponent, one argues analogously.

In the following we present numerical results for a specific random linear oscillator of the form (11.6.1). We have to specify the random process $\theta_t(\omega)$. Here we start with an Ornstein–Uhlenbeck process

(11.6.3) $$d\eta_t = -\alpha \eta_t + \sigma dW_t \text{ in } \mathbb{R}^1,$$

where α and σ are positive numbers. The system noise is given by
$$\theta_t(\omega) = \xi_t = \sin \eta_t,$$
resulting in an Ornstein–Uhlenbeck process on the circle $\mathbb{S}^1$. The stochastic differential equation (11.6.3) is solved numerically using the Euler scheme. The resulting random linear oscillator equation of the form (11.6.1) with noise magnitude ρ given by

(11.6.4) $$\ddot{x} + 2b\dot{x} + (1 + \rho \cdot \xi_t)x = 0,$$

is solved numerically using an explicit 4-th order Runge–Kutta scheme. Figure 11.1 presents the level curves of the maximal Lyapunov exponent in dependence on b and ρ.

Note that for $\rho = 0$, i.e., for the unperturbed system, and for $b > 0$ the (maximal) real part of the eigenvalues is negative, hence the system is stable. But as the damping b increases beyond 1, the system experiences overdamping and the (maximal) Lyapunov exponent increases. The $\lambda_{\max} = 0$ level curve in Figure 11.1 separates the stable region (below the curve) from the unstable region for the random oscillator. It is remarkable that

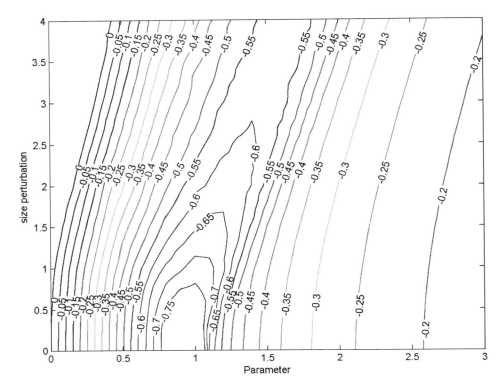

Figure 11.1. Level curves of the maximal Lyapunov exponents for the random linear oscillator (11.6.4) in dependence on the damping parameter b (on the horizontal axis) and the perturbation size ρ (on the vertical axis).

for $b > 1$ the (maximal) Lyapunov exponent for increasing noise size ρ first decreases, then increases and and at (finite) ρ-level reaches 0. Figure 11.2 shows the maximal Lyapunov exponent as a function of the damping b and the noise magnitude ρ.

11.7. Exercises

Exercise 11.7.1. Show hat for two orthogonal projections P and Q on $\mathbb{R}^d$ the following statements are equivalent:

(i) $P[\mathbb{R}^d] \subset Q[\mathbb{R}^d]$,
(ii) $PQ = P$,
(iii) $QP = P$.

Exercise 11.7.2. Let $P, Q : \mathbb{R}^d \to \mathbb{R}^d$ be two orthogonal projections. Show that $\|P - Q\| \leq 1$.

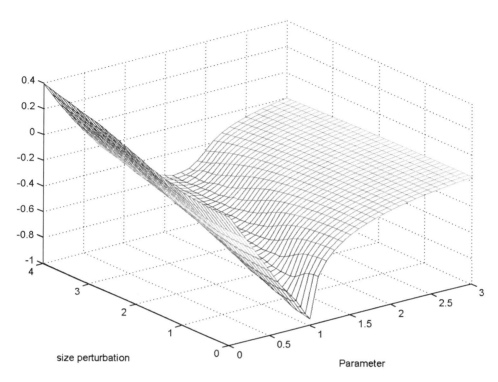

Figure 11.2. The maximal Lyapunov exponents for the random linear oscillator (11.6.4) in dependence on the damping parameter b and the perturbation size ρ.

Exercise 11.7.3. Let $\Omega = \mathbb{S}^1$ be parametrized by $\psi \in [0,1)$ with the Borel σ-algebra $\mathcal{B}$ and define P as the uniform distribution on $\mathbb{S}^1$, i.e., P coincides with Lebesgue measure. Define θ as the shift $\theta(t,\tau) = (t+\tau)(\mathrm{mod}\,1)$ for $t \in \mathbb{R}, \tau \in \Omega$. Show that $\theta : \mathbb{R} \times \Omega \to \Omega$ is an ergodic metric dynamical system.

Exercise 11.7.4. Let (X_1, μ_1) and (X_2, μ_2) be probability spaces and consider measure preserving maps $T_1 : X_1 \to X_1$ and $T_2 : X_2 \to X_2$ which are isomorphic. That is, there are null sets N_1 and N_2 and a measurable map $H : X_1 \setminus N_1 \to X_2 \setminus N_2$ with measurable inverse H^{-1} such that
$$H(T_1(x_1)) = T_2(H(x_1)) \text{ for all } x_1 \in X_1 \setminus N_1.$$
Show that T_1 is ergodic if and only if T_2 is ergodic.

Exercise 11.7.5. Consider the unit interval $X = [0,1]$ and let $T(x) = x + \theta \pmod{1}$ where $\theta \in [0,1]$. Show that the Lebesgue measure is an invariant probability measure which is not ergodic for T if $\theta \in \mathbb{Q}$. (Note that T is ergodic if $\theta \notin \mathbb{Q}$.)
Hint: By Lemma 10.1.3 it suffices to show that every invariant function

f in the Hilbert space $L_2([0,1], \mathbb{C})$ is constant (this lemma also holds for complex-valued functions by considering the real and imaginary parts separately.). Using a Fourier expansion in terms of the complete orthonormal system $e^{2\pi i n x}, n \in \mathbb{Z}$, one finds that it suffices to show this claim for each base function for which it is easily seen.

Exercise 11.7.6. Consider a metric space Ω with a measure μ on the Borel σ-algebra $\mathcal{B}(\Omega)$ and let $T : \Omega \to \Omega$ be a measurable map preserving μ. Let $E \in \mathcal{B}(\Omega)$ be a set with finite measure $\mu(E) < \infty$. Show

$$\mu\left(\left\{\omega \in \Omega \setminus E \,\middle|\, \begin{array}{c} \text{there is } n \in \mathbb{N} \text{ with } T^n(\omega) \in E \text{ and} \\ (T^m(\omega) \in E \text{ implies } T^k(\omega) \in E \text{ for all } k \geq m) \end{array}\right\}\right) = 0.$$

Hint: Define $n(y) := \min\{n \in \mathbb{N} \,|\, T^{-n}(y) \notin E\}$ for $y \in E$ and $E_j := \{y \in E \,|\, n(y) = j \text{ and } T^k(y) \in E \text{ for all } k \in \mathbb{N}\}$. Then show first that $\mu(E_j) = 0$ for all j.

Exercise 11.7.7. Prove the following ergodic theorem for irreducible permutations:

Let σ be an irreducible permutation of a finite set $X = \{x_1, x_2, \ldots, x_N\}$, i.e., for every $x \in X$ the orbit $\mathcal{O}(x) = \{x, \sigma(x), \sigma^2(x), \ldots\}$ has exactly N different elements (equivalently, every point is periodic with period N). Then for every function $f : X \to \mathbb{R}$,

$$\lim_{n \to \infty} \frac{1}{n} \sum_{k=0}^{n-1} f(\sigma^k(x)) = \frac{1}{N}(f(x_1) + \cdots + f(x_N)).$$

11.8. Orientation, Notes and References

Orientation. This chapter describes the linear algebra associated with linear dynamical systems (in discrete and continuous time) with random coefficients. The Multiplicative Ergodic Theorem shows that in the ergodic case there are Lyapunov decompositions of the state space into subspaces, which are invariant under the dynamical system and which are characterized by the Lyapunov exponents in forward and backward time. The Lyapunov exponents are constants while the subspaces are random, hence they depend on the base point, just as in the periodic case considered in Chapter 7. These assertions hold almost surely, i.e., with probability one. Hence nothing can be said about points in the base space which are in a null set. Thus part of the base space is excluded form the analysis. The null set, naturally, depends on the considered probability measure on the base space, and, in fact, it may be quite large by other measures or by topological considerations, such as genericity.

Comparing the Multiplicative Ergodic Theorem, Theorem 11.1.6, and the corresponding result for topological linear flows, Theorem 9.1.5, one

11.8. Orientation, Notes and References

sees that continuity is replaced by measurability and chain transitivity is replaced by ergodicity of the base flow. The Multiplicative Ergodic Theorem is stronger, since the Lyapunov exponents are numbers, not intervals, and they exist as limits; it is weaker, because the assertions only hold in a set of full measure, not for every point in the base space.

Notes and references. A comprehensive exposition of several versions of the Multiplicative Ergodic Theorem as well as far reaching further theory are contained in Ludwig Arnold's monograph [6]. Our proof in this chapter is distilled from his proofs of various versions of the MET. Ludwig Arnold, in turn, is carefully working out the arguments proposed in Goldsheid and Margulis [53]. We restrict the analysis to invertible systems on the time domain $\mathbb{Z}$. The restriction to the invertible case, in particular, considerably shortens the proof of the deterministic MET. There are many other proofs of the MET; see [6] for a discussion. For an extension of the MET to infinite dimensional systems, see, in particular, Lian and Lu [95]. Theorem 11.5.1 is due to Furstenberg and Kesten [49].

Corollary 11.1.12 determines the almost sure stability behavior of random linear differential equations. One could, of course, also look at the pth moments, $p \geq 0$, of the trajectories and try to determine their stability behavior. Results in this direction were first obtained in Arnold, Kliemann and Oeljeklaus [7] for Markovian background noise $\theta_t\omega$, compare Arnold [6] for a complete overview. Topological properties of random systems are analyzed in Cong [32].

The proof of the MET presented here uses singular value decompositions of the system matrices (dynamics) $(\Phi(n,\omega)^\top \Phi(n,\omega))^{1/2n}$. Of course, singular values are relevant in many areas; we refer, in particular, to Hinrichsen and Pritchard [68, Section 4.3] for further properties in the context of systems and control theory, and to Harville [63, Section 21.12] in the context of statistics. For basic facts on singular values and their computation we refer to the Handbook of Linear Algebra [71, Section 5.6 and Chapters 24 and 58].

To define appropriate metrics on Grassmannians and flag manifolds one needs some properties regarding orthogonal projections, compare Sections 11.2 and 11.3. Basic references for these properties are Horn and Johnson [72] and Kato [74].

Our version of the MET in continuous time, Theorem 11.1.6, is formulated for random dynamical systems, for which the underlying metric dynamical system is generated, e.g., via Kolmogorov's construction, compare Example 11.1.4. Including stochastic differential equations in this setup requires some work, since stochastic differential equations 'naturally' lead to 2-parameter flows $\varphi_{s,t}(\omega) : \mathbb{R}^d \to \mathbb{R}^d$, compare Kunita [91]. The process to generate random dynamical systems for Stratonovich-type stochastic differential equations is described in Arnold [6, Section 2.3]. What needs to be shown is that $\varphi_{0,t}(\omega)$ is a cocycle over the metric dynamical system that models the driving noise of the stochastic differential equation. The key idea is that of a semimartingale helix combining both metric dynamical systems and filtered probability spaces: For a metric dynamical system $(\Omega, \mathcal{F}, P, (\theta_t)_{t \in \mathbb{R}})$ a map $F : R \times \Omega \to \mathbb{R}^d$ is called a helix if $F(t+s, \omega) = F(t, \theta_s(\omega)) + F(s, \omega)$ for all $s, t \in \mathbb{R}$ and $\omega \in \Omega$. If our probability space admits also a filtration of sigma algebras $(\mathcal{F}_s^t)_{s \leq t}$ one can define a semimartingale helix as a helix such that $F_s(t, \omega) := F(t, \omega) - F(s, \omega)$ is a $\mathcal{F}_s^t$-semimartingale. Under certain smoothness conditions, such semimartingale helixes define (unique) random dynamical systems over the metric dynamical system $(\Omega, \mathcal{F}, P, (\theta_t)_{t \in \mathbb{R}})$. One obtains for the (classical) Stratonovich-type stochastic differential equations that they define random dynamical systems over the filtered metric system that describes Brownian motion This construction also works vice versa, so that certain (filtered) dynamical systems generate (unique) semimartingale helixes. This mechanism was first described by Arnold and Scheutzow in [8].

In general, Lyapunov exponents for random linear systems are difficult to compute explicitly—numerical methods are usually the way to go; see, e.g., Talay [132], Grorud and Talay [57], or the engineering oriented monograph Xie [144]. The approach to numerical computation of Lyapunov exponents in Section 11.6 is based on Verdejo, Vargas, and Kliemann [134]. We refer to Kloeden and Platen [82] for a basic text on numerical methods for stochastic differential equations.

Bibliography

[1] D. Acheson, *From Calculus to Chaos.* An Introduction to Chaos, Oxford University Press 1997.

[2] J.M. Alongi and G.S. Nelson, *Recurrence and Topology,* Amer. Math. Soc., 2007.

[3] E. Akin, *The General Topology of Dynamical Systems*, Amer. Math. Soc., 1993.

[4] H. Amann, *Ordinary Differential Equations*, Walter de Gruyter, 1990.

[5] H. Amann, J. Escher, *Analysis. II,* Birkhäuser Verlag, Basel, 2008.

[6] L. Arnold, *Random Dynamical Systems,* Springer-Verlag, 1998.

[7] L. Arnold, W. Kliemann, and E. Oeljeklaus, *Lyapunov exponents of linear stochastic systems*, Springer Lecture Notes in Mathematics No. 1186 (1986), 85-125.

[8] L. Arnold and M. Scheutzow, *Perfect cocycles through stochastic differential equations,* Probab. Theory Relat. Fields 101 (1995), 65-88.

[9] D.K. Arrowsmith and C.M. Place, *An Introduction to Dynamical Systems*, Cambridge University Press, 1990.

[10] Z. Artstein and I. Bright, *Periodic optimization suffices for infinite horizon planar optimal control,* SIAM J. Control Optim. 48 (2010), 4963-4986.

[11] J. Ayala-Hoffmann, P. Corbin, K. McConville, F. Colonius, W. Kliemann, and J. Peters, *Morse decompositions, attractors and chain recurrence*, Proyecciones Journal of Mathematics 25 (2006), 79-109.

[12] V. Ayala, F. Colonius, and W. Kliemann, *Dynamic characterization of the Lyapunov form of matrices*, Linear Algebra and Its Applications 402 (2005), 272-290.

[13] V. Ayala, F. Colonius, and W. Kliemann, *On topological equivalence of linear flows with applications to bilinear control systems,* J. Dynamical and Control Systems 17 (2007), 337-362.

[14] V. Ayala and C. Kawan, *Topological conjugacy of real projective flows*, J. London Math. Soc. 90 (2014), 49-66.

[15] L. Barreira and Ya.B. Pesin, *Lyapunov Exponents and Smooth Ergodic Theory*, University Lecture Series, vol. 23, Amer. Math. Soc., 2002.

[16] L. BARREIRA AND Y.B. PESIN, *Nonuniform Hyperbolicity*, Cambridge University Press, 2007.

[17] H. BAUMGÄRTEL, *Analytic Perturbation Theory for Matrices and Operators*, Birkhäuser, 1985.

[18] M.V. BEBUTOV, *Dynamical systems in the space of continuous functions*, Dokl. Akad. Nauk SSSR 27 (1940), 904-906

[19] W.M. BOOTHBY, *An Introduction to Differentiable Manifolds and Riemannian Geometry*, Academic Press, New York, 1975.

[20] C.J. BRAGA BARROS AND L.A.B. SAN MARTIN, *Chain transitive sets for flows on flag bundles*. Forum Mathematicum 19 (2007), 19-60.

[21] I.U. BRONSTEIN AND A.YA KOPANSKII, *Smooth Invariant Manifolds and Normal Forms*, World Scientific, 1994.

[22] A. BRUCKNER, J. BRUCKNER, AND B. THOMPSON, *Real Analysis*, Prentice Hall, 1997.

[23] B. F. BYLOV, R. E. VINOGRAD, D. M. GROBMAN, AND V. V. NEMYTSKII, *Theory of Lyapunov Exponents*, Nauka, 1966 (in Russian).

[24] L. CESARI, *Asymptotic Behavior and Stability Problems in Ordinary Differential Equations*, Springer-Verlag, 1971, Third edition.

[25] F. CHATELIN, *Eigenvalues of Matrices*, SIAM 2012, Revised edition.

[26] C. CHICONE, *Ordinary Differential Equations with Applications*, Springer-Verlag, 1999.

[27] C. CHICONE AND YU. LATUSHKIN, *Evolution Semigroups in Dynamical Systems and Differential Equations*, Mathematical Surveys and Monographs, 70, American Mathematical Society, Providence, RI, 1999.

[28] G.H. CHOE, *Computational Ergodic Theory*, Springer-Verlag, 2005.

[29] F. COLONIUS AND W. KLIEMANN, *The Dynamics of Control*, Birkhäuser Boston, 2000.

[30] F. COLONIUS AND W. KLIEMANN, *Morse decomposition and spectra on flag bundles*, J. Dynamics and Differential Equations 14 (2002), 719-741.

[31] F. COLONIUS AND W. KLIEMANN, *Dynamical systems and linear algebra*, in: Handbook of Linear Algebra (L. Hogben, ed.), CRC Press, Boca Raton, 2nd ed., 2014, 79-1 – 79-23.

[32] N.D. CONG, *Topological Dynamics of Random Dynamical Systems*, Oxford Mathematical Monographs, Clarendon Press, 1997.

[33] N.D. CONG, *Structural stability and topological classification of continuous time, linear hyperbolic cocycles*, Random Comp. Dynamics 5 (1997), 19-63.

[34] C. CONLEY, *Isolated Invariant Sets and the Morse Index*, Reg. Conf. Ser. Math., vol. 38, Amer. Math. Soc., Providence, RI (1978).

[35] C. CONLEY, *The gradient structure of a flow I*, Ergodic Theory Dyn. Sys. 8 (1988), 11-26.

[36] W.A. COPPEL, *Dichotomies in Stability Theory*, Springer-Verlag, 1978.

[37] H. CRAUEL AND F. FLANDOLI, *Attractors for random dynamical systems*, Probab. Theory Related Fields, 100 (1994), 1095-1113.

[38] B.D. CRAVEN, *Lebesgue Measure and Integral*, Pitman, 1982.

[39] J. DALECKII AND M. KREIN, *Stability of Solutions of Differential Equations in Banach Space*, Amer. Math. Soc., 1974.

[40] N. DUNFORD AND J.T. SCHWARTZ, *Linear Operators, Part I: General Theory*, Wiley-Interscience, 1977.

[41] R. DURRETT, *Probability: Theory and Examples*, 4th ed., Cambridge University Press, Cambridge, 2010.

[42] R. EASTON, *Geometric Methods for Discrete Dynamical Systems*, Oxford University Press, 1998.

[43] S. ELAYDI, *An Introduction to Difference Equations*. Third edition, Undergraduate Texts in Mathematics. Springer-Verlag, New York, 2005.

[44] D. ELLIOTT, *Bilinear Control Systems: Matrices in Action*, Springer-Verlag, 2009.

[45] R. FABBRI, R. JOHNSON, AND L. ZAMPOGNI, *Nonautonomous differential systems in two dimensions*, Chapter 2 in: Handbook of Differential Equations, Ordinary Differential Equations, volume 4, F. Batelli and M. Fečkan, eds., Elsevier 2008.

[46] H. FEDERER, *Gometric Measure Theory*, Springer, New York, 1969, Reprint 2013.

[47] N. FENICHEL, *Persistence and smoothness of invariant manifolds for flows*, Indiana Univ. Math., 21 (1971), 193-226.

[48] R. FERES, *Dynamical Systems and Semisimple Groups. An Introduction*, Cambridge University Press, 1998.

[49] H. FURSTENBERG AND H. KESTEN, *Products of random matrices*, Ann. Math. 31 (1960), 457-469.

[50] G. FLOQUET, *Sur les équations différentielles linéaires à coefficients périodiques*, Ann. École Norm. Sup. 12 (1883), 47-88.

[51] P. GÄNSSLER AND W. STUTE, *Wahrscheinlichkeitstheorie*, Springer-Verlag, 1977.

[52] I. GOHBERG, S. GOLDBERG, AND N. KRUPNIK, *Traces and Determinants of Linear Operators*, Birkhäuser, 2000.

[53] I.Y. GOLDSHEID AND G.A. MARGULIS, *Lyapunov indices of a product of random matrices*. Russian Mathematical Surveys 44:5(1989), 11-71.

[54] M. GOLUBITSKY AND M. DELLNITZ, *Linear Algebra and Differential Equations*, Brooks Cole Pub. Co., 1999.

[55] W.H. GREUB, *Multilinear Algebra*, Springer-Verlag, Berlin 1967, 2nd ed., 1978.

[56] D.M. GROBMAN, *On homeomorphisms of systems of differential equations*, Doklady AN SSSR 128 (1959), 880-881.

[57] A. GRORUD AND D. TALAY, *Approximation of Lyapunov exponents of nonlinear stochastic differential equations*, SIAM Journal on Applied Mathematics, 56(2), (1996), 627-650.

[58] L. GRÜNE, *Numerical stabilization of bilinear control systems*, SIAM Journal on Control and Optimization 34 (1996), 2024-2050.

[59] L. GRÜNE, *A uniform exponential spectrum for linear flows on vector bundles*, J. Dyn. Diff. Equations, 12 (2000), 435-448.

[60] J. GUCKENHEIMER AND P. HOLMES, *Nonlinear Oscillations, Dynamical Systems, and Bifurcation of Vector Fields*, Springer-Verlag, 1983.

[61] W. HAHN, *Stability of Motion*, Springer-Verlag, 1967.

[62] P. HARTMAN, *A lemma in the theory of structural stability of differential equations*, Proc. Amer. Math. Soc. 11 (1960), 610-620.

[63] D.A. HARVILLE, *Matrix Algebra from a Statistician's Perspective*, Springer-Verlag, New York, 1997.

[64] E. HEIL, *Differentialformen und Anwendungen auf Vektoranalysis, Differentialgleichungen, Geometrie*, Bibliographisches Institut, 1974.

[65] U. HELMKE AND J. B. MOORE, *Optimization and Dynamical Systems*, Springer-Verlag, 1994.

[66] S. HILGER, *Lineare Systeme Periodischer Differentialgleichungen*, Diplomarbeit, Universität Würzburg, 1986.

[67] G.W. HILL, *On the part of the motion of the lunar perigee, which is a function of the mean motions of the sun and the moon*, Acta Math. 8 (1886), 1-36.

[68] D. HINRICHSEN AND A.J. PRITCHARD, *Mathematical Systems Theory*, Springer-Verlag, 2005.

[69] M.W. HIRSCH AND S. SMALE, *Differential Equations, Dynamical Systems, and Linear Algebra*, Academic Press, New York, 1974.

[70] M.W. HIRSCH, S. SMALE AND R.L. DEVANEY, *Differential Equations, Dynamical Systems and an Introduction to Chaos*, Elsevier, 2004.

[71] L. HOGBEN (ed.), *Handbook of Linear Algebra*, 2nd ed., CRC Press, Boca Raton, 2014.

[72] R.A. HORN AND C.R. JOHNSON, *Matrix Analysis*, 2nd. ed., Cambridge University Press, Cambridge, 2013.

[73] K. JOSIĆ AND R. ROSENBAUM, *Unstable solutions of nonautonomous linear differential equations*, SIAM Review 50 (2008), 570-584.

[74] T. KATO, *Perturbation Theory for Linear Operators*, Springer-Verlag, 1984.

[75] A. KATOK AND B. HASSELBLATT, *Introduction to the Modern Theory of Dynamical Systems*, Cambridge University Press, 1995.

[76] C. KAWAN AND T. STENDER, *Lipschitz conjugacy of linear flows*, J. London Math. Soc. 80 (2009), 699-715.

[77] C. KAWAN AND T. STENDER, *Growth rates for semiflows on Hausdorff spaces*, J. Dynamics and Differential Equations 24 (2012), 369-390.

[78] R.Z. KHASMINSKII, *Necessary and sufficient conditions for the asymptotic stability of linear stochastic systems*, Theory of Probability and its Applications 11 (1967), 144-147.

[79] R.Z. KHASMINSKII, *Stochastic Stability of Differential Equations*, Sijthoff and Noordhoff, Alphen 1980. (Translation of the Russian edition, Nauka, Moscow 1969).

[80] J.F.C. KINGMAN, *Subadditive ergodic theory*, Ann. Prob. 1 (1973), 883-904.

[81] P. KLOEDEN AND M. RASMUSSEN, *Nonautonomous Dynamical Systems*, Mathematical Surveys and Monographs 176, American Mathematical Society, 2011.

[82] P. KLOEDEN AND E. PLATEN, *Numerical Solution of Stochastic Differential Equations*, Springer-Verlag, 2000.

[83] H.-J. KOWALSKY, *Lineare Algebra*, 4. Auflage, DeGruyter, Berlin 1969.

[84] H.-J. KOWALSKY AND G.O. MICHLER, *Lineare Algebra*, 12. Auflage, DeGruyter, Berlin 2003.

[85] U. KRAUSE AND T. NESEMANN, *Differenzengleichungen und Diskrete Dynamische Systeme*, 2. Auflage, De Gruyter, Berlin/Boston, 2011.

[86] U. KRENGEL, *Ergodic Theorems*, De Gruyter, 1985.

[87] N. H. KUIPER, *The topology of the solutions of a linear differential equation on $\mathbb{R}^n$*, Manifolds, Tokyo 1973 (Proc. Internat. Conf., Tokyo, 1973), pp. 195-203. Univ. Tokyo Press, Tokyo, 1975.

[88] N.H. KUIPER, *Topological conjugacy of real projective transformations*, Topology 15 (1976), 13-22.

[89] N.H. KUIPER, J. W. ROBBIN, *Topological classification of linear endomorphisms*, Invent. Math. 19 (1973), 83-106.

[90] M. KULCZYCKI, *Hadamard's inequality in inner product spaces*, Universitatis Iagellonicae Acta Mathematica, Fasc. XL (2002), 113-116.

[91] H. KUNITA, *Stochastic Flows and Stochastic Differential Equations*, Cambridge University Press, 1990.

[92] N. N. LADIS, *The topological equivalence of linear flows*, Differ. Equations 9 (1975), 938-947.

[93] E. LAMPRECHT, *Lineare Algebra 2*, Birkhäuser, 1983.

[94] G.A. LEONOV, *Strange Attractors and Classical Stability Theory*, St. Petersburg University Press, St. Petersburg, Russia, 2008.

[95] Z. LIAN AND K. LU, *Lyapunov exponents and invariant manifolds for random dynamical systems in a Banach space*, Memoirs of the AMS, 2010.

[96] F. LOWENTHAL, *Linear Algebra with Linear Differential Equations*, John Wiley & Sons, 1975.

[97] A.M. LYAPUNOV, *The General Problem of the Stability of Motion*, Comm. Soc. Math. Kharkov (in Russian), 1892. Problème Géneral de la Stabilité de Mouvement, Ann. Fac. Sci. Univ. Toulouse 9 (1907), 203-474, reprinted in Ann. Math. Studies 17, Princeton (1949), in English Taylor & Francis 1992.

[98] W. MAGNUS AND S. WINKLER, *Hill's Equation*, Dover Publications 1979. Corrected reprint of the 1966 edition.

[99] M. MARCUS, *Finite Dimensional Multilinear Algebra, Part I and II*, Marcel Dekker, New York 1973 and 1975.

[100] L. MARKUS AND H. YAMABE, *Global stability criteria for differential systems*, Osaka Math. J. 12 (1960), 305-317.

[101] W.S. MASSEY, *A Basic Course in Algebraic Topology*, Springer-Verlag, 1991.

[102] R. MENNICKEN, *On the convergence of infinite Hill-type determinants*, Archive for Rational Mechanics and Analysis, 30 (1968), 12-37.

[103] R. MERRIS, *Multilinear Algebra*, Gordon Breach, Amsterdam, 1997.

[104] K.R. MEYER AND G.R. HALL, *Introduction to Hamiltonian Dynamical Systems and the N-Body Problem*, Springer-Verlag, 1992.

[105] P.D. MCSWIGGEN, K. MEYER, *Conjugate phase portraits of linear systems*, American Mathematical Monthly, 115 (2008), 596-614.

[106] R.K. MILLER, *Almost periodic differential equations as dynamical systems with applications to the existence of a.p. solutions*, J. Diff. Equations, 1 (1965), 337-345.

[107] K. MISCHAIKOW, *Six lectures on Conley index theory*. In: Dynamical Systems, (CIME Summer School 1994), R. Johnson editor, Lecture Notes in Math., vol. 1609, Springer-Verlag, Berlin, 1995, pp. 119-207.

[108] E. OJA, *A simplified neuron model as a principal component analyzer*, J. Math. Biol. 15(3) (1982), 267-273.

[109] V.I. OSELEDETS, *A multiplicative ergodic theorem. Lyapunov characteristic numbers for dynamical systems*, Trans. Moscow Math. Soc. 19 (1968), 197-231.

[110] M. PATRÃO AND L.A.B. SAN MARTIN, *Semiflows on topological spaces: chain transitivity and semigroups*, J. Dyn. Diff. Equations 19 (2007), 155-180.

[111] M. PATRÃO AND L.A.B. SAN MARTIN, *Morse decompositions of semiflows on fiber bundles,* Discrete and Continuous Dynamical Systems 17 (2007), 113-139.

[112] L. PERKO, *Differential Equations and Dynamical Systems,* Springer-Verlag, 1991.

[113] H. POINCARÉ, *Sur les déterminants d'ordre infini,* Bulletin de la Société Mathematiques de France, 14 (1886), 77-90.

[114] M. RASMUSSEN, *Morse decompositions of nonautonomous dynamical systems,* Trans. Amer. Math. Soc., 359 (2007), 5091-5115.

[115] M. RASMUSSEN, *All-time Morse decompositions of linear nonautonomous dynamical systems,* Proc. Amer. Math. Soc., 136 (2008), 1045-1055.

[116] M. RASMUSSEN, *Attractivity and Bifurcation for Nonautonomous Dynamical Systems,* Lecture Notes in Mathematics, vol. 1907, Springer-Verlag, 2007.

[117] C. ROBINSON, *Dynamical Systems,* 2nd edition, CRC Press, 1998.

[118] K.P. RYBAKOWSKI, *The Homotopy Index and Partial Differential Equations,* Springer-Verlag, 1987.

[119] R.J. SACKER AND G.R. SELL, *Existence of dichotomies and invariant splittings for linear differential systems I,* J. Diff. Equations, 13 (1974), 429-458.

[120] D. SALAMON, *Connected simple systems and the Conley index of isolated invariant sets,* Trans. Amer. Math. Soc., 291 (1985), 1-41.

[121] D. SALAMON, E. ZEHNDER, *Flows on vector bundles and hyperbolic sets,* Trans. Amer. Math. Soc., 306 (1988), 623-649.

[122] L. SAN MARTIN AND L. SECO, *Morse and Lyapunov spectra and dynamics on flag bundles,* Ergodic Theory and Dynamical Systems 30 (2010), 893-922.

[123] B. SCHMALFUSS, *Backward Cocycles and Attractors of Stochastic Differential Equations.* In International Seminar on Applied Mathematics – Nonlinear Dynamics: Attractor Approximation and Global Behaviour (1992), V. Reitmann, T. Riedrich, and N. Koksch, eds., Technische Universität Dresden, pp. 185-192.

[124] J. SELGRADE, *Isolated invariant sets for flows on vector bundles,* Trans. Amer. Math. Soc. 203 (1975), 259-390.

[125] G. SELL AND Y. YOU, *Dynamics of Evolutionary Equations,* Springer-Verlag, 2002.

[126] M.A. SHAYMAN, *Phase portrait of the matrix Riccati equation,* SIAM J. Control Optim. 24 (1986), 1-65.

[127] R. SHORTEN, F. WIRTH, O. MASON, K. WULFF, C. KING, *Stability criteria for switched and hybrid systems,* SIAM Review 49 (2007), 545-592.

[128] C.E. SILVA, *Invitation to Ergodic Theory,* Student Mathematical Library, Vol. 42, Amer. Math. Soc., 2008.

[129] E.D. SONTAG, *Mathematical Control Theory,* Springer-Verlag, 1998.

[130] J.M. STEELE, *Kingman's subadditive ergodic theorem,* Ann. Inst. Poincaré, Prob. Stat. 26 (1989), 93-98.

[131] J.J. STOKER, *Nonlinear Vibrations in Mechanical and Electrical Systems,* John Wiley & Sons, 1950 (reprint Wiley Classics Library 1992).

[132] D. TALAY, *Approximation of upper Lyapunov exponents of bilinear stochastic differential systems,* SIAM Journal on Numerical Analysis, 28(4), (1991), 1141-1164.

[133] G. TESCHL, *Ordinary Differential Equations and Dynamical Systems,* Graduate Studies in Math. Vol. 149, Amer. Math. Soc., 2012.

[134] H. VERDEJO, L. VARGAS AND W. KLIEMANN, *Stability of linear stochastic systems via Lyapunov exponents and applications to power systems,* Appl. Math. Comput. 218 (2012), 11021-11032.

[135] R.E. VINOGRAD, *On a criterion of instability in the sense of Lyapunov of the solutions of a linear system of ordinary differential equations,* Doklady Akad. Nauk SSSR (N.S.) 84 (1952), 201-204.

[136] A. WACH, *A note on Hadamard's inequality,* Universitatis Iagellonicae Acta Mathematica, Fasc. XXXI (1994), 87-92.

[137] P. WALTERS, *An Introduction to Ergodic Theory,* Springer-Verlag, 1982.

[138] M.J. WARD, *Industrial Mathematics,* Lecture Notes, Dept. of Mathematics, Univ. of British Columbia, Vancouver, B.C., Canada, 2008 (unpublished).

[139] S. WIGGINS, *Normally Hyperbolic Invariant Manifolds in Dynamical Systems,* Springer-Verlag, 1994.

[140] S. WIGGINS, *Introduction to Applied Nonlinear Dynamical Systems and Applications,* Springer-Verlag, 1996.

[141] F. WIRTH, *Dynamics of time-varying discrete-time linear systems: spectral theory and the projected system,* SIAM J. Control Optim. 36 (1998), 447-487.

[142] F. WIRTH, *A converse Lyapunov theorem for linear parameter-varying and linear switching systems,* SIAM J. Control Optim. 44 (2006), 210-239.

[143] W.M. WONHAM, *Linear Mulivariable Control: a Geometric Approach,* Springer-Verlag, 1979.

[144] W.-C. XIE, *Dynamic Stability of Structures,* Cambridge University Press, Cambridge, 2006.

Index

adapted norm, 35, 40
almost periodic function, 154
alternating k-linear map, 83
alternating product, 84
attractor, 159, 168
    and Morse decomposition, 162
    neighborhood, 159
attractor-repeller pair, 160

base component, 114, 121
base flow, 114
base space, 114, 121
basic set, 60
bilinear control system, 201
binary representation, 65
Birkhoff's ergodic theorem, 214
bit shift, 65
Blaschke's theorem, 55
Borel-Cantelli Lemma, 256

center subbundle
    linear periodic difference equation, 135
    linear periodic differential equation, 143
    random linear system, 229, 232
    robust linear system, 205
    topological linear system, 175
center subspace
    linear autonomous difference equation, 24
    linear autonomous differential equation, 16

linear periodic difference equation, 135
linear periodic differential equations, 143
random linear system, 229, 232
chain, 50, 60
    concatenation, 186
    jump time, 51, 52, 61, 66
    total time, 51
chain component, 51, 58, 60, 62, 165
    and Lyapunov space, 72, 76, 92, 134, 142
    in Grassmannian
        linear autonomous differential equation, 92
    in projective bundle
        linear periodic difference equation, 134
        linear periodic differential equation, 142
        topological linear flow, 173, 182
    in projective space
        linear autonomous difference equation, 75
        linear autonomous differential equation, 72
chain exponent, 172
chain limit set, 164
chain reachability, 51
chain recurrent point, 51, 60
chain recurrent set, 51, 58, 60, 165
    and connectedness, 58, 61
    and limit sets, 53

279

chain transitive set, 51, 60
  and time reversal, 53, 61
  for time shift, 200
  maximal, 51, 58, 60
characteristic number, 99
cocycle, 114, 120
  2-parameter, 102, 118, 131
  linear periodic difference equation, 131
  linear topological system, 170
  random linear, 225, 231
conjugacy
  and chain transitivity, 54, 63
  and fixed points, 32
  and limit sets, 54
  and Morse decompositions, 156
  and periodic solutions, 32
  and structural stability, 34
  dynamical systems, 32, 39
  in projective space, 79
  linear, 33
  linear autonomous differential equations, 33
  linear contractions, 41
  linear difference equations, 43
  projective flows, 75, 77
  smooth, 32, 33, 39
Conley index, 66
connected component, 57
control system, 117
  bilinear, 171, 197
  linear, 171
cycles, 156
cylinder sets, 226

dynamical system
  continuous, 30, 38
  linear skew product, 120, 121, 231
    bilinear control system, 201
    periodic difference equation, 131
    robust linear system, 201
  metric, 117, 121, 224
  random linear, 117, 121, 230
  topological linear, 170

eigenspace, 6
  complex generalized, 6
  random, 258
  real, 9
  real generalized, 9, 26
eigenvalue
  stability, 16, 24, 109, 128, 137

equilibrium, 31
equivalence of flows, 45
ergodic flow, 227
ergodic map, 212
  uniquely, 221
evaluation map, 116
exponential growth rate
  finite time, 172
  linear autonomous difference equation, 20
  linear autonomous differential equation, 12
  linear periodic difference equation, 128, 131, 132
  linear periodic differential equation, 137, 139
  linear topological flow, 172, 173, 192
  of a function, 121, 123
  volume, 87, 88, 90
exterior power, 238
exterior product, 85

Fatou's lemma, 219
Fenichel's uniformity lemma, 175
fiber, 170
Fibonacci numbers, 78
fixed point, 31
  asymptotically stable, 16, 23
  exponentially stable, 16, 24
  stable, 16, 23
  unstable, 16, 24
flag manifold, 95, 240
  metric, 240
flag of subspaces, 240, 247
  linear autonomous difference equation, 22
  linear autonomous differential equation, 15, 71
  linear periodic difference equation, 133
  linear periodic differential equation, 141
  random, 258
Floquet exponent, 128, 137
Floquet multiplier, 128, 137
Floquet space, 132, 139
Floquet theory, 127, 174, 228
flow
  continuous, 30
  ergodic, 227
  linear skew product, 114
  structurally stable, 34

Fourier expansion, 148, 213
fundamental domain, 41
fundamental solution
   linear autonomous difference
      equation, 18
   linear autonomous differential
      equation, 4
   nonautonomous linear difference
      equation, 118
   nonautonomous linear differential
      equation, 101
Furstenberg-Kesten Theorem, 252

generalized eigenspace
   complex, 6
   real, 9, 26
generalized eigenvector, 10
generator, 38, 121
Goldsheid-Margulis metric, 240, 247
gradient-like system, 66
Grassmannian
   linear autonomous differential
      equation, 87
   metric, 85, 87, 239, 240
Gronwall's lemma, 103

Hadamard inequality, 85, 94
Hamiltonian system, 143, 152
Hausdorff distance, 56
Hill's equation, 145
homoclinic structures, 156
hyperbolic matrix, 34, 43
hyperellipsoid, 263

integrability condition, 227, 231
invariance of domain theorem, 38
invariant set, 156
   isolated, 156
   metric dynamical system, 227
   minimal, 54

Jordan curve theorem, 209
Jordan normal form, 129
   and smooth conjugacy, 33, 39
   complex, 6
   real, 7
Jordan subspace, 73, 77
jump times, 51, 52, 61, 66

kinematic similarity transformation, 142
kinetic energy, 152
Kingman's subadditive ergodic
   theorem, 217

Kolmogorov's construction, 226

limit set, 47, 49, 59
   and chain component, 58
   and chain transitivity, 53, 61
   and time reversal, 50
linear autonomous differential equation
   on Grassmannian, 87
linear contraction, 40
linear expansion, 40
linear oscillator
   autonomous, 18
   periodic, 144
   with periodic restoring force, 145, 150
   with random restoring force, 263, 265
   with uncertain damping, 206
   with uncertain restoring force, 204, 207
linearized differential equation, 117
Liouville formula, 105
locally integrable matrix function, 100
Lyapunov exponent
   average, 263
   formula on the unit sphere, 122, 203
   linear autonomous difference
      equation, 20
   linear autonomous differential
      equation, 12
   linear nonautonomous difference
      equation, 119
   linear nonautonomous differential
      equation, 106
   linear periodic difference equation, 131, 132
   linear periodic differential equation, 139, 140, 142
   linear random system
      numerical computation, 270
   linear skew product flow, 115
   linear topological flow, 172, 173, 192, 196
   random linear system
      continuous time, 228
      discrete time, 232
   random system, 263
Lyapunov space
   and volume growth rate, 88
   linear autonomous difference equation
      in projective space, 76
   linear autonomous differential
      equation, 13, 21
      in projective space, 72

linear periodic difference equation, 132
linear periodic differential equation, 139
linear topological systems, 173
nonautonomous linear differential equation, 111
random linear system
  continuous time, 228
  discrete time, 232
Lyapunov transformation, 122, 133, 142

Mathieu's equation, 145, 151
  stability diagram, 150
maximal ergodic theorem, 213
measure preserving map, 212
metric space
  complete, 47
  connected, 49
minimal invariant set, 54
monodromy matrix, 137
Morse decomposition, 156
  and attractor sequence, 162
  finest, 157, 165, 173, 182
  order, 157
Morse sets, 156
Morse spectral interval, 173, 188
  boundary points, 191, 196
Morse spectrum, 172
  and Lyapunov exponents, 192
  for time reversed flow, 187
  periodic, 187
Multiplicative Ergodic Theorem
  continuous time, 227
  deterministic, 242
  discrete time, 231
multiplicity
  Lyapunov exponent, 111

Newton method, 150
normal basis, 108

Oja's flow, 79
one-sided time set, 30, 39
Ornstein–Uhlenbeck process, 265
Oseledets space
  continuous time, 228
  discrete time, 232
Oseledets' Theorem
  continuous time, 227
  discrete time, 231

parallelepiped, 83, 106

pendulum, 152
  damped, 145
  inverted, 151
  with oscillating pivot, 150
periodic function, 200
Plücker embedding, 85
polar coordinates, 48
polar decomposition, 78
potential energy, 152
principal component analysis, 79
principal fundamental solution, 101, 118
  linear periodic difference equation, 130
  linear periodic differential equation, 136
probability measure
  ergodic, 212, 215, 227
  invariant, 211
projection
  orthogonal, 235, 240
projective bundle, 171
  metric, 171
projective flow, 70, 75
projective space, 69
  metric, 69, 178

quasi-periodic function, 154

random linear differential equation, 225, 229
repeller, 159
  complementary, 160
  neighborhood, 159
Riccati equation, 68, 81
robust linear system, 117, 197, 201
rotation, 76, 109

Selgrade bundle, 173
Selgrade's theorem, 182
semi-dynamical system, 168
semiconjugacy, 70
semimartingale helix, 270
semisimple eigenvalue, 17
shift, 200
similarity of matrices, 32, 39, 234
simple vector, 84, 85
singular value, 228, 232, 237, 253, 263
singular value decomposition, 237
  exterior power, 238
skew-component, 115, 170
solution
  Carathéodory, 100, 124

Index

existence, 30
linear autonomous difference
    equation, 18
linear autonomous differential
    equation, 4
periodic, 31
solution formula
    in Jordan block, 11, 12, 19, 20
    in projective space, 71, 73, 77
    on the sphere, 78, 203
    scalar differential equation, 101
solution map
    continuity, 5, 18, 104, 119
    continuity with respect to
        parameters, 104, 119
spectrum
    eigenvalues of a matrix, 6
    Lyapunov spectrum, 204
    Lyapunov spectrum for a random
        system, 228
    Morse spectrum, 172, 204
    uniform growth spectrum, 195
stability
    and eigenvalues, 16, 24, 109, 152
    asymptotic, 16, 24, 135, 143, 177
    exponential, 16, 24, 135, 143, 177, 205
    almost sure, 229, 232
stability diagram
    linear oscillator with uncertain
        restoring force, 205, 207
    Mathieu equation, 150
    random linear oscillator, 265
stability radius, 206
stable fixed point, 16, 23
stable subbundle, 175, 177
    linear periodic difference equation,
        135
    linear periodic differential equation,
        143
    random linear system, 229, 232
    robust linear system, 205
stable subspace
    linear autonomous difference
        equation, 24
    linear autonomous differential
        equation, 16
    linear periodic difference equation,
        135
    linear periodic differential equation,
        143
    random linear system, 229, 232

stochastic differential equation, 265
stochastic linear differential equation,
    226, 269
subadditive ergodic theorem, 217
subadditive sequence, 244
subbundle, 171
    exponentially separated, 189, 190

Theon of Smyrna, 78
theorem
    Arzelá-Ascoli, 115
    Banach's fixed point, 102
    Birkhoff's Ergodic, 214
    Blaschke's, 55
    Furstenberg-Kesten, 252
    invariance of domain, 38
    Jordan curve, 209
    Kingman's subadditive ergodic, 217
    Lebesgue's on dominated
        convergence, 104
    maximal ergodic, 213
    Multiplicative Ergodic, 227, 231
        deterministic, 242
    Oseledets', 227, 231
    Poincaré-Bendixson, 65
    Selgrade's, 182
time shift, 115, 200
time-n map, 38
time-t map, 30
time-one map, 120
time-reversed equation, 13
    linear difference equation, 21
time-reversed flow, 60, 187, 194
    and chain transitivity, 53, 61
    and limit sets, 50
time-varying perturbation, 117
topology
    uniform convergence, 115
type number, 123

uniform growth spectrum, 195
uniquely ergodic map, 221
unstable subbundle, 175
    linear periodic difference equation,
        135
    linear periodic differential equation,
        143
    random linear system, 229, 232
    robust linear system, 205
unstable subspace
    linear autonomous difference
        equation, 24

linear autonomous differential
    equation, 16
  linear periodic difference equation,
    135
  linear periodic differential equation,
    143
  random linear system, 229, 232

variation-of-constants formula, 103
vector bundle, 170
  fiber, 170
  subbundle, 171
volume, 83
volume growth rate, 87, 88, 90, 106, 111

Whitney sum, 173, 182
Wronskian, 105

DATE DUE

PRINTED IN U.S.A.